Advances in Geographical and Environmental Sciences

Advances in Geographical and Environmental Sciences synthesizes series diagnostigation and prognostication of earth environment, incorporating challenging interactive areas within ecological envelope of geosphere, biosphere, hydrosphere, atmosphere and cryosphere. It deals with land use land cover change (LUCC), urbanization, energy flux, land-ocean fluxes, climate, food security, ecohydrology, biodiversity, natural hazards and disasters, human health and their mutual interaction and feedback mechanism in order to contribute towards sustainable future. The geosciences methods range from traditional field techniques and conventional data collection, use of remote sensing and geographical information system, computer aided technique to advance geostatistical and dynamic modeling.

The series integrate past, present and future of geospheric attributes incorporating biophysical and human dimensions in spatio-temporal perspectives. The geosciences, encompassing land-ocean-atmosphere interaction is considered as a vital component in the context of environmental issues, especially in observation and prediction of air and water pollution, global warming and urban heat islands. It is important to communicate the advances in geosciences to increase resilience of society through capacity building for mitigating the impact of natural hazards and disasters. Sustainability of human society depends strongly on the earth environment, and thus the development of geosciences is critical for a better understanding of our living environment, and its sustainable development.

Geoscience also has the responsibility to not confine itself to addressing current problems but it is also developing a framework to address future issues. In order to build a 'Future Earth Model' for understanding and predicting the functioning of the whole climatic system, collaboration of experts in the traditional earth disciplines as well as in ecology, information technology, instrumentation and complex system is essential, through initiatives from human geoscientists. Thus human geoscience is emerging as key policy science for contributing towards sustainability/survivality science together with future earth initiative.

Advances in Geographical and Environmental Sciences series publishes books that contain novel approaches in tackling issues of human geoscience in its broadest sense-books in the series should focus on true progress in a particular area or region. The series includes monographs and edited volumes without any limitations in the page numbers.

All submitted proposals are carefully reviewed by the series editor.

Publication Ethics and Publication Malpractice Statement: https://www.springernature.com/gp/policies/book-publishing-policies

Ethical, Conflict-of-Interest, and Editorial Policy Statement: https://www.springer.com/us/authors-editors/journal-author/journal-author-helpdesk/before-you-start/before-you-start/1330#c14214

Mery Biswas · Arun Mondal ·
Bedhas Ujjwal Mandal
Editors

Himalayan Flood Hazards

Analyses on Climate Change and Fluvial Hydrology

Springer

Editors
Mery Biswas
Department of Geography
Presidency University
Kolkata, West Bengal, India

Arun Mondal
Department of Geography
University of Allahabad
Prayagraj, Uttar Pradesh, India

Bedhas Ujjwal Mandal
Department of Geography
Shyampur Siddheswari Mahavidyalaya
Howrah, West Bengal, India

ISSN 2198-3542 ISSN 2198-3550 (electronic)
Advances in Geographical and Environmental Sciences
ISBN 978-981-92-0701-5 ISBN 978-981-92-0702-2 (eBook)
https://doi.org/10.1007/978-981-92-0702-2

This Springer imprint is published by the registered company Springer Nature Singapore Pte Ltd.
The registered company address is: 152 Beach Road, #21-01/04 Gateway East, Singapore 189721, Singapore

Contents

Chapter 1
Flood Inundation Mapping in Ramganga River Basin Using Synthetic Aperture Radar (SAR) Based Sentinel-1 Data in 2018

Arun Mondal, Anupam Pandey, Surendra Kumar Chandniha, and Sananda Kundu

Abstract Remote sensing data is a widely used technique in real-time flood monitoring and assessing the effects of the flood. The study focuses on the extension of the flood water at pre-flood, during flood, and post-flood events using Synthetic Aperture Radar (SAR) images, which are derived from the sensor of Sentinel-1 satellite in VV polarisation. The study also emphasizes the change of river direction during flood and helps to identify the most vulnerable zones of the study area. Identification of flooded areas and their temporal changes is calculated by the 'Modified Normalised Differential Water Index' (MNDWI) formula using the 'threshold method'. Optical imagery of Sentinel-2 is used to understand the temporal changes of the river path in a better way. Daily precipitation data of the study area is acquired from 'Tropical Rainfall Measuring Mission (TRMM)' to assess the contribution of precipitation to the flood that occurred in August-September 2018. The study covers some selected parts of the middle-lower Ramganga river basin in Uttar Pradesh. The results indicate that the peak flood condition occurred on 02 September 2018 (54.51 Sq. Km.) due to heavy precipitation on 24th Aug 2018 (166.41 mm) and 28th Aug 2018 (46.82 mm). The findings show a negative relationship between the distance of surrounding lands from the river channel and the duration of the flood. In the study, it is also observed that the width of the river channel increases at the time of flood, which indicates the necessity of sustainable planning for the development of that area. Overall, the results help to achieve the sustainable goals, making the human and human settlements more inclusive, resilient, safe, and sustainable.

A. Mondal · A. Pandey
Department of Geography, University of Allahabad, Prayagraj, India

S. K. Chandniha
BRSM College of Agricultural Engineering and Technology and Research Station, IGKV, Mungeli, India

S. Kundu (✉)
Department of Geography, Manipur University, Imphal, India
e-mail: skundu@manipuruniv.ac.in

M. Biswas et al. (eds.), *Himalayan Flood Hazards*, Advances in Geographical and Environmental Sciences, https://doi.org/10.1007/978-981-92-0702-2_1

Keywords Flood inundation · Sentinel-1 · TRMM · Using synthetic aperture radar (SAR) · MNDWI

1.1 Introduction

Disaster is a natural or man-made phenomenon that interrupts the normal lifestyle of humans and decreases the standard of living of individuals. The occurrences of disasters become more intense and frequent due to climate change. Disasters damage the property of more than $500 billion every year (As Natural Disasters Rise 2017). Worldwide, 78% of the total population is affected by the natural hazards, and the flood is one of the most frequent (46%) catastrophic events among all natural hazards (Kugler and De Groeve 2007). A flood is a situation when the water overflows from its channel and spreads over the dry area in the surroundings (Anusha and Bharathi 2019; Agnihotri et al. 2019). India is one of the severely flood-prone countries in the world, with one-eighth of its total geographical area (approximately 40 million hectares) prone to flood (NDMA 2008; Manjusree et al. 2012).

Floods, compounded by continuous heavy precipitation and discharge of high amounts of water from dams, increase the intensity of damages of flood effected areas (economic &structural damages) (Pramanick et al. 2021; Anusha and Bharathi 2020; Mondal et al. 2018a, b; Kundu et al. 2017a, b). The duration of flood can last several months, which makes a deep adverse impact on ecological balances, arable land, human lives, crops, infrastructures, and overall sustainable development of that place (Pramanick et al. 2021; Muttarak and Dimitrova 2019; Ray et al. 2019). Flood can be categorised as riverine flood (mainly due to high precipitation) and glacial flood (due to expansion of glacial lake (Anusha and Bharathi 2020). Riverine flood extents over a large area, but the destructive power of the Glacial Lake Outburst Flood (GLOF) is more intense (Bajracharya et al. 2007). In India, the number of riverine floods is relatively more and it mainly occurs in the months of monsoon (June–September). The flood can be caused by different reasons, like heavy precipitation, inadequate discharge rate, and river system, etc. (Mondal et al. 2017a, b; Mondal et al. 2016; Manjusree et al. 2012). Flood also challenges the food security of the country, damaging the crop in the monsoon (kharif crop) and winter season (rabi crop) (Ray et al. 2019). The Central Water Commission, the principal agency in India, forecasts the flood and also fixes the warning level and danger level at different locations (Agnihotri et al. 2019).

Floods happen in a very irregular way, and it is very difficult to understand the hydraulic character as well as the geographical extension of the flood-affected areas (Agnihotri et al. 2019). The flood extension-related data can be collected from field survey (in-place data or in-situ data collection, etc.) or remote sensing technologies (aerial and satellite imageries) (Mondal et al. 2022; Anusha and Bharathi 2020; Clement et al. 2018; Jensen 2009). Collecting the data using the field survey technique at the time of flood is very difficult because the communication systems (road networks and telecommunications, etc.) are damaged (Manjusree et al. 2012; Mondal et al. 2012). So, remote sensing technologies, which are used by many researchers, are a very useful technique to demarcate the boundary of flood-affected areas because gauge stations can't provide such an areal extent-related data (Bhatt and Rao 2016; Klemas 2015; Manjusree et al. 2012). Synthetic Aperture Radar (SAR) sensors are considered more effective than optical sensors (visible/infrared) in flood mapping because SAR sensors can capture the spatial information (depth and extension of flood water, etc.) during day-night times and in all weather conditions (Mukhopadhyay et al. 2014; Klemas 2015; Manjusree et al. 2012; Voigt et al. 2007; Sanyal and Lu 2004). The backscatter value of SAR is used to identify the flood-affected area because the backscatter value of the water is very low and the flood-affected area appears with a darker tone on the imagery than its surroundings' land surface (Di Baldassarre et al. 2011).

India, a riverine country, faces several floods in different parts of the country every year. In the recent decade, many devastating floods occurred in different parts of India. Bihar is one of the most suffering states of floods where several consecutive floods occurred in July–September 2020, 2019 (Koshi and Gandak River basin), 2017 (Koshi river basin, north Bihar) etc., due to heavy precipitation and over bank flow (Kumar et al. 2022; Parida et al. 2022; Tripathi et al. 2020; Tripathi and Tiwari 2019a; Kundu et al. 2015a, b). The death of more than 200–500 people and huge damage to infrastructures were reported in different parts of the Ganga basin (especially, Patna district in Bihar) in the 2019 flood (Khan et al. 2022). Another devastating flood, which devastated a huge portion of East and North-East India (Bihar, West Bengal, Assam), occurred in the Ganga–Brahmaputra basin in July–September 2020 (Mondal et al. 2025; Pandey et al. 2022). In northern India, Punjab is one of the flood-prone states in India where several districts were inundated in 2019, causing several damages to infrastructure, crops, and huge economic losses in Punjab (Sahoo et al. 2021; Tripathi et al. 2021). Southern India, which is located in the tropical plateau region (Eastern and Western Ghats), receives heavy precipitation through conventional and cyclonic rainfall. The flood that occurred in Kerala in 2018 devastated the area of more than 100,000 sq. hectors (Pramanick et al. 2022; Tiwari et al. 2020). The flood, which occurred in Mumbai in June 2015, killed about 1200 people (Lal et al. 2020; Gupta and Nair 2011). The recurrent floods hit Chennai city several times (2005, 2010, and 2015), causing huge economic damage (Gupta and Nair 2011). The study of 'Dhar and Nandargi (1995)' indicates that the maximum number of floods mainly occur in northern, southern, and peninsular India.

The previous studies on floods indicate that SAR images (Sentinel 1, Radarsat 1) were widely used by numerous scholars to identify the flood-affected area (Pramanick et al. 2022; Devrani et al. 2022; Sahoo et al. 2021; Tiwari et al. 2020; Desh and Sar 2020; Agnihotri et al. 2019; Tripathi and Tiwari 2019b). Along with SAR images, optical remote sensing data (Sentinel 2, MODIS NRT) were also used by many researchers (Khan et al. 2022; Kumar et al. 2022; Parida et al. 2022; Tripathi et al. 2021; Anusha and Bharti 2020; Tripathi et al. 2020). Many scholars conducted a ground survey to check the validity of the research (Pandey et al. 2022; Kumar 2019). In the SAR images, single polarisation in VV orientation (Tiwari et al. 2020; Anusha and Bharti 2020) and dual polarisation in VV-VH and VH-VB orientation (Kumar et al. 2022; Tripathi et al. 2021; Parida et al. 2022; Tripathi et al. 2020; Tripathi and Tiwari 2019a, b; etc.) were used by different scholars to get better reflectance value of the flooded area. SRTM DEM (Khan et al. 2022; Kumar et al. 2022; Tripathi and Tiwari 2019a, b, etc.), as well as ALOS Pulsar (Parida et al. 2022; Tiwari et al. 2020, etc.) and ASTER DEM (Tripathi et al. 2021; Desh and Sar 2020, etc.), were used by several scholars in their studies. According to Devrani et al. (2022), the spatial resolution of satellite imagery influenced the river morphometry of the basin area. Most of the researchers collected precipitation data from Tropical Rainfall Measuring Mission (TRMM) (Pandey et al. 2022; Parida et al. 2022; Sahoo et al. 2021; Tiwari et al. 2020, etc.) and India Meteorological Department (IMD) (Kumar et al. 2022; Anusha and Bharti 2020, etc.) websites. Based on the value of different indices like MNDWI, NDWI, etc., different researchers used various methods like Threshold classification method (Tripathi et al. 2021; Tiwari et al. 2020; Anusha and Bharti 2020; Tripathi et al. 2020, etc.), supervised classification method (Devrani et al. 2022), and unsupervised classification method (Khan et al. 2022; Anusha and Bharti 2020; Tripathi et al. 2020) to identify and separate the inundated flood area from its surrounding uninundated area.

However, some studies were done on Ramganga river basin which were mainly focused on the zonation of flood-prone area based on various hazard forming factors and vulnerable factors (Mukherjee and Deb et al. 2023), planning for sustainable use of flooded water through 'Underground Transfer of Floods for Irrigation' (Reddy et al. 2020) and flood frequency analysis (Mukherjee and Bilas 2014). In the very small area of the lower Ramganga river, a study of flood inundation was conducted by Agnihetri et al. (2019). All the above-mentioned studies indicate the limited number of publications on flood-related analysis in the Ramganga basin. But those are essential to take proper measures for the sustainable planning of that area. Our study area, which is located in the middle to lower segment of the Ramganga river basin, is unique because flood extension and duration-related studies using SAR and optical data have not been done ever in the middle-lower segment of the Ramganga river. The present study is conducted to (i) estimate the extension of the flood water at pre-flood, during flood, and post-flood events, analysing the SAR images, (ii) understand the changing nature of the course of the Ramganga River during flood, and (iii) assess the waterlogging condition of different places of the study area during the flood period.

1.2 Study Area

The Ramganga River is one of the most important left bank tributaries of the Ganga River. The total area of the Ramganga basin is about 30,635.1 km^2, and the total length of the river is about 649.11 km (Mukherjee and Deb 2024). The river originates from a spring called Ramnali located in Uttarakhand at an elevation of 2926 m (Mukherjee et al. 2017). The present study area is a part of the lower Ramganga basin. The study area extends from 27°41′21″ N to 27°57′7″ N and from 79°22′27″

E to 79°38′14″ E (Fig. 1.1). It covers the Badaun and Shahjahanpur districts of Uttar Pradesh, India. The study area covers an area of 403.81 Km^2 and the length of the river is about 28.54 km. The average temperature of the study area is around 25.7 °C. The average maximum temperature of the summer months is about 40 °C, and the average minimum temperature of the winter months is around 10 °C (Agnihotri et al. 2019). The average rainfall is about 90–100 cm/year. The rainfall in the study area mainly occurs during the summer months (July–September), whereas the rest of the year remains mostly dry (Agnihotri et al. 2020). The soil types in the study area are Khadar, Bhangar, and Bhur. The Khadar soil occupies an active channel belt and active floodplain zone, whereas the Bhangar soil occupies an upland terrace surface, and the Bhur soil is found sparsely along the river tract.

1.3 Materials and Methods

The Sentinel-1 Synthetic Aperture Radar (SAR) and Sentinel-2 data set of the European Space Agency (ESA) obtained from Copernicus Open Access Hub were utilised in this present study to analyse the spatial and temporal distribution of flood and the duration of flood. Table 1.1 represents a detailed description of the images. Alongside these data sets, TRMM daily precipitation data were also utilised to provide the spatial distribution of rainfall over the Ramganga River and surrounding areas. The methodological flow chart of the study is shown in Fig. 1.2.

The Sentinel-1 data were obtained from two satellites in the same orbital plane, Sentinel-1A and Sentinel-1B. The images of Sentinel- 1 were acquired for the 21 August-26 September period. Sentinel-1 is a C-band system developed by the ESA. Its data are very useful for flood monitoring due to their frequent revisit time (Kundu et al. 2022) and their capability to capture data at day/night irrespective of the weather conditions (Anusha and Bharathi 2019). These images were obtained in the Interferometric Wide Swath (IW) sensor mode. The sentinel images were with VV and VH polarisations of high resolution (HR) Level-1 Ground range detected (GRD) processing level (Pramanick et al. 2021). For the present study, a sigma calibration was performed, which creates a new image with calibrated backscatter coefficients. After that, a filtration was applied on calibrated images using the Refined Lee filter to remove the noise from the images using a speckle filtering process (Borah et al. 2018). Then, the SRTM data was used for Doppler terrain correction. After that, the process of binarisation was applied using a threshold value. The threshold method has been a popular method used to extract the flooded area. This requires careful identification of the threshold backscatter values (Kundu et al. 2022). Identification of the flood-inundated area of this study was carried out with this method.

Sentinel 2 images of 21 June and 14 September were used for the pre- and post-flood analysis. There are several methods available to map the water bodies from the Sentinel 2, like band density slicing (Rahman and Thakur 2018), supervised and unsupervised classification (Huang et al. 2014), spectral water index (Li et al. 2013), etc. Among these algorithms, the spectral water index is preferred more. Its

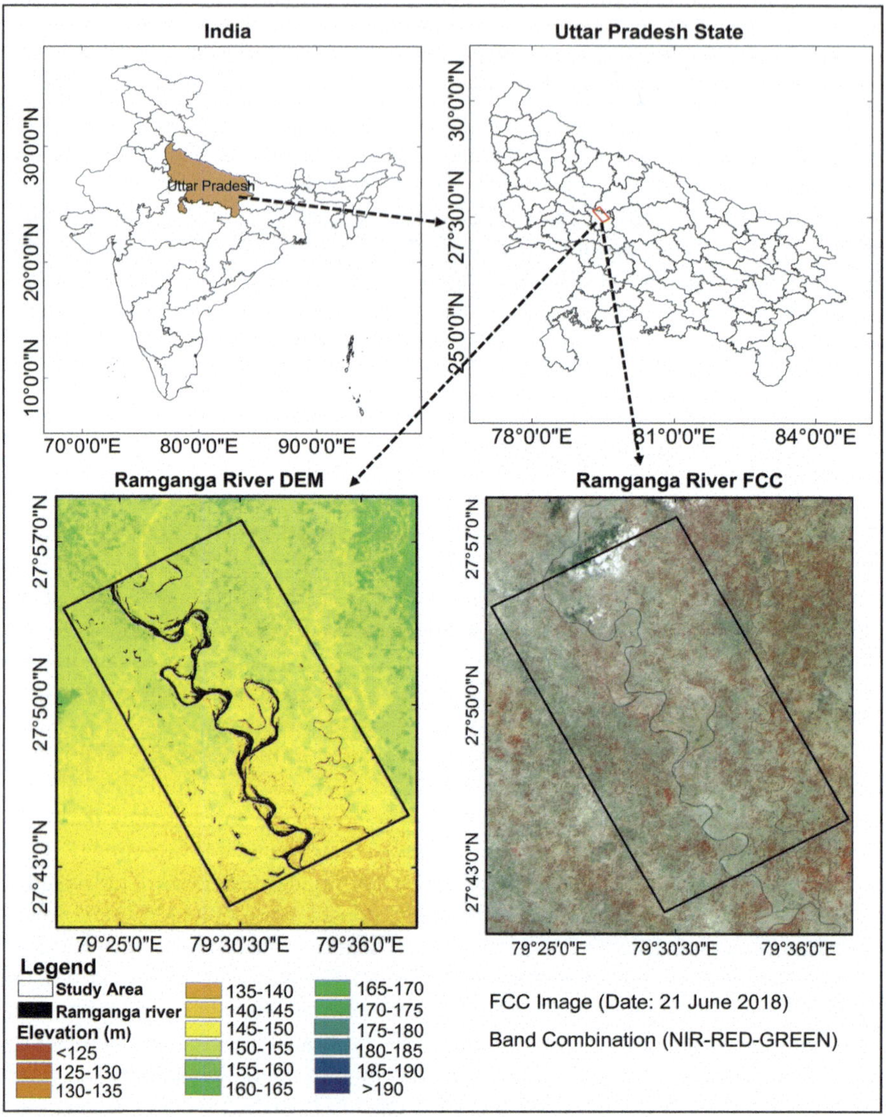

Fig. 1.1 Ramganga River was shown in the Uttar Pradesh state, DEM, and FCC image

computational cost is low, user-friendly, and reliable. Due to this, it is considered the best (RYU et al. 2002). In our present study to map the flood inundation area, Modified Normalised Differential Index (MNDWI) was used. Green band (Band 3) and SWIR (Band 11) of Sentinel-2 with spatial resolutions 10 m and 20 m were used, and the 10-m green band was increased to 20 m to calculate the MNDWI.

Table 1.1 Data source

Sl. No	Image ID	Date	Track ID
Sentinel-1 image			
1	S1A_IW_GRDH_1SDV_20180809T003608_ 20180809T003633_023162_02842B_A8C2	09-Aug-2018	165
2	S1A_IW_GRDH_1SDV_20180818T124618_ 20180818T124647_023301_0288BA_2216	18-Aug-2018	129
3	S1A_IW_GRDH_1SDV_20180821T003609_ 20180821T003634_023337_0289D5_4EF2	21-Aug-2018	165
4	S1A_IW_GRDH_1SDV_20180830T124619_ 20180830T124648_023476_028E47_FE33	30-Aug-2018	129
5	S1A_IW_GRDH_1SDV_20180902T003610_ 20180902T003635_023512_028F64_63D3	02-Sep-2018	165
6	S1A_IW_GRDH_1SDV_20180911T124619_ 20180911T124648_023651_0293E3_B406	11-Sep-2018	129
7	S1A_IW_GRDH_1SDV_20180914T003610_ 20180914T003635_023687_029500_54F5	14-Sep-2018	165
8	S1A_IW_GRDH_1SDV_20180923T124619_ 20180923T124648_023826_029995_DEBC	23-Sep-2018	129
Sentinel-2 image			
9	S2A_MSIL1C_20180621T051651_N0206_R062_ T44RLR_20180621T081647	21-June-201	99,999
10	S2B_MSIL1C_20180914T051639_N0206_R062_ T44RLR_20180914T091348	14-Sep-2018	99,999

$$\text{MNDWI}_{\text{20m}} = \frac{\rho_3^{20m} - \rho_{11}}{\rho_3^{20m} + \rho_{11}}$$

where, ρ_{11} = TOA reflectance of 11 (SWIR) Band, ρ_3^{20m} = TOA reflectance of the increased 3 Band with a spatial resolution of 20 m. The value of ρ_3^{20m} is calculated as the average value of the corresponding 2 × 2 ρ3 values.

Tropical Rainfall Measuring Mission (TRMM) daily precipitation data was used to evaluate rainfall over the region Ramganga and the surrounding area for a period of 18 August to 27 September 2018. TRMM data has the global coverage of 50°N and 50°S from 1998 to 2015, with a spatial resolution of 0.25° × 0.25° (roughly 26 × 28 km in India). The TRMM products come from different sources. Different sensors like radar, infrared, and microwave are used on the TRMM core observatory. In addition to this, microwave and infrared data obtained from multiple geostationary satellites and ground-based precipitation measurements are also incorporated. (Mondal et al. 2018a, b).

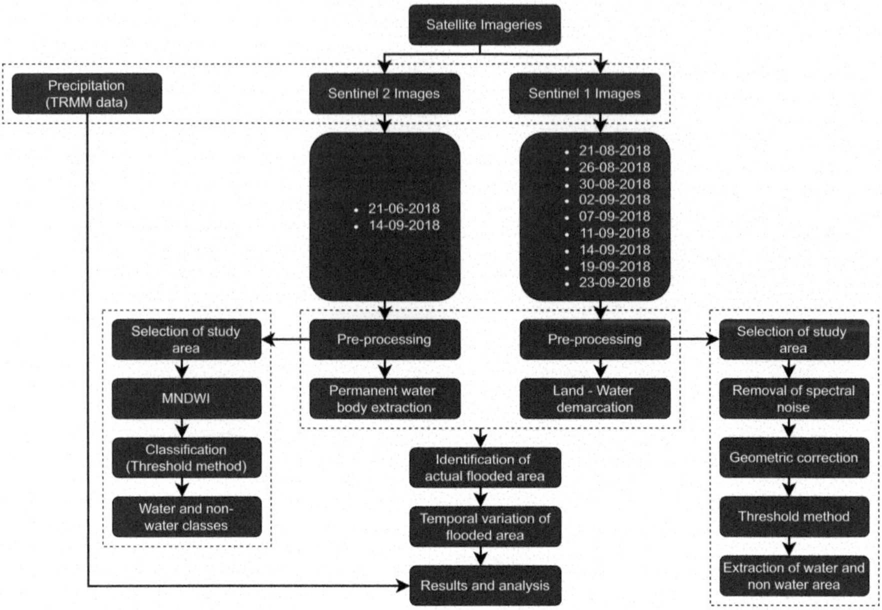

Fig. 1.2 Methodological flow chart

1.4 Extraction of Flood Extent and Duration

Flood extent shows the area under flood and places where damage could occur. It is very important to delineate a specific flood boundary for an accurate assessment of the flooded area (Kundu et al. 2014). As stated earlier, the threshold method was used to delineate the flooded and non-flooded areas from the pre-processed images of Sentinel-1. The permanent water body was marked from the Sentinel-2 images. A permanent water body was subtracted from the water of the Sentinel-1, which gives the actual flooded area. On the other hand, flood duration indicates the period of water that stays on the land. Flood duration is one of the indicators by which we can know the flood loss (damage of crops, settlements, etc.). Sentinel-1 images of 26, 30 August, and 2, 7, and 11 September 2018 were used to estimate the flood duration. The areas where flood water stayed for 1 day were considered as 1-day flood, where flood water concentrated for consecutive 2 to 5 days, 6 to 10 days, and 11 to 16 days were considered areas were flooded for 2 to 5 days, 6–10 days, and 11–16 days.

1.5 Results

1.5.1 Flood Extent Using Temporal SAR Images

Table 1.2 shows the detailed statistics of backscatter values (minimum, maximum, mean, and standard deviation) of the Sentinel-1 images, based on which flood inundation maps were prepared. The minimum, maximum, and mean backscatter values of 21 August image were −46.64, 11.3 and −10.76; 26 August image were −38.80, 7.05, and −9.56; 30 August image were −36.62, 8.50, and −11.82; 2 September image were −33.34, 14.46, and −11.95; 7 September image were −41.79, 8.12, and −10.20; 11 September image were −31.84, 7.94, and −10.62;14 September image were −31,11.07, and −10.83, 19 September image were −45.24,8.83, and −9.60; and 23 September image values were −31.12, 7.75 and −10.10. From the table, it's clearly visible that the highest minimum backscatter value was observed on 21 August. Whereas the maximum backscatter value was observed on 2 September, the highest mean backscatter value was observed on 26 August. Based on these backscatter values, the threshold method was applied to estimate the temporal flood inundated area. The threshold method is a very simple and useful method used to identify flooded and non-flooded areas using Sentinel-1 images. This method was applied by considering the backscatter (dB) values of the images. Values that were less than or equal to the threshold limit were marked as water pixels (assigned a value 1), and the rest of the pixels were marked as land pixels (assigned a value 2). (Table 1.3) shows the threshold limit of all the images that were used to delineate water and non-water areas in the present study.

Figure 1.3 and Table 1.4 depicts the temporal distribution of flood inundated area in the study area from 21 August to 23 September 2018, with peak flood on 30 August and 02 September when the actual flooded area was 52.54 Km^2 (12.70%) and 54.51 Km^2 (13.17%). While the least area (4.32 Km^2) under flood was observed on 19 September. After 02 September, due to the decrease in rainfall, the actual flooded

Table 1.2 Backscatter values of Sentinel-1 images from 21 August 2018 to 23 September 2018

Date	Min	Max	Mean	SD
21 August 2018	−46.64	11.3	−10.76	3.85
26 August 2018	−38.8	7.05	−9.56	3.4
30 August 2018	−36.62	8.5	−11.82	4.81
02 September 2018	−33.34	14.46	−11.95	5.03
07 September 2018	−41.79	8.12	−10.2	4.15
11 September 2018	−31.84	7.94	−10.62	3.16
14 September 2018	−31.03	11.07	−10.83	3.36
19 September 2018	−45.24	8.83	−9.6	3.4
23 September 2018	−31.12	7.75	−10.1	2.9

Table 1.3 Threshold values for water and non-water area demarcation of Sentinel-1 images from 21 August 2018 to 23 September 2018

Date	Min	Max	Threshold for water demarcation
21 August 2018	−46.64	11.3	−18
26 August 2018	−38.8	7.05	−16.5
30 August 2018	−36.62	8.5	−16.5
02 September 2018	−33.34	14.46	−17
07 September 2018	−41.79	8.12	−16
11 September 2018	−31.84	7.94	−17
14 September 2018	−31.03	11.07	−17
19 September 2018	−45.24	8.83	−16
23 September 2018	−31.12	7.75	−16

area decreased by 7.72% from 02 to 07 September. The actual flooded area on 07 September was 22.54 Km^2 (5.45%), which decreased to 5.21 Km^2 (1.26%) on 11 September. Again on 14 September, some increase in the extent of the flooded area (8.09 Km^2) was observed compared to 11 September. Meanwhile, on 23 September 2018, the study area had 5.89 km^2 (1.42%) under flood water.

Figure 1.4 shows the temporal inundation area of 21 June and 14 September 2018. Modified Normalised Difference Water Index (MNDWI) was used to extract flooded areas from Sentinel-2 images of 21 June and 14 September 2018. The threshold method was applied to MNDWI values to extract flooded and non-flooded areas. The minimum and maximum MNDWI values of the 21 June image were −0.39 and 0.095, whereas the threshold value that was used to demarcate flooded and non-flooded areas was −17. On the other hand, the minimum and maximum MNDWI values of the 14 September 2018 image were −0.45 and 0.56, where the threshold value + 0.15 was used to demarcate flooded and non- flooded areas (Table 1.5). From the figure, it's clearly visible that there was no flood on 21 June 2018. Whereas, on 14 September, flood water extended along the course of the river in the study area.

Figure 1.5 displays the change in the course of the Ramganga River from the lean period (21 June 2018) to the peak flood (14 September 2018). From the false colour composite (FCC) images, it can be said that from 21 June 2018 to 14 September 2018, the width of the river expanded tremendously, and the areas mainly covered with rural settlements and vegetation were inundated on 14 September 2018.

Figure 1.6 and Table 1.6 shows the flood duration in the study area, which ranges from 1 day to 11- 16 days. Most areas (25.02 Km^2) in the study area remained underwater for 2 to 5 days, and less area (5.54 Km^2) remained under flood for 11- 16 days. Meanwhile, a large section of the study area (322.88 km^2) remained unaffected by the flood. The areas close to the main river channel had a flood duration of 11–16 days. Whereas, areas away from the main river channel remained underwater for 1 day or 2–5 days.

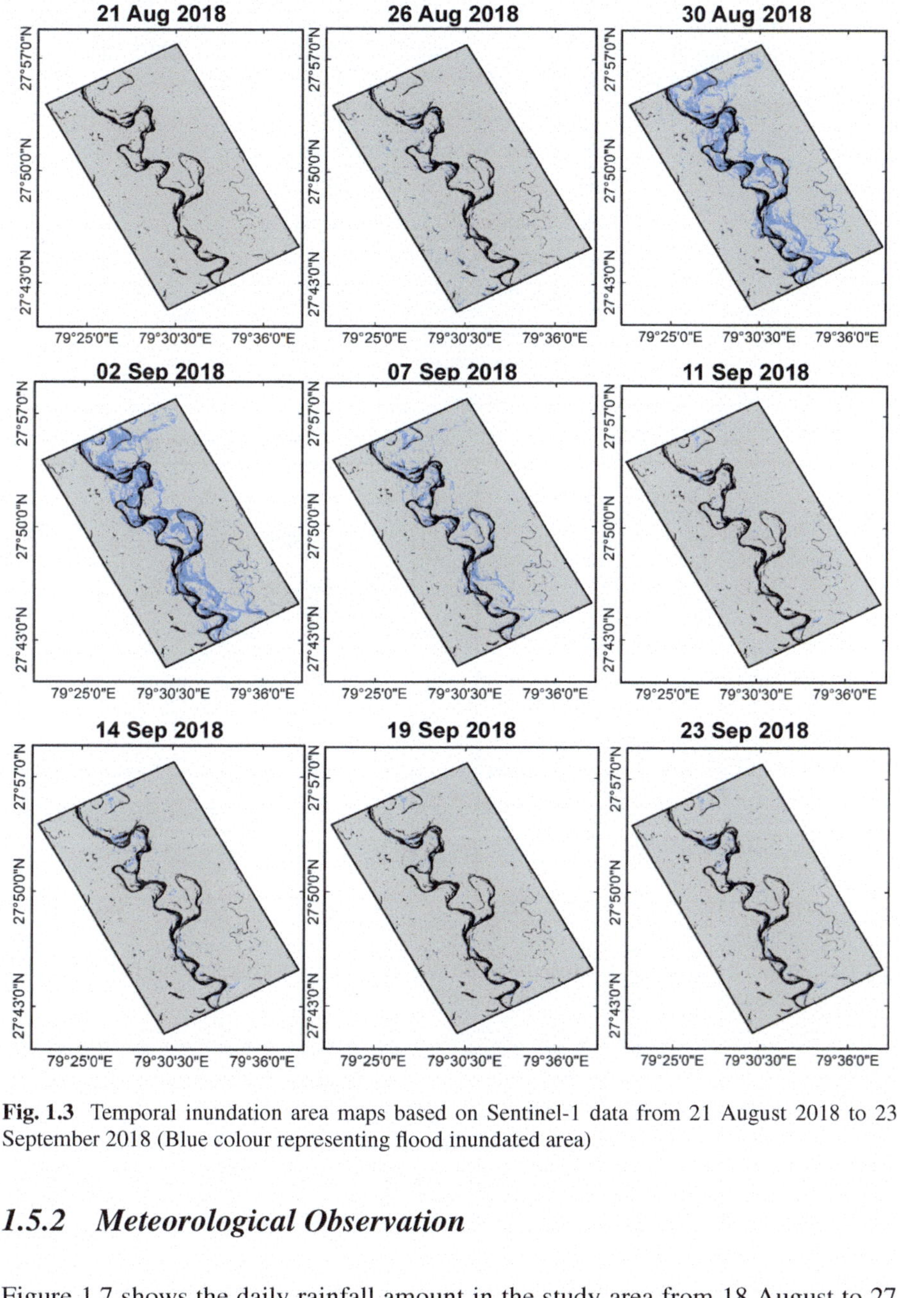

Fig. 1.3 Temporal inundation area maps based on Sentinel-1 data from 21 August 2018 to 23 September 2018 (Blue colour representing flood inundated area)

1.5.2 Meteorological Observation

Figure 1.7 shows the daily rainfall amount in the study area from 18 August to 27 September 2018. The highest rainfall (210.9 mm) occurred on 24 August. Whereas, the rainfall amount on other dates was very less, except for 23 August and 1, 5, and 9 September when rainfall was more than 25 mm. More of the rainfall was observed

Table 1.4 Actual flooded area excluding permanent water bodies from 21 August 2018 to 23 September 2018

Date	Actual flooded area (Sq. Km.)
21 August 2018	0
26 August 2018	7.67
30 August 2018	52.54
02 September 2018	54.51
07 September 2018	22.54
11 September 2018	5.21
14 September 2018	8.09
19 September 2018	4.32
23 September 2018	5.89

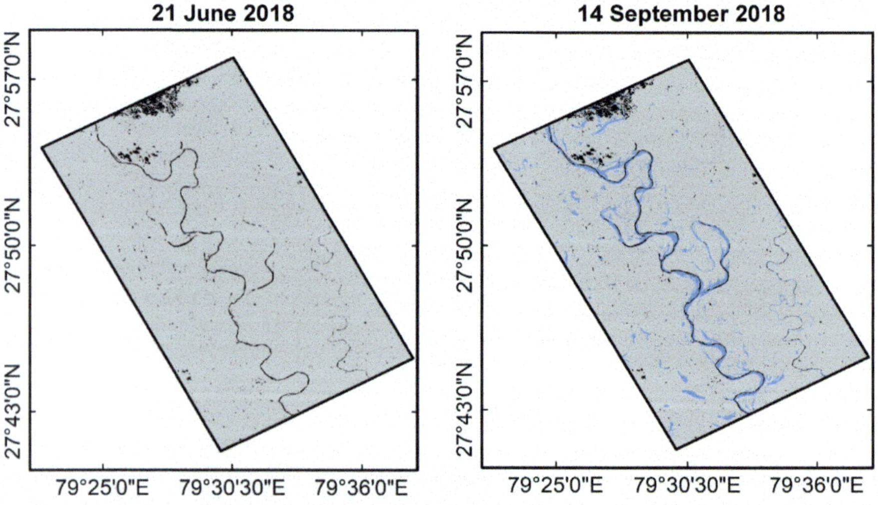

Fig. 1.4 Temporal inundation area maps based on Sentinel-2 data of 21 June 2018 and 14 September 2018 (Blue colour representing flood inundated area)

Table 1.5 Threshold values for water and non-water area demarcation of Sentinel-2 images of 21 June 2018 and 14 September 2018

Date	MNDWI Min	MNDWI Max	Threshold for water demarcation
21 June 2018	−0.39	0.095	−17
14 September 2018	−0.45	0.56	+0.15

from 20 August to 5 September, and this was the period when more of the areas were affected by floods. Due to the rise in water level in the Ramganga River.

Figure 1.8 shows the spatial distribution of daily precipitation (TRMM) over the Ramganga River and surrounding areas from 21 August to 23 September 2018. On

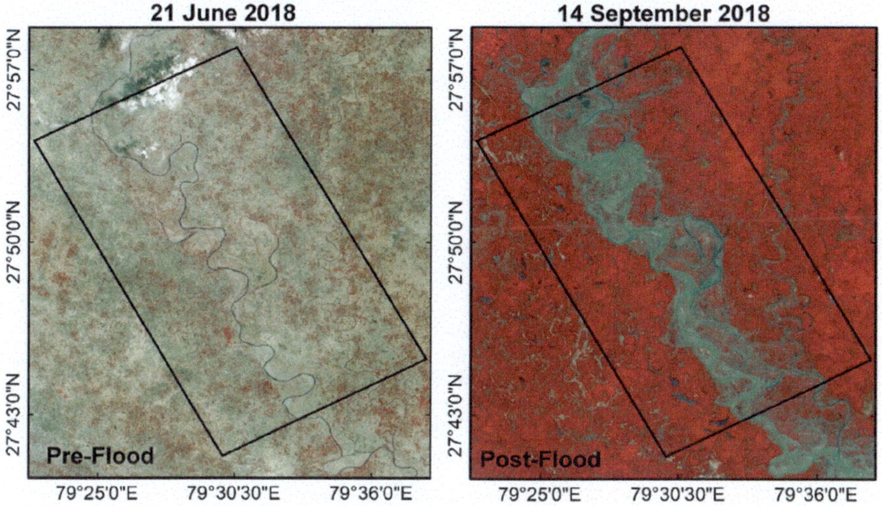

Fig. 1.5 Sentinel-2 image of 21 June 2018 and 14 September 2018 showing changes in the course of the Ramganga River during flood

most of the dates observed, rainfall varied from 0 to 3 mm, except for 26, 30 August, and 23 September. On 26 and 30 August, the study area experienced major rainfall events in all parts of the region, and the amount of rainfall varied from 0 to > 15. This was the reason for the peak flood during this time. On the other hand, on 23 September, the area also experienced more than 15 mm of rainfall, and the majority of the rainfall occurred in the eastern and southeastern parts of the region.

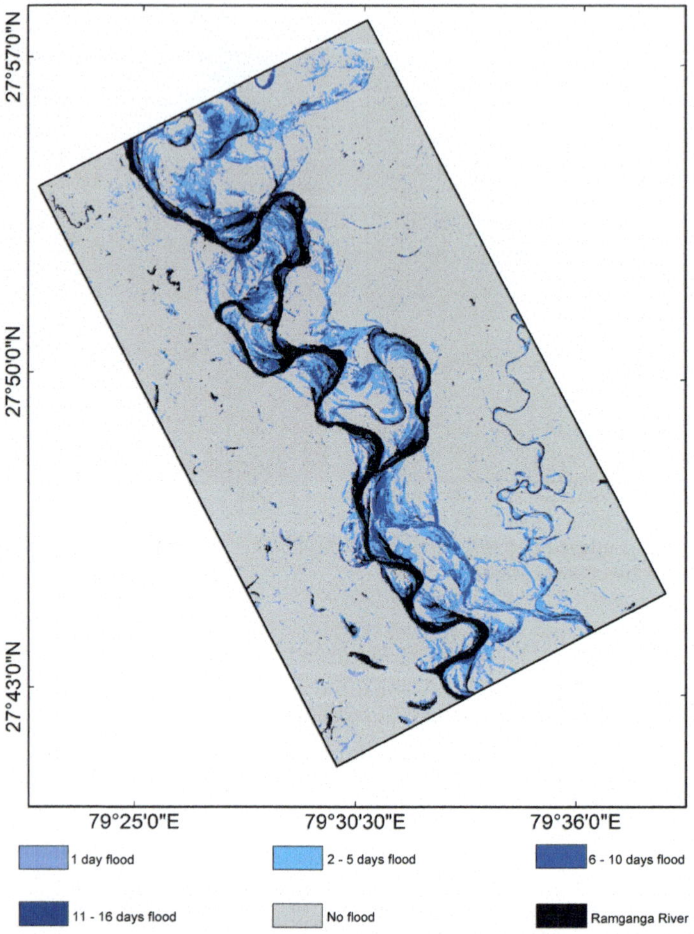

Fig. 1.6 Flood duration map of the Ramganga study area

Table 1.6 Actual flooded area excluding permanent water bodies on different days

No. of days of flood	Actual flooded area (Sq. Km.)
1 day water	19.18
2–5 days water	25.02
6 to 10 days water	16.13
11–16 days water	5.54
No flood	322.88

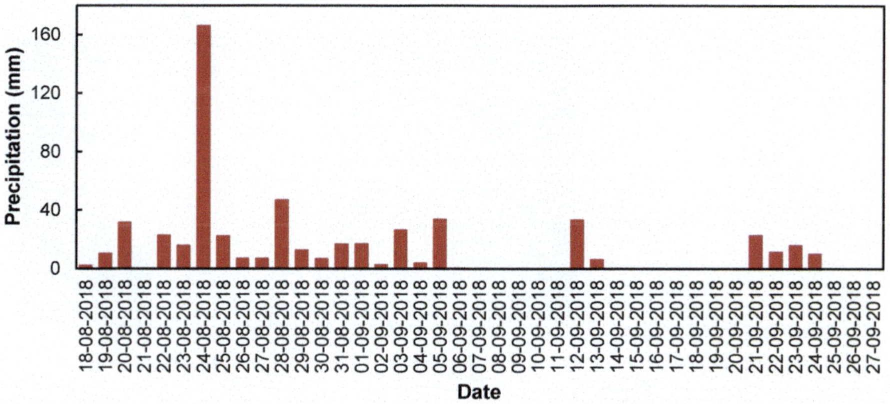

Fig. 1.7 Amount of daily precipitation in the study area region of the Ramganga river from 18 August 2018 to 27 September 2018

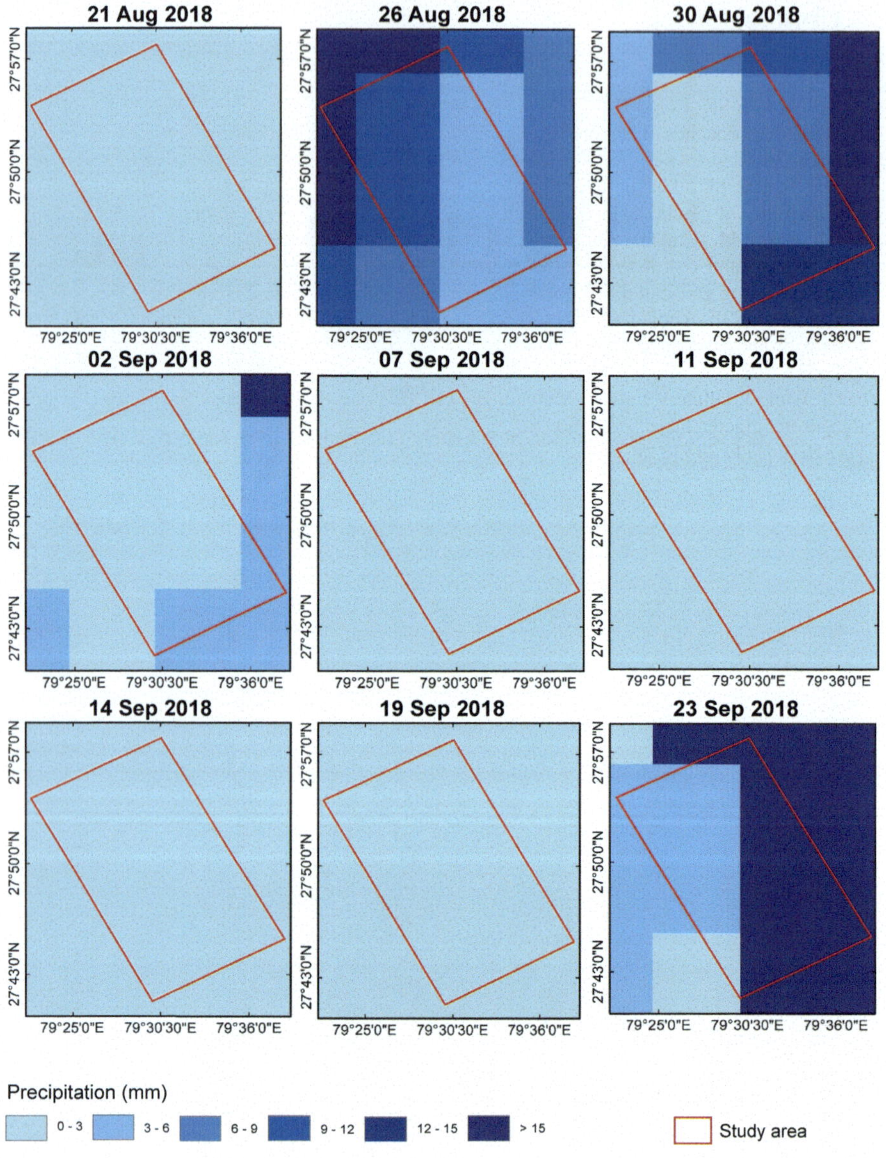

Fig. 1.8 Spatial distribution of daily precipitation over Ramganga and surrounding areas from TRMM observations (21 August –23 September 2018)

1.6 Conclusion

In this study, the flooded area and its temporal changes have been analysed using SAR (Sentinel-1) and optical imagery (Sentinel-2). The area, which is located in the middle-lower course of the Ganga-Ramganga doab region, receives excess water supply through heavy precipitation (especially in the monsoon season) and downstream flow of flooded water. This makes that place a severely flooded area (Kundu et al. 2015a; Mondal et al. 2026). The study of the flood, which occurred in August–September 2018, reveals various outcomes. Those are-

1. The lowest mean backscatter value and the highest anomaly of backscatter value are observed in two images when the flood was in its peak conditions (30 August 2018 and 02 September 2018). So, it depicts the relationship between the backscatter value and occurrence of floods.
2. The threshold value of SAR images (Sentinel-1), which varies from -16 to -18 dB, is very low to differentiate the flooded and non-flooded areas in VV polarisation.
3. The occurrence of the flood starts after 21 Aug 2018. The analysis reveals that the flood was at its peak condition from 30 August 2018 to 07 September 2018. The flood water inundated the land surfaces were increased very quickly from 7.67 Sq. Km. to 52.54 Sq. Km. on 30 August 2018. It reached its peak condition on 02 September 2018, when 54.51 Sq. Km. The areas were waterlogged. After 07 September 2018, the flood water moved at the downstream reach of the river, and the flooded area started to decrease.
4. From the analysis of the flood duration map, it is observed that the duration of flood decreased from the river channel to the outer areas as well as S-E to N-W direction.
5. The shifting of the river flow and its increasing width towards the riverbank during the flood increases the susceptibility of that place, which needs to protect.
6. TRMM daily precipitation data shows the impact of precipitation on the areal extension of flood. Two waves of the flood mainly occurred in the study area. The highest precipitation occurred on 24th Aug 2018, and because of this, the flooded area was suddenly increased to 45.59 Sq. km. The second-highest precipitation occurred on 28–08-2018, increasing the extension of the flood to 54.51 Sq. Km.

Finally, the article proves the importance of geospatial technology in flood mapping and monitoring. Due to the absence of gauge station and presence of heavy clouds, collecting the data using remote sensing technology is one of the scientific and cost-effective ways. This data can be used in hydrological modelling. During a flood, the changing nature of river width, flow path, and bank condition helps the planner to take proper measures at the time of constructing the road, bridge, rail line, and others. So, the study provides the input of the data that needs to be implemented in the policy-making to save and properly use the resources of that place.

References

Agnihotri AK, Ohri A, Mishra S (2020) Channel planform dynamics of lower Ramganga River, Ganga basin, GIS and remote sensing analyses. Geocarto Int 35(9):934–953. https://doi.org/10.1080/10106049.2018.1552323

Agnihotri AK, Ohri A, Gaur S, Shivam Das N, Mishra S (2019) Flood inundation mapping and monitoring using SAR data and its impact on Ramganga River in Ganga basin. Environ Monit Assess 191(12). https://doi.org/10.1007/s10661-019-7903-4

Anusha N, Bharathi B (2020) Flood detection and flood mapping using multi-temporal synthetic aperture radar and optical data. Egyptian J Rem Sens Space Sci 23(2):207–219. https://doi.org/10.1016/j.ejrs.2019.01.001

As Natural Disasters Rise, Countries Call for Action on Resilient Crisis Recovery Planning (2017). URL <http://www.worldbank.org/en/news/feature/2017/06/06/as-natural-disasters-rise-countries-call-for-action-on-resilient-crisis-recovery-planning> (accessed 20 July 2018)

Bajracharya B, Shrestha AB, Rajbhandari L (2007) Glacial lake outburst floods in the Sagarmatha region. Mt Res Dev 27(4):336–344

Bhatt CM, Rao GS (2016) Ganga floods of 2010 in Uttar Pradesh, north India: a perspective analysis using satellite remote sensing data. Geomat Nat Haz Risk 7(2):747–763

Borah SB, Sivasankar T, Ramya MNS, Raju PLN (2018) Flood inundation mapping and monitoring in Kaziranga National Park, Assam using Sentinel-1 SAR data. Environ Monit Assess 190(9):520

Clement MA, Kilsby CG, Moore P (2018) Multi-temporal synthetic aperture radar flood mapping using change detection. J Flood Risk Manag 11(2):152–168

Dash P, Sar J (2020) Identification and validation of potential flood hazard area using GIS-based multi-criteria analysis and satellite data-derived water index. J Flood Risk Manag 13(3):e12620

Devrani R, Srivastava P, Kumar R, Kasana P (2022) Characterization and assessment of flood inundated areas of lower Brahmaputra River Basin using multitemporal Synthetic Aperture Radar data: a case study from NE India. Geol J 57(2):622–646

Di Baldassarre G, Schumann G, Brandimarte L, Bates P (2011) Timely low resolution SAR imagery to support floodplain modelling: a case study review. Surv Geophys 32:255–269

Gupta AK, Nair SS (2011) Urban floods in Bangalore and Chennai: risk management challenges and lessons for sustainable urban ecology. Curr Sci 1638–1645

Huang C, Chen Y, Wu J (2014) Mapping spatio-temporal flood inundation dynamics at large river basin scale using time-series flow data and MODIS imagery. Int J Appl Earth Obs Geoinf 26:350–362

Jensen JR (2009) Remote sensing of the environment: an earth resource perspective 2/e. Pearson Education India

Khan A, Govil H, Khan HH, Thakur PK, Yunus AP, Pani P (2022) Channel responses to flooding of Ganga River, Bihar India, 2019 using SAR and optical remote sensing. Adv Space Res 69(4):1930–1947

Khan MYA, Daityari S, Chakrapani GJ (2016) Factors responsible for temporal and spatial variations in water and sediment discharge in Ramganga River, Ganga Basin India. Environ Earth Sci 75(4):1–18. https://doi.org/10.1007/s12665-015-5148-2

Klemas V (2015) Remote sensing of floods and flood-prone areas: an overview. J Coastal Res 31(4):1005–1013

Kugler Z, De Groeve T (2007) The global flood detection system. JRC Sci Tech Rep 1–45

Kumar H, Karwariya SK, Kumar R (2022) Google earth engine-based identification of flood extent and flood-affected paddy rice fields using Sentinel-2 MSI and Sentinel-1 SAR data in Bihar state, India. J Indian Soc Rem Sens 50(5):791–803

Kumar R (2019) Flood inundation and hazard mapping of 2017 floods in the Rapti River basin using Sentinel-1A synthetic aperture radar images. Appl Chall Geospat Technol Pot Fut Trends 77–98

Kundu S, Aggarwal SP, Kingma N, Mondal A, Khare D (2015a) Flood monitoring using microwave remote sensing in a part of Nuna river basin, Odisha, India. Nat Hazards 76:123–138

Kundu S, Khare D, Mondal A, Mishra PK (2015b) Analysis of spatial and temporal variation in rainfall trend of Madhya Pradesh, India (1901–2011). Environ Earth Sci 73(12):8197–8216
Kundu S, Khare D, Mondal A (2017a) Landuse change impact on sub-watersheds prioritization by analytical hierarchy process (AHP). Eco Inform 42:100–113
Kundu S, Khare D, Mondal A (2017b) Future changes in rainfall, temperature and reference evapotranspiration in the central India by least square support vector machine. Geosci Front 8(3):583–596
Kundu S, Lakshmi V, Torres R (2022) Flood depth estimation during hurricane harvey using sentinel-1 and UAVSAR data. Rem Sens 14(6). https://doi.org/10.3390/rs14061450
Lal P, Prakash A, Kumar A, Srivastava PK, Saikia P, Pandey AC,... Khan ML (2020) Evaluating the 2018 extreme flood hazard events in Kerala, India. Rem Sens Lett 11(5):436–445
Li W, Du Z, Ling F, Zhou D, Wang H, Gui Y, Zhang X (2013) A comparison of land surface water mapping using the normalized difference water index from TM, ETM+ and ALI. Rem Sens 5(11):5530–5549
Manjusree P, Prasanna Kumar L, Bhatt CM, Rao GS, Bhanumurthy V (2012) Optimization of threshold ranges for rapid flood inundation mapping by evaluating backscatter profiles of high incidence angle SAR images. Int J Disast Risk Sci 3(2):113–122. https://doi.org/10.1007/s13753-012-0011-5
Mondal A, Khare D, Kundu S (2016) Change in rainfall erosivity in the past and future due to climate change in the central part of India. Int Soil Water Conserv Res 4(3):186–194
Mondal A, Khare D, Kundu S (2017a) Uncertainty analysis of soil erosion modelling using different resolution of open-source DEMs. Geocarto Int 32(3):334–349
Mondal A, Khare D, Kundu S (2018a) A comparative study of soil erosion modelling by MMF USLE and RUSLE. Geocarto Int 33(1):89–103
Mondal A, Khare D, Kundu S, Mukherjee S, Mukhopadhyay A, Mondal S (2017b) Uncertainty of soil erosion modelling using open-source high resolution and aggregated DEMs. Geosci Front 8(3):425–436
Mondal A, Kundu S, Mukhopadhyay A (2012) Rainfall trend analysis by Mann-Kendall test: a case study of north-eastern part of Cuttack district, Orissa. Int J Geol Earth Environ Sci 2(1):70–78
Mondal A, Lakshmi V, Hashemi H (2018b) Intercomparison of trend analysis of multisatellite monthly precipitation products and gauge measurements for river basins of India. J Hydrol 565:779–790
Mondal A, Le MH, Lakshmi V (2022) Land use, climate, and water change in the Vietnamese Mekong Delta (VMD) using earth observation and hydrological modeling. J Hydrol Reg Stud 42:101132
Mondal A, Pandey A, Mondal S, Kundu S (2025) Impact assessment of glacial lake outburst floods (GLOFs) in the Himalaya using satellite imageries over the Last 25 Years (2000–2024): a comprehensive review. Society and Environment, Remote Sensing Applications, p 101845
Mondal A, Pandey A, Mondal S, Kundu S (2025) Impact assessment of glacial lake outburst floods (GLOFs) in the Himalaya using satellite imageries over the last 25 years (2000–2024): a comprehensive review. Remote Sens Appl: Soc Environ 101845. https://doi.org/10.1016/j.rsase.2025.101845
Mukherjee R, Deb P (2024) Application of GIS-based analytical hierarchy process for assessment and mapping of flood risk zone in the lower Ramganga River Basin, Western Gangetic Plain, India. Environ Dev Sustain 26(3):6163–6193. https://doi.org/10.1007/s10668-023-02957-z
Mukhopadhyay A, Mondal A, Mukherjee S, Khatua D, Ghosh S, Mitra D, Ghosh T (2014) Forest cover change prediction using hybrid methodology of geoinformatics and Markov chain model: a case study on sub-Himalayan town Gangtok India. J Earth Syst Sci 123(6):1349–1360
Muttarak R, Dimitrova A (2019) Climate change and seasonal floods: potential long-term nutritional consequences for children in Kerala India. BMJ Glob Health 4(2):e001215
NDMA (2008) Management of floods. URL < https://indiawris.gov.in/wiki/doku.php?id=flood_management>.

Pandey AC, Kaushik K, Parida BR (2022) Google Earth Engine for large-scale flood mapping using SAR data and impact assessment on agriculture and population of Ganga-Brahmaputra basin. Sustainability 14(7):4210

Parida BR, Tripathi G, Pandey AC, Kumar A (2022) Estimating floodwater depth using SAR-derived flood inundation maps and geomorphic model in kosi river basin (India). Geocarto Int 37(15):4336–4360

Pramanick N, Acharyya R, Mukherjee S, Mukherjee S, Pal I, Mitra D, Mukhopadhyay A (2022) SAR based flood risk analysis: a case study Kerala flood 2018. Adv Space Res 69(4):1915–1929. https://doi.org/10.1016/j.asr.2021.07.003

Rahman MR, Thakur PK (2018) Detecting, mapping and analysing of flood water propagation using synthetic aperture radar (SAR) satellite data and GIS: A case study from the Kendrapara District of Orissa State of India. Egypt J Rem Sens Space Sci 21:S37–S41

Ray K, Pandey P, Pandey C, Dimri AP, Kishore K (2019) On the recent floods in India. Curr Sci 117(2):204–218

Ryu JH, Won JS, Min KD (2002) Waterline extraction from Landsat TM data in a tidal flat: a case study in Gomso Bay Korea. Rem Sens Environ 83(3):442–456

Sahoo S, Arora M, Pateriya B (2021) Mapping of flood inundated area and impact assessment using Sar data over Ghaggar river basin of Punjab, India

Sanyal J, Lu XX (2004) Application of remote sensing in flood management with special reference to monsoon Asia: a review. Nat Hazards 33:283–301

Tiwari V, Kumar V, Matin MA, Thapa A, Ellenburg WL, Gupta N, Thapa S (2020) Flood inundation mapping-Kerala 2018; Harnessing the power of SAR, automatic threshold detection method and Google Earth Engine. PLoS ONE 15(8):e0237324

Tripathi A, Tiwari RK (2019a) Utilization of spaceborne C-band SAR data for analysis of flood impact on agriculture and its management. Int Arch Photogramm Remote Sens Spat Inf Sci 42:521–524

Tripathi A, Tiwari RK (2019b) C-band SAR Interferometry based flood inundation mapping for Gorakhpur and adjoining areas. In: 2019 international conference on computer, electrical and communication engineering (ICCECE). IEEE, pp 1–6

Tripathi A, Attri L, Tiwari RK (2021) Spaceborne C-band SAR remote sensing–based flood mapping and runoff estimation for 2019 flood scenario in Rupnagar, Punjab India. Environ Monit Assess 193(3):110

Tripathi G, Pandey AC, Parida BR, Kumar A (2020) Flood inundation mapping and impact assessment using multi-temporal optical and SAR satellite data: a case study of 2017 Flood in Darbhanga district, Bihar, India. Water Resour Manage 34:1871–1892

Voigt S, Kemper T, Riedlinger T, Kiefl R, Scholte K, Mehl H (2007) Satellite image analysis for disaster and crisis-management support. IEEE Trans Geosci Remote Sens 45(6):1520–1528

Chapter 2
Trend of Glacial Lake Outburst Floods (GLOFs) in the Central Himalayan Region Over the Last 25 years: A Review-Based Approach

Arun Mondal, Anupam Pandey, Souvik Mondal, Partha Paul, and Sananda Kundu

Abstract The frequency and intensity of different hazards are increasing due to climate change. Two main climatic parameters, named temperature and precipitation, have a deep impact on maintaining the hydro-ecological balance of the Earth. Continuous melting of glaciers in the polar and high mountain areas forms numerous glacial lakes and increases their size. This process, combined with high seasonal precipitation, makes the mountain area one of the GLOF-prone zones. This study focused on a systematic analysis of GLOF-related articles, published from 2000 to 2024, in the Central Himalayan region. A total of 68 articles have been selected for analysing the flood locations, intensities, frequencies, and damages of the study area. The record of GLOF indicates that Eastern Central or Nepal Himalaya experienced the highest number of GLOF events specially Sagarmatha region. The analysis also shows that maximum frequencies of GLOF events occurred in the season of Monsoon (June to September) and Summer (April to June). The study also denotes the importance of remote sensing techniques for demarcating GLOF-affected areas, identifying the pre and post GLOF situations, and marking potential hazardous lakes. The study will provide a clear conceptual framework and development of thinking of researchers about the GLOF-related studies in the Central Himalaya from 2000 to 2024 by analysing the recent trends and methods of previous researchers, and will help to find out the research gap for further analysis.

Keywords Hazard · Climate change · Damages · Glacial lake outburst flood (GLOF) · Himalaya

---

A. Mondal · A. Pandey · S. Mondal · P. Paul
Department of Geography, University of Allahabad, Prayagraj, India

S. Kundu (✉)
Department of Geography, Manipur University, Imphal, India
e-mail: skundu@manipuruniv.ac.in

M. Biswas et al. (eds.), *Himalayan Flood Hazards*, Advances in Geographical and Environmental Sciences, https://doi.org/10.1007/978-981-92-0702-2_2

2.1 Introduction

Climate change is a worldwide phenomenon, making a huge influence on Energy-matter circulation (Chen et al. 2024; Byers et al. 2017). All natural processes are affected by it, such as cycle of erosion, biological cycle, water cycle, and others. The change of two important parameters, temperature and rainfall, creates a huge change in the environment (Dubey et al. 2024; Sattar et al. 2020). The frequency of natural disasters are also increasing. The ice of polar and high mountain areas decreases, and the area of glacier lakes increases with the increase in average Global temperature. The retreating rate of the Mount Everest region (10–59 ma^{-1}) denotes the evidence of increasing temperature on glacial retreats (Bajracharya and Mool 2009). This is a common phenomenon all over the Central Himalaya. This unusual areal growth of glacial lakes often create flood like hazard situation and destroys the local infrastructures (roads, houses, built-up lands, etc.), vegetation cover, and arable lands. This flood is called 'Glacial Lake outburst Flood' (GLOF) (Satter et al. 2022; Hu et al. 2022a, b; Ruiz-Villanueva et al. 2017; Das and Murugesh Prabhu 2012). Because of these catastrophic events, millions of dollars of property and several numbers of human casualties are reported every year (Gouli et al. 2023; Yang et al. 2023; Zheng et al. 2021).

Although increasing mean Global temperature is an important factor of occurring the GLOF, other intrinsic (expansion of glacial lake, high volume of water, unstable lateral moraine, and others) and extrinsic factors (heavy rainfall, earthquake, avalanches, landslide, glacier calving, etc.) also play an important role in GLOF (Mondal et al. 2025; Jiang et al. 2024; Sattar et al. 2021a, b; Ruiz-Villanueva et al. 2017; Kumar et al. 2021; Rawat et al. 2022). The destruction of GLOF depends upon the amount of water released from the glacier lake. The presence of silt, gravel, and bloke of rocks makes the GLOF more destructive, and the amount of damage becomes high (Akbar et al. 2024; Chowdhury et al. 2022).

Third pole region is the 2nd glaciated area in the World after the Polar region (Mondal et al. 2023; Wangchuk et al. 2022). Central Himalaya is one of the important glaciated areas in the Himalayan Mountains, and the glaciers of the Sagarmatha region are located in this region. From the historical period, this place has witnessed several GLOF events (Veh 2019; Higaki and Sato 2012). Unique geographical location and undulating hog-back topography make this place one of the GLOF-prone zones in the World. The devastating power of Nare (3rd September 1977) and Dig Tsho (4th August 1985) GLOFs were so high that several mini and Namche hydropower plants were destroyed, several roads and buildings were damaged, and the deaths of more than hundreds of people were recorded (Osti and Egashira 2009; Chen et al. 2013). The impact attracted the attention of researchers to analysis GLOF factors and possible GLOF events in future. Such types of GLOF events also been recorded in the years 1997, 1998, 2011, 2016, etc.

GLOF is considered a mesoscale natural hazard by climatologists and hydrologists (Islam and Patel 2022; Fischer et al. 2021; Zhang et al. 2019; Prakash et al. 2013; Dey 2002). In the Central Himalayan region, researchers mainly focused on the mapping of glacial lakes, identification of hazardous lakes, case studies on GLOF events, and others (Singhal et al. 2022; Satter et al. 2020; Veh et al. 2019; Haritashya et al. 2018; Hoy et al. 2016). There is a lack of systematic records of techniques and methods used by researchers in their research articles (Das 2024; Rahman 2023; Ashraf et al. 2021; Ahmed et al. 2021; Chen et al. 2014; Mondal et al. 2012). There are also some review articles, but proper documentation of GLOF details, importance of satellite imageries, publisher and journal-wise publication of articles, and contribution of authors and organisations is not provided in detail.

The study covers all published articles between 2000 and 2024 that are related to GLOF events in the Central Himalaya. The article is used to analyze the collaboration and association within researchers and research organisations. In this study, used data, methods, and trends of previously published articles are analysed. Assessment of infrastructural and economic damages of GLOFs helps to understand the devastating power of GLOFs and alarms the government and local people to take a proper mitigation plan in the Central Himalayan region.

2.2 Study Area

The Himalayan Mountain range is 2500 km-long uneven and undulating orogenic landform, resulting from the collision between the Eurasian and Indian plates. The formation of the Himalayan Mountain started at about 55 million years ago in the Eocene epoch (Mohanty et al. 2023a, b; Wang et al. 2015; Liu et al. 2014). The movement of the Indian plateau is about 5 cm yr^{-1} in a northern direction. The range of the Himalayan Mountain spans over six countries, named India, China, Nepal, Bhutan, Pakistan and Afghanistan (Shijin and Shitai 2015; Khullar 2006). Several scholars divide the Himalayan Mountain region in different categories. Sing (1971) divide the whole Himalayan region into three mesoregions on the basis of topography (Khullar 2006). Those are – Western Himalaya, Central Himalaya and Eastern Himalaya. According to Sing (1971) Central Himalaya consists of Uttarakhand Himalaya and Nepal Himalaya (Khullar 2006). Its length and width are approx. 800 and 400 km, respectively (Nie et al. 2013). The average temperature in the month July varies between 5°–30 °C and it dips to 0°–4 °C in the month January (Khullar 2006). The precipitation mainly occurs in month of Monsoon. Winter is the dry season but in the high-altitude precipitation occurs in the form of snowfall. Among the different types of soils, Podzol soil is the main type of soils (Goswami and Goyal 2021; Bazai et al. 2021; Sharma et al. 2018). The rivers (Kosi, Gandaki and Karnali), originated from Central Himalaya are perennial in nature. The geographical location of the Central Himalaya is important, as it acts as a barrier to flow the cold wind to the southern slope in the winter season (Fischer et al. 2021; Mohanty and Maiti 2021; Harrison et al. 2018) (Fig. 2.1).

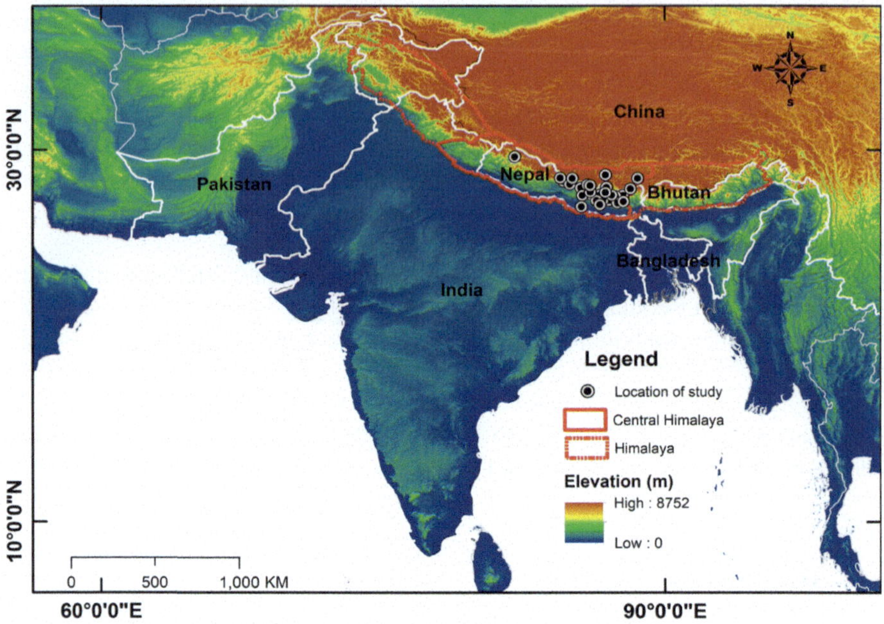

Fig. 2.1 Location map of the study area. The red dotted line indicates the outline of the Himalayan Mountain range, and solid red line indicates the Central Himalayan region. Black dots on the diagram indicate the places where GLOF-related studies have been done from 2000 to 2024. GTOPO30 is used to prepare a DEM of the study area

2.3 Data Collection and Methods

'Web of Science' database, an important source of data, played an important role in collecting the GLOF-related articles in the Central Himalayan region from 2000 to 2024. The database was used to analyze the detailed quantitative and qualitative properties of GLOF-related published articles. The total number of selected articles was 68 in numbers. The hotspot zone, where the highest number of GLOFs occurs, was identified in the Central Himalaya. Then, year-wise published articles, published articles in different journals and publications indicate the growth, trend, and growing importance of GLOF related researches in the Central Himalayan region from 2000 to 2024. The bibliometric analysis, used to show the coordination and collaboration among researchers and research organisations, played an important role in the GLOF related analysis. For the analysis of the methods of previously published articles, it is categorised into two main sub-sections (Primary data source and Secondary data source). The use of remote sensing data in the GLOF relates studies were arranges properly. The information of structural and economic damages, caused by various GLOFs, were collected from those 68 articles. The flow-chart of the study is given in Fig. 2.2.

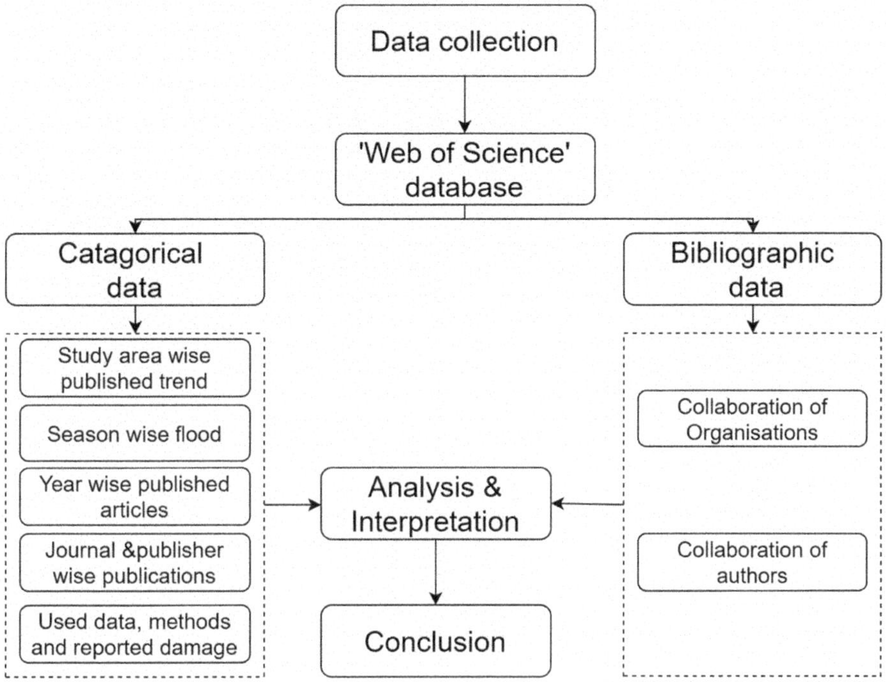

Fig. 2.2 Flow-chart of the whole study

2.4 Results

The result of the analysis is discussed in different sections and sub-sections. Those are discussed.

2.4.1 Flood Occurrence Place

Table 2.1 presents the list of GLOF events that occurred in the Central Himalayan region. The data show that GLOF events are mainly concentrated in the Eastern Nepal Himalayan region. Dudh Koshi river basin and Arun River basin, specially Mount Everest region, are the main hotspot areas where maximum number of GLOFs occur. On the other hand, the occurrence of GLOF events in western Nepal is not high in number.

Table 2.1 List of some reported GLOF events in the Central Himalayan region

SI. No.	Date	Lake name	Location	Damage	References
1	03.09.1977	Nare	Dudh Koshi, Nepal	Damaged mini-hydroplant, roads, bridges, cultivation fields, etc	Osti and Egashira (2009), Cenderelli and Wohl (2002)
2	04.08.1985	Dig Tsho	Dudh Kosi River, Nepal	Destroyed Namche hydropower plant completely, damaged 14 bridges, and several buildings	Chen et al. (2013)
3	27.08.1985	Jinco	Arun River, Nepal	Damaged eight villages, livestock, farm land, roads, bridges, etc	Chen et al. (2013)
4	7.1997	Chubung	Dudh Kosi River, Nepal	Damaged houses and cultivation fields at Beding village	Chen et al. (2013)
5	03.09.1997	Nare Drangka	Dudh Kosi River, Nepal		Chen et al. (2013)
6	03.09.1998	Sabai Tsho	Nepal	Caused damage worth 2 million dollars and the loss of lives	Khadka et al. (2021)
7	Aug- Sep 2011	Ghunsa2	Nepal	Damage to properties in the downstream area	Byers et al. (2020)
8	2015, 2016	Unknown	Mahalangur region, Nepal	No record	Khadka et al. (2022)
9	05.07.2016	Gongbatongsha Lake	Nepal	Destroyed the Bhotekoshi hydropower plant and the surrounding area	Dubey et al. (2024)
10	17.04.2017	Langmale Lake	Barun Valley, Nepal	Damaged bridges, buildings, and farmland	Khadka et al. (2021)
11	16.08.2024	Upper and lower Tso Mutse	Mahalangur region, Nepal		Khatri (2025)

2.4.2 Flood Occurrence in Different Seasons

Table 2.2 shows the season-wise occurrence of GLOF events in the Central Himalayan region. The data show that climatic factors are one of the important factors of GLOF events. Most of the GLOF events occurred in the month of the Monsoon (72.73%) and Summer (9.09%). In the months of Summer, glaciers started to melt and increase its size, and in the months of Monsoon, high precipitation acts

Table 2.2 Season-wise occurrence of some important GLOFs in the Central Himalayan Mountain

Seasons	Frequency	Percentage
Monsoon	8	72.73
Post-monsoon	1	9.09
Summer	1	9.09
Not mention	1	9.09

as a triggering factor of occurring the GLOF events. Thus, the frequency of GLOF events is higher in the Monsoon season than in the summer season.

2.4.3 Analysis of Published Articles

A detailed analysis of previously published articles from 2000 to 2024 is discussed.

2.4.3.1 Year-Wise Published Articles

Figure 2.3 indicates the year-wise number of publications in the Central Himalayan region from 2000 to 2024. A few numbers of articles were published before 2012, except in 2007 and 2009, when 2 and 3 articles were published, respectively. After 2012, the number of publications increased, and highest number of publications was found in the year 2019, when 8 number of articles were published.

2.4.3.2 Published Articles in Journals

Figure 2.4 shows the journal-wise publication of GLOF-related articles in the Central Himalayan region from 2000 to 2024. In the Central Himalaya, total 68 number of GLOF related articles were published between 2000 and 2024 in 37 different journals. Remote Sensing is the top journal where total 5 GLOF related articles were published in last 25 years (2000–2025). Other important journals are 'Hydrology and Earth System Sciences' (4 articles), 'Natural Hazards' (4 articles), 'Natural Hazards and Risk' (3 articles), 'Annals of Glaciology' (3 articles) etc.

2.4.3.3 Publisher-Wise Number of Articles

Figure 2.5 indicates the top 10 names of the publisher which provide the highest number of articles about GLOF-related studies in the Central Himalaya from 2000 to 2024. The names of the top 5 publishers are 'Taylor and Francis' (11 articles), 'Springer' (9 articles), 'Elsevier' (8 articles), 'Copernicus' (8 articles), and 'MDPI' (5 articles).

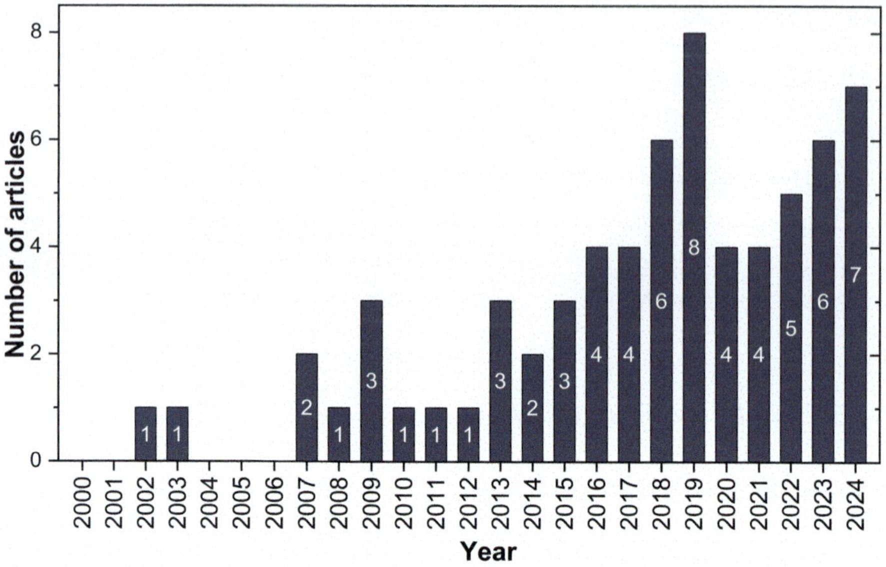

Fig. 2.3 Year-wise publication of GLOF-related articles in the Central Himalayan region between 2000 and 2024

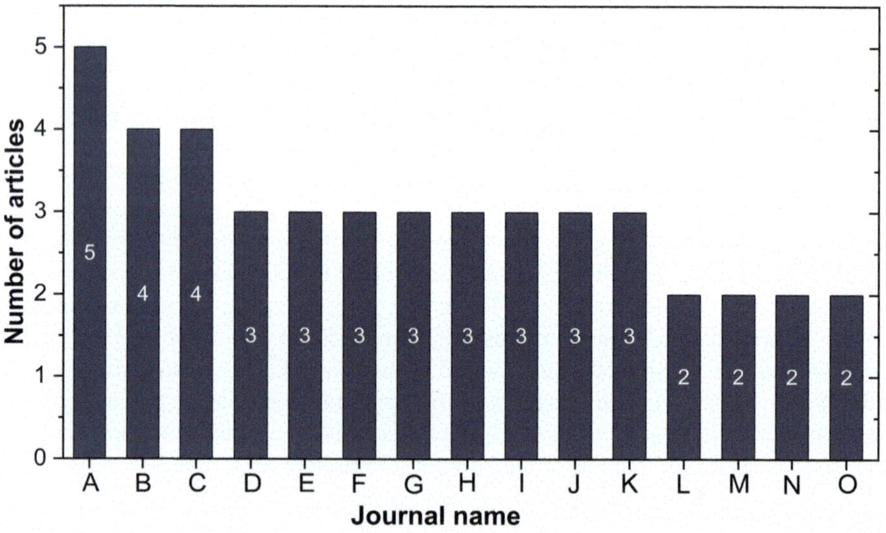

Fig. 2.4 Journal-wise publication of articles on GLOF-related studies in the Central Himalayan region between 2000 and 2024. In the horizontal scale, the name of the journal is written where A = Remote Sensing, B = Hydrology and Earth System Sciences, C = Natural Hazards, D = Hydrological Processes: An International Journal, E = Geomatics, Natural Hazards and Risk, F = Annals of Glaciology, G = Geomorphology, H = Frontiers in Earth Science, I = Science of The Total Environment, J = Natural Hazards and Earth System Sciences, K = Geocarto International, L = Mountain Research and Development, M = Earth Surface Processes and Landforms, N = Natural Hazards and Risk, O = The Cryosphere

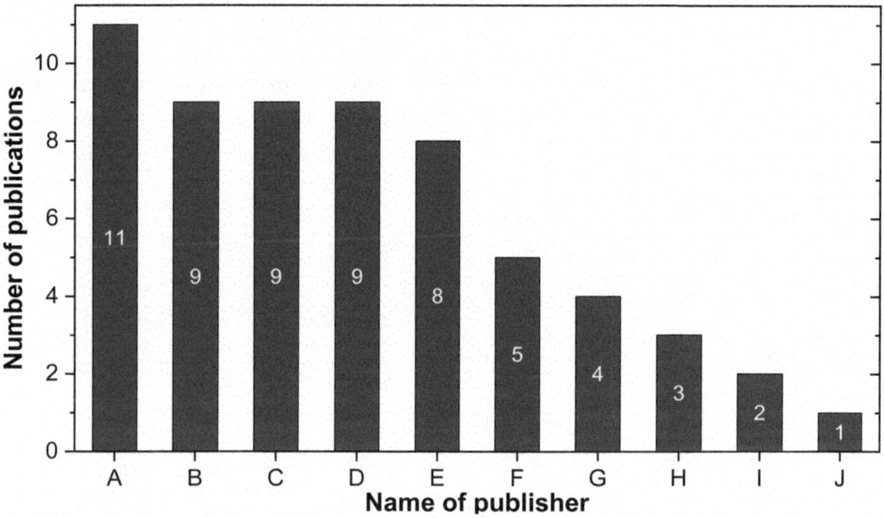

Fig. 2.5 The number of related papers published in different publications from 2000 to 2024 on the Central Himalayan GLOFs. The horizontal scale indicates the name of publishers where A = Taylor and Francis, B = Springer, C = Elsevier, D = Copernicus, E = Copernicus, F = MDPI, G = John Wiley and Sons, H = Cambridge University Press, I = Frontiers Media S.A, J = International Mountain Society

2.4.3.4 Contribution of Research Organisations

The contribution of research organisations indicates the collaboration of researchers from different research institutes in the field of research. The total link strength of any research institute or organisation is important to understand its influence in research (Fig. 2.6). Total 172 organisations were focused on GLOF-related studies in the Central Himalayan region from 2000 to 2024. Some of those are 'University of Chinese Academy Science' (4 articles, 15 total link strength), 'Planetary Science Institute' (2 articles, 11 total link strength), 'Institute of environmental science and Geography' (2 articles, 9 total link strength), 'Geophysical Institute' (2 articles, 6 total link strength) and 'Centre for Research in Water Resource' (2 articles, 5 total link strength).

2.4.3.5 Contribution of Authors

Figure 2.7 shows the collaboration and contribution of research scholars about GLOF-related studies in the Central Himalayan region between 2000 and 2024. Total 191 authors shared their knowledge by publishing numerous articles regarding GLOFs in the Central Himalaya. Total 38 authors (19.89%) were published two or more than two articles while 15 authors (7.85%) were published three or more than three articles, and only seven authors (3.66%) were contributed four or more than

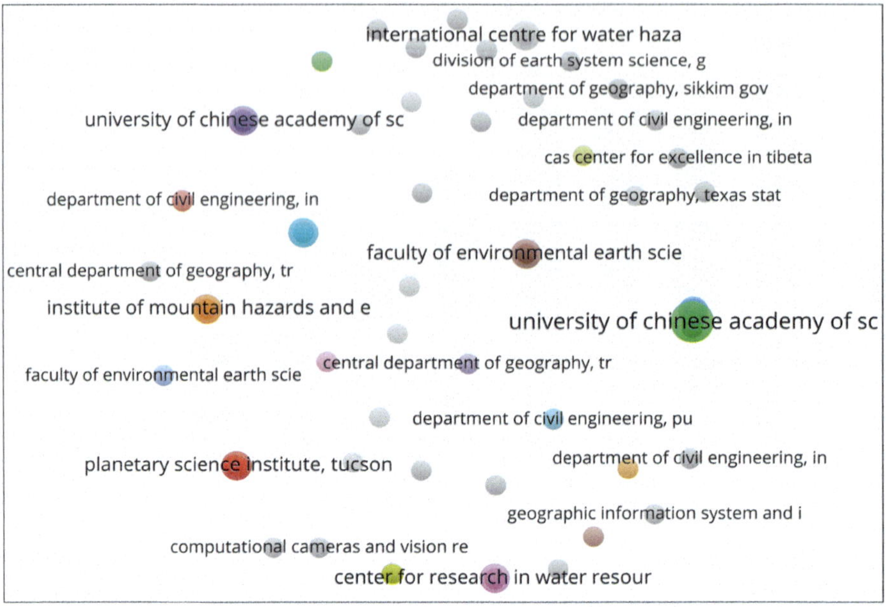

Fig. 2.6 The cooperation network analysis among various research organisations on the Central Himalayan GLOF events from 2000 to 2024

four articles. The name of top 5 authors with higher number of published articles were 'Byers, Alton C.' (6 articles, 25 total link strength), 'Watanabe' (6 articles, 19 total link strength), 'Zhang, Guoqing' (5 articles, 22 total link strength), 'Satter, Ashim' (4 articles, 18 total link strength) and 'Nie, Yong' (4 articles, 17 total link strength).

2.4.4 Data Used

For the GLOF-related study in the Central Himalayan region (2000–2024), researchers collected data from various primary and secondary sources. Those are discussed below.

2.4.4.1 Primary Data Source

Primary data source indicates the data that is collected from the field survey. As the terrain of the Himalayan region is very rugged, the scope of the primary survey to collect the data is limited (Thompson et al. 2020; Sherpa et al. 2019; Dey 2002). But the collected information of the previous GLOF-related published articles from 2000 to 2024 indicates that 21 articles out of 68 articles used primary data to analyze

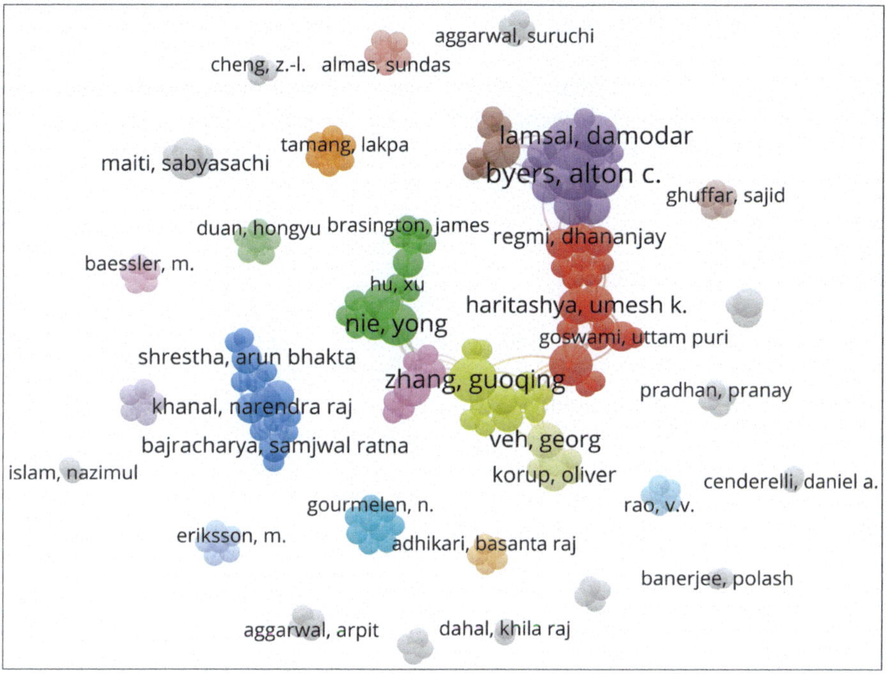

Fig. 2.7 The cooperation network analysis among researchers of various research organisations on the Central Himalayan GLOF events from 2000 to 2024

the GLOF-related study (Wang et al. 2024; Nie et al. 2020; Thompson et al. 2020; Dahal and Hagelman III 2011). Interviews with local people, field measurements, and questionnaires were the best techniques to collect the primary data.

Primary sourced data provide ground-level and accurate information related to GLOF-related studies in the Himalayan region. Expansion of any glacial lakes and their vulnerability to GLOF can be assessed by the field survey (Fischer et al. 2021; Washakh et al. 2019; Kropáček et al. 2015). Many researchers focused on the seasonal growth of glacial lakes, bathometric analysis, and people's perception regarding GLOFs (Byers et al. 2019; Lala et al. 2017; Khanal et al. 2015; Liu et al. 2014; Cenderelli and Wohl 2003). Seasonal growth and volume of water were assessed for different glacial lakes like 'Imja Tsho' (Nie et al. 2020), 'Guangxieco Lake' (Liu et al. 2014), and 'Chamlang South Glacier' (Lamsal et al. 2016a, b). Bathometric analysis was conducted by Lala et al. (2018) to simulate the GLOF event of 'Imja Tsho' lake in Nepal. Primary survey was also used to understand the perception of GLOF susceptibility in the downstream area of GLOF (Thompson et al. 2020; Dahal and Hagelman III 2011).

2.4.4.2 Secondary Data Sources

Secondary data is the information that is collected from another published journals, government reports, and websites. In the GLOF-related studies of the Central Himalaya, 64 out of 68 articles used secondary data sources between 2000 and 2024. Researchers used different types of data. Those are-

Ground-Based Observed Data

Researchers collected topographical, meteorological, and river gauge information from different secondary data sources. Researchers used topographical data of Survey of India to analyze the topography of the study area (Goswami and Goyal 2021; Fischer et al. 2021; Aggarwal et al. 2017; Bolch et al. 2008). Meteorological data (precipitation and temperature) were collected from different agencies like the Indian Meteorological Department (IMD) and the China Meteorological Department (Goswami and Goyal 2021; Fischer et al. 2021; Veh et al. 2019; Nie et al. 2017).

Space-Based Observed Data

Remote sensing technique proves itself as an important tool to analyze the GLOF-related research in the Himalayan region (Mondal et al. 2026a; Rinzin et al. 2021; Cook et al. 2018; Watanabe et al. 2009; Osti et al. 2013). Satellite imageries of different agencies were used by the researchers for analyzing the GLOF events in detail and accurately (Mohanty et al. 2023a, b; Khadka et al. 2021; Byers et al. 2020; Nie et al. 2017; Lamsal et al. 2016a, b; Westoby et al. 2014; Shrestha et al. 2010). In the Central Himalayan region, 59 out of 68 published articles used satellite imagery from 2000 to 2024. Aerial photography also plays an important role to analysis the GLOF events.

Figure 2.8 indicates the use of different satellite imagery in GLOF-related studies in the central Himalayan region between 2000 and 2024. The data show that researchers preferred Landsat satellite imagery over others to analyze the GLOF events in the Central Himalayan region (66 out of 119 times, 55.46%). For the analysis of GLOF events in the Central Himalaya, the aerial photography is also used, but at low frequencies. The analysis shows that the use of satellite imageries was started in the year 2007. The imagery of Corona and Hexagon satellites was relevant in past and also useful in present-day GLOF studies. The use of imageries of Sentinel satellite started in 2019 and its sensor (active sensors) makes it so effective that it can capture imagery of before and after GLOF events in any weather conditions.

Optical remote sensing (Banerjee and Bhuiyan 2023; Sharma et al. 2018; Mir et al. 2018; Sattar et al. 2019; Aggarwal et al. 2017; Nie et al. 2013; Aggarwal et al. 2016; Komori et al. 2012; Osti and Egashira 2009; Bajracharya et al. 2007b, a) and microwave remote sensing (Rinzin et al. 2021; Lamsal et al. 2016a, b; Byers et al. 2013; Bolch et al. 2008) are very important technique to analysis the GLOF events

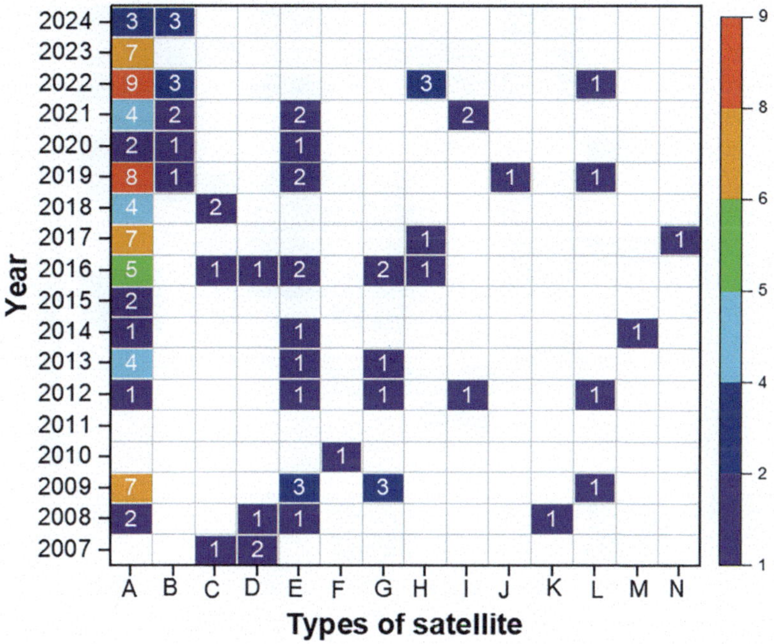

Fig. 2.8 Types of satellite data used by the researchers on the GLOF-related studies in the Himalaya from 2000 to 2024. The horizontal scale indicates the name of satellites where A = Landsat, B = Sentinel-2, C = IRS, D = IKNOS, E = Corona, F = Quick Bird, G = ALOS, H = Resourcesat, I = SPOT, J = Planet Scope, K = Terra, L = KH-9 Hexagon, M = GeoEye satellite, N = Cartosat

because microwave remote sensing is not depended on weather condition and take imagery of any place in any time. The detailed analysis of previous GLOF-related articles (2000–2024) indicates that researchers used 18 active remote sensing satellite imagery (Sentinel, ALOS, and Terra satellite) and the remaining 101 satellite images are passive remote sensing (Landsat, Corona, Quick Bird, etc.).

As the Landsat satellite imagery was used frequently for the analysis of GLOF events, Fig. 2.9 shows the contribution of different sensors of Landsat to the analysis of GLOF events in the Central Himalaya from 2000 to 2024. The data indicates that two sensors (TM and ETM+), each used 16 times, played a crucial role in providing the imagery of before and after GLOF events, seasonal variation of glacial lakes, and others. The sensors, OLI and TIRS, started in use from the year 2016 to 2019, respectively, and gave a great contribution in the field of research.

Figure 2.10 is used to indicate different types of Digital Elevation Models (DEMs) for the GLOF-related analysis in the Central Himalaya between 2000 and 2024. SRTM DEM, ASTER DEM, and ALOS Palsar DEM were some notable DEMs, which are used frequently in the GLOF-related study. SRTM DEM was used 23 times, followed by ASTER DEM and ALOS Palsar DEM, each used 8 times. The use of SRTM DEM in GLOF-related studies in the Central Himalaya started from the year 2010 and continuously increases its frequency (except the years 2011, 2012,

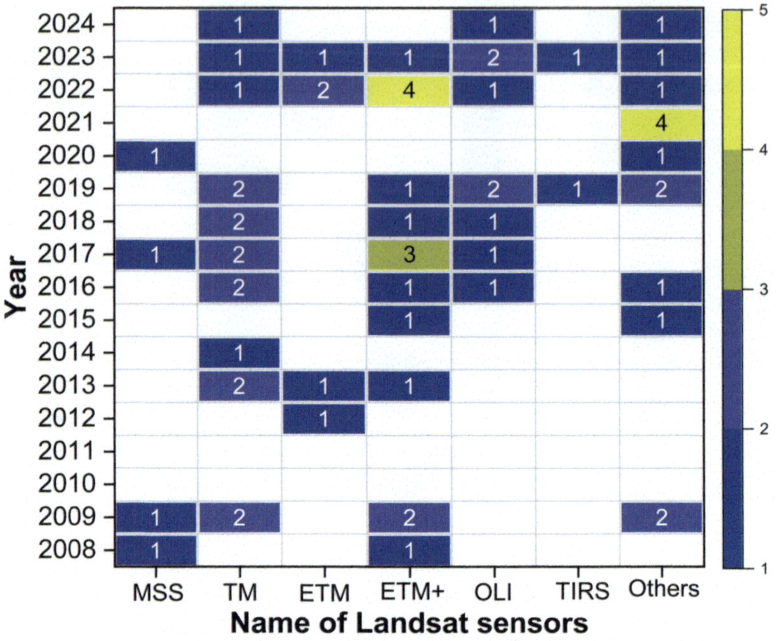

Fig. 2.9 Different types of Landsat sensors used on GLOF-related studies in the Central Himalaya from 2000 to 2024

2014, and 2016). ASTER DEM and ALOS Palsar DEM were introduced in the year 2008 but were used frequently from the year 2013 to 2015, respectively.

2.4.5 Method Used

Figure 2.11 illustrates different methods that were used by the previous researchers in their articles from 2000 to 2024. The result shows the importance and influence of the remote-sensing technique in GLOF-related studies. Total 59 articles out of 68 articles had been used remote-sensing technique. Only six articles out of all had been used field survey method in their research articles.

Different methods were used to analyze the GLOF events in the Himalayan region. In the GLOF-related studies, one of the most important tasks is to find out the location of glacial lakes and its expansion. NDWI (Normalized Difference Water Index) is a popular method, used by several scholars, to differentiate waterbodies from its surroundings (Kayastha and Maskey et al., 2024; Hu et al., 2022a, b; Sattar et al. 2019; Chand and Watanabe, 2019; Somos-Valenzuela et al. 2015; Shrestha et al. 2010). For the identification of hazardous lakes, areal extent of lakes is important. Normalized Difference Snow/Ice Index (NDSII) was used to identify the extent of the glacial lake (Gouli et al. 2023; Sattar et al. 2022). In the present decade, HEC-RAS

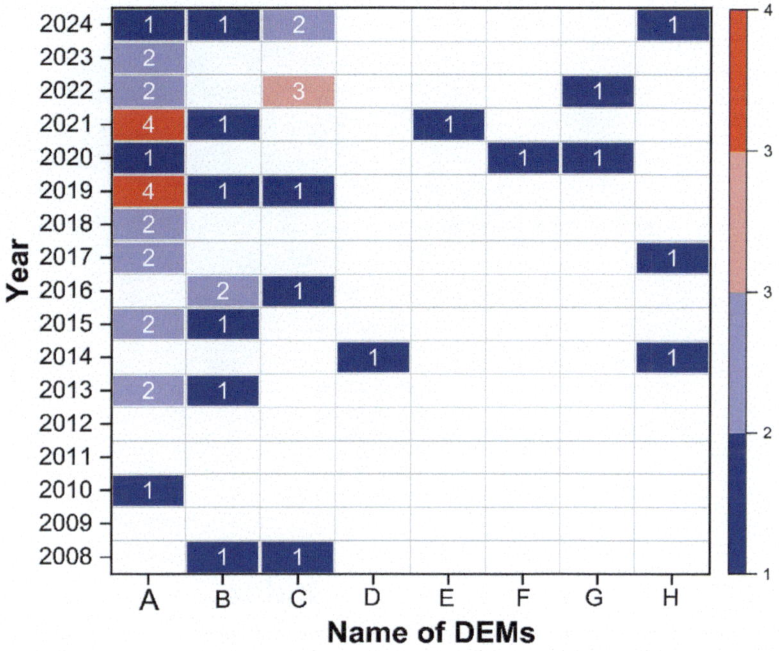

Fig. 2.10 Different types of DEMs were used in GLOF-related studies in the Himalayan region between 2000 and 2024. The horizontal scale indicates the name of satellites where A = SRTM DEM, B = ASTER DEM, C = ALOS Palsar DEM, D = 'Structure from Motion' (SfM) DEM, E = CartoDEM, F = Unmanned Aerial Vehicle DEM (UAV DEM), G = High Mountain Asia DEM (HMA DEM), H = Others

1D and 2D unsteady flow models were used for simulating the discharge of water at the time of GLOF (Kayastha and Maskey 2024; Sattar et al. 2022). The Markov model was also used to predict the future changes of the GLOF (Hu et al. 2022a, b).

2.4.6 Climate Change and Occurrence of GLOFs in the Central Himalaya

Climate change is a Global phenomenon, and warming the temperature of the Earth is its Universal impact (Kayastha and Maskey 2024; Pandey et al. 2022; Kundu et al. 2017a, b; Gurung et al. 2010). Although rising temperatures cause changes in the ecological, biological, and hydrological balance of the Earth, Cryosphere is affected more due to its high susceptibility property to heat (Chowdhury et al. 2022; Hagg et al. 2021; Scapozza et al. 2019; Mondal et al. 2018a, b). The glaciers of mountain areas and third pole region (the Himalayan Mountain) are melting quickly than the Glaciers of Polar regions because of the thin ice layer in the High Mountain areas. The average temperature of the Earth increases from 0.74 to 0.8 °C in last 100 years

Fig. 2.11 Different methods were used by the researchers to analyze the Himalayan GLOF-related study from 2000 to 2024

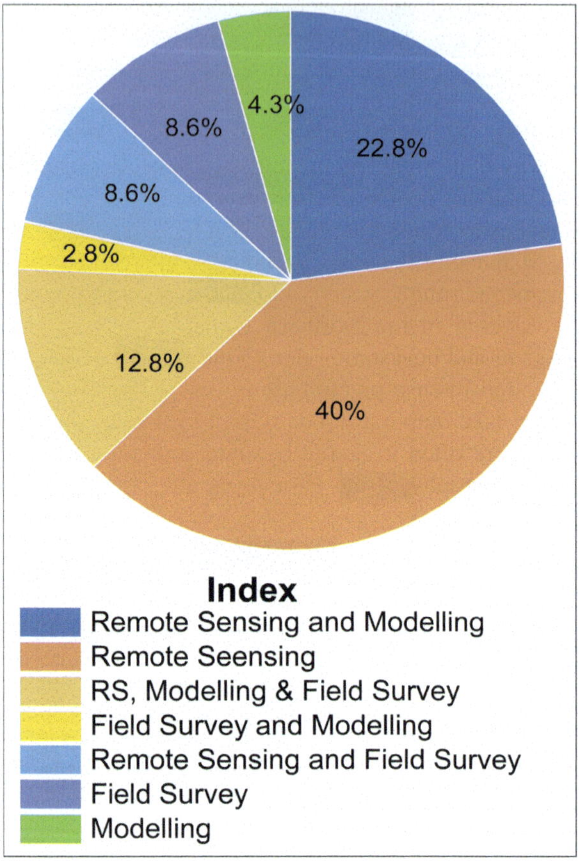

(1901–2000) (Vishwas et al. 2024; Sattar et al. 2021b; Mukhtar et al. 2021). On the other hand, temperature of Central Himalaya (Nepal Himalya) is increased 2.58 °C in 43 years (from 1971 to 2014) and the value is greater than the increasing temperature rate of N-W Himalaya, that is 1.6 °C in last century (Gouli et al. 2023; Dorji 2021; Mukhopadhyay et al. 2014). This highly growing temperature increases the frequency and size of glacial lakes. In the Nepal Himalaya, the glacial lake surface has been increased at the rate of 25% from the year 1987 to 2017 (Vishwas et al. 2024; Gouli et al. 2023; Mondal et al. 2022). This increasing size of those lakes is responsible of occurring various natural hazards, one of which is GLOF. The anomaly of temperature and precipitation makes the Central Himalayan region as one of the GLOF prone zones in the World. The glaciers of High Mountain area retreats at the rate of 0.04% per year but the retreating rate of Central Himalayan region exceeds the retreating rate of High Mountain area (Yu et al. 2024; Hu et al. 2022a, b; Goswami and Goyal 2021; Sattar et al. 2020b; Schmidt et al. 2020; Mondal et al. 2017a, b). Mahalangur or mount Everest region, the highest glaciated area in per square kilo-meter in Nepal, is the prime location of GLOFs and the glaciers of that place is decreasing at the rate

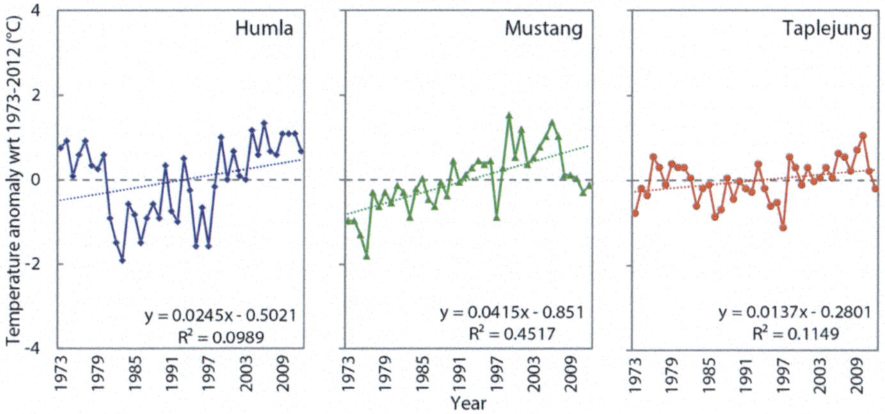

Fig. 2.12 Temperature anomalies of Humla (Western Nepal), Mustang (Central Nepal), and Taplejung (Eastern Nepal) districts in Nepal from 1973 to 2009. *Source* Uprety et al. (2017)

of (−0.27)– (±0.06) % from the year 1962 to 2011 (Rinzin et al. 2021; Schmidt et al. 2020; Kundu et al. 2015). The continuous growth of glacial lakes caused to occur several GLOF events in the Central or Nepal Himalaya. The records shows that approx 35 to 40 GLOF events have been recorded in the Himalayan region in last 40 years (upto the year 2017) (Ahmed et al. 2022; Zhang et al. 2019; Mondal et al. 2016; Shrestha et al. 2010). Nepal, a highly undulation and unstable region, experienced more than 25 GLOF events upto the year 2010 (Saha et al. 2024; Rinzin et al. 2021; Mal et al. 2021; Shrestha et al. 2010). Due to the high industrialisation and global warming, the frequency of occurrences of GLOFs is increased. Sagarmatha or Mount Everest region is the hotspot zone of GLOF in The Nepal Himalaya. The region faces total seven GLOF events (1977, 1985, 1998, 2015, 2016, 2017 and 2024), of which four occurred after the year 2000 (Yang et al. 2023; Khadka et al. 2021; Zhang et al. 2019; Rounce et al. 2016). Not only frequencies of GLOFs, but also downstream impacts of GLOF have been increased due to the climate change and increasing economic involvement of human (Fig. 2.12).

2.4.7 Damage

Damage is an inevitable effect of any destructive event. In the Himalayan or mountain area, GLOF can damage some part of the mountain area, but the amount of destruction depends on the place where it happens (Mondal et al. 2026b; Dubey et al. 2024; Mohanty et al. 2023a, b; Satter et al. 2022; Veh et al. 2019; Bajracharya 2009; Bajracharya and Mool 2009; Bajracharya et al. 2007b, a). If the human involvement and their economic activities increase, the impact or damage caused by GLOFs increases (Hu et al. 2022a, b). As the Himalayan region is very rugged surface, the details of GLOF events are poorly recorded (Qayyum et al. 2020; Khadka et al. 2018; Haritashya et al. 2018; Shrestha et al. 2017). The Damage of the GLOF can be classified into two major sub-sections. Those are-

2.4.7.1 Infrastructure Damages

Infrastructure damage indicates the destruction of built-up areas or man-made structures like houses, roads, bridges, hydropower facilities, and others. On 3rd September 1997 and 4th August 1985, two GLOF events, named 'Nare GLOF' and 'Dig Tsho GLOF', respectively, destroyed many hydropower plants (Namche Hydropower plant and others), more than 14 bridges, and several buildings in the Dudh Koshi basin, Nepal (Osti and Egashira 2009; Cenderelli and Wohl 2002; Chen et al. 2013). The devastating nature of GLOFs attracted the attention of researchers for GLOF-related analysis. The GLOF event in the Arun River basin (27th August 1985) damaged eight villages and several infrastructures (Chen et al. 2013). The GLOF events of the Bhote Kosi river valley in 1998 damaged the hill-slope and hampered the normal life of the downstream people. From the twentieth century, the devastating power of GLOFs has increased. On 5th July 2016, a devastating GLOF event occurred and destroyed Bhotekoshi hydropower plant and its surrounding areas (Dubey et al. 2024). Another devastating GLOF occurred on 16th August 2024 in the Sagarmatha region, which destroyed several downstream infrastructure, including houses, schools, and other valuable man-made features (Khatri 2025).

2.4.7.2 Economic Damages

Economic damage means the financial losses due to the GLOF events. Every year, Nepal suffers millions of dollars of economic losses due to the destruction of GLOF events. On 4th August 1985, a major GLOF event, named Dig Tsho GLOF, damaged the property of about 2.6 million USD (Chen et al. 2013). Another GLOF event devastated the property of downstream area of about 2 million USD, which occurred on 3rd September 1998 in the Nepal Himalaya (Khadka et al. 2021). During the GLOF events, physical structures and agricultural lands were damaged, and caused the losses of millions of dollars every year (Dubey et al. 2024; Byers et al. 2020).

The deaths of civilians were another impact of GLOF. The Dig Tsho glacier lake outburst flood (4th August 1985) recorded heavy losses of human life (Chen et al. 2013). A similar type of results also recorded at the time of the Sabai Tsho GLOF on 3rd September 1998 (Khadka et al. 2021). The recorded data of GLOF indicates that GLOF damaged the vegetation, arable land, and infrastructures in the downstream area.

2.5 Conclusion

The contribution of researchers in GLOF-related studies in the Central Himalayan region has been examined from 2000 to 2024. The unique geophysical properties and frequently occurrence of GLOFs in the Nepal Himalaya attract the attention of researchers. The data shows that increasing temperature and anomaly of precipitation increase the size and number of glacial lakes, which triggers the GLOF phenomenon in the Nepal Himalaya. GLOF occurring frequencies in the Nepal Himalaya indicates that Sagarmatha or Mount Everest region is the GLOF hotspot zone due to high retreating of glaciers and thinning of ice layer. Melting of glaciers and size of glacial lakes increase in Monsoon and summer seasons, and the record of GLOF events provide similar types of result. In the western side of the Central Himalaya, the observed GLOF events are very few in number. In the GLOF related studies, the number of published articles were increased after the year 2013. Taylor and Francis, Springer and Elsevier are some important publishers of publishing the articles in various journals. The analysis of GLOF related articles in the Central Himalaya from 2000 to 2024, indicates the application and importance of remote sensing technique of GLOF analysis. Remote sensing technique provides the imagery of lake expansion, before and after GLOF events, affected areas and others. The occurrence of GLOF events in the Central Himalaya increases in number after the year 2000, Sagarmatha region is its inevitable example. Not only frequency, but also the infrastructure and economic damages caused by GLOFs increase. Increasing economic activity and human involvement in the Central Himalayan region increase the amount of damage, destroying several hydropower plant, road, buildings and other build-up facilities. In the view of above observations, it is concluded that more attention is needed to analysis the factors of GLOF events and implement early warning system in the GLOF prone zone to save the life of downstream residents.

References

Aggarwal A, Jain SK, Lohani AK, Jain N (2016) Glacial lake outburst flood risk assessment using combined approaches of remote sensing, GIS and dam break modelling. Geomat Nat Haz Risk 7(1):18–36

Aggarwal S, Rai SC, Thakur PK, Emmer A (2017) Inventory and recently increasing GLOF susceptibility of glacial lakes in Sikkim, Eastern Himalaya. Geomorphology 295:39–54

Ahmed R, Rawat M, Wani GF, Ahmad ST, Ahmed P, Jain SK, Farooq M (2022) Glacial lake outburst flood hazard and risk assessment of Gangabal Lake in the Upper Jhelum Basin of Kashmir Himalaya using geospatial technology and hydrodynamic modeling. Rem Sens 14(23):5957

Ahmed R, Wani GF, Ahmad ST, Sahana M, Singh H, Ahmed P (2021) A review of glacial lake expansion and associated glacial lake outburst floods in the Himalayan region. Earth Syst Environ 5(3):695–708

Akbar S, Wang J, Ullah A, Latif Y, Muhammad S (2024) The growth and emergence of potentially dangerous glacier lakes in Astore Basin, Western Himalaya during 1993–2021. Geomat Nat Haz Risk 15(1):2353838

Ashraf A, Iqbal MB, Mustafa N, Naz R, Ahmad B (2021) Prevalent risk of glacial lake outburst flood hazard in the Hindu Kush–Karakoram–Himalaya region of Pakistan. Environ Earth Sci 80(12):451

Bajracharya B, Shrestha AB, Rajbhandari L (2007a) Glacial lake outburst floods in the Sagarmatha region. Mt Res Dev 27(4):336–344

Bajracharya SR (2009) Glacial lake outburst floods risk reduction activities in Nepal. In: Asia-Pacific Symposium on new technologies for prediction and mitigation of sediment disasters, Japan Society of Erosion Control Engineering (JSECE), Tokyo, Japan, pp 18–19

Bajracharya SR, Mool P (2009) Glaciers, glacial lakes and glacial lake outburst floods in the Mount Everest region Nepal. Annal Glaciol 50(53):81–86

Bajracharya SR, Mool PK, Shrestha BR (2007b) Impact of climate change on Himalayan glaciers and glacial lakes: case studies on GLOF and associated hazards in Nepal and Bhutan, vol 12. International Centre for Integrated Mountain Development, Kathmandu

Banerjee P, Bhuiyan C (2023) Glacial lakes of Sikkim Himalaya: their dynamics, trends, and likely fate—a timeseries analysis through cloud-based geocomputing, and machine learning. Geomat Nat Haz Risk 14(1):2286903

Bazai NA, Cui P, Carling PA, Wang H, Hassan J, Liu D,... Jin W (2021) Increasing glacial lake outburst flood hazard in response to surge glaciers in the Karakoram. Earth-Sci Rev 212:103432

Bolch T, Buchroithner MF, Peters J, Baessler M, Bajracharya S (2008) Identification of glacier motion and potentially dangerous glacial lakes in the Mt. Everest region/Nepal using spaceborne imagery. Nat Hazards Earth Syst Sci 8(6):1329–1340

Byers AC III, Byers EA, McKinney DC, Rounce DR (2017) A field-based study of impacts of the 2015 earthquake on potentially dangerous glacial lakes in Nepal Himalaya. J Assoc Nepal Himal Stud 37(2):7

Byers AC, Chand MB, Lala J, Shrestha M, Byers EA, Watanabe T (2020) Reconstructing the history of glacial lake outburst floods (GLOF) in the Kanchenjunga conservation area East Nepal: an interdisciplinary approach. Sustainability 12(13):5407

Byers AC, McKinney DC, Somos-Valenzuela M, Watanabe T, Lamsal D (2013) Glacial lakes of the hinku and hongu valleys, makalu Barun national park and buffer zone Nepal. Nat Hazards 69:115–139

Byers AC, Rounce DR, Shugar DH, Lala JM, Byers EA, Regmi D (2019) A rockfall-induced glacial lake outburst flood, Upper Barun Valley Nepal. Landslides 16:533–549

Cenderelli DA, Wohl EE (2003) Flow hydraulics and geomorphic effects of glacial-lake outburst floods in the Mount Everest region, Nepal. Earth Surf Process Landform J Br Geomorphol Res Group 28(4):385–407

Chand MB, Watanabe T (2019) Development of supraglacial ponds in the Everest Region, Nepal, between 1989 and 2018. Rem Sens 11(9):1058

Chen F, Zhang M, Zhao H, Guan W, Yang A (2024) Pakistan's 2022 floods: spatial distribution, causes and future trends from Sentinel-1 SAR observations. Rem Sens Environ 304:114055

Chen N, Hu GS, Deng W, Khanal N, Zhu YH, Han D (2013) On the water hazards in the trans-boundary Kosi River basin. Nat Hazard 13(3):795–808. https://doi.org/10.5194/nhess-13-795-2013

Chen W, Doko T, Liu C, Ichinose T, Fukui H, Feng Q, Gou P (2014) Changes in Rongbuk lake and Imja lake in the Everest region of Himalaya. Int Arch Photogramm Remote Sens Spat Inf Sci 40:259–266

Chowdhury A, De SK, Sharma MC, Debnath M (2022) Potential glacial lake outburst flood assessment in a changing environment, Chhombo Chhu Watershed, Sikkim Himalaya India. Geocarto Int 37(27):15627–15655

Cook KL, Andermann C, Gimbert F, Adhikari BR, Hovius N (2018) Glacial lake outburst floods as drivers of fluvial erosion in the Himalaya. Science 362(6410):53–57

Dahal KR, Hagelman R III (2011) People's risk perception of glacial lake outburst flooding: a case of Tsho Rolpa Lake Nepal. Environ Haz 10(2):154–170

Das BK, Murugesh Prabhu TS (2012) Impacts of climate change: Glacial lake outburst floods (GLOFs). Climate Change in Sikkim
Das S (2024) The effects of climatic alteration on the glacial lake flash flood in the Himadri range of Nepal and adjacent Indian region. Idealistic J Adv Res Progr Spectr (IJARPS) eISSN–2583–6986 3(10):218–224
Dey D (2002) Watershed conservation and management of glacial lake outburst flood; combating climate change in Himalayan environment. South Asian Forum for Environment, Calcutta, India
Dorji K (2021) Adaptive human settlement planning for glacier lake outburst floods in Bhutan. Soc Sci Asia 7(4):9–16
Dubey S, Sattar A, Gupta V, Goyal MK, Haritashya UK, Kargel JS (2024) Transboundary hazard and downstream impact of glacial lakes in Hindu-Kush Karakoram Himalayas. Sci Total Environ 914:169758
Fischer M, Korup O, Veh G, Walz A (2021) Controls of outbursts of moraine-dammed lakes in the greater Himalayan region. Cryosphere 15(8):4145–4163
Goswami UP, Goyal MK (2021) Assessment of glacial lake development and downstream flood impacts of critical glacial lake. Nat Haz 109(1):1027–1046
Gouli MR, Hu K, Khadka N, Talchabhadel R (2023) Hazard assessment of a pair of glacial lakes in Nepal Himalaya: unfolding combined outbursts of Upper and Lower Barun. Geomat Nat Haz Risk 14(1):2266219
Gurung DR, Bajracharya S, Shrestha BR, Pradhan P (2010) Wi-Fi network at Imja Tsho (lake), Nepal: an early warning system (EWS) for glacial lake outburst flood (GLOF). Grazer Schr Geogr Raumforsch 45:321–326
Hagg W, Ram S, Klaus A, Aschauer S, Babernits S, Brand D, Pappas T (2021) Hazard assessment for a glacier lake outburst flood in the Mo Chu River Basin Bhutan. Appl Sci 11(20):9463
Haritashya UK, Kargel JS, Shugar DH, Leonard GJ, Strattman K, Watson CS,... Regmi D (2018) Evolution and controls of large glacial lakes in the Nepal Himalaya. Rem Sens 10(5):798
Harrison S, Kargel JS, Huggel C, Reynolds J, Shugar DH, Betts RA, Vilímek V (2018) Climate change and the global pattern of moraine-dammed glacial lake outburst floods. Cryosphere 12(4):1195–1209
Higaki D, Sato G (2012) Erosion and sedimentation caused by glacial lake outburst floods in the Nepal and Bhutan Himalayas. Glob Environ Res 16(1):71–76
Hoy A, Katel O, Thapa P, Dendup N, Matschullat J (2016) Climatic changes and their impact on socio-economic sectors in the Bhutan Himalayas: an implementation strategy. Reg Environ Change 16:1401–1415
Hu J, Yao X, Duan H, Zhang Y, Wang Y, Wu T (2022a) Temporal and spatial changes and GLOF susceptibility assessment of glacial lakes in Nepal from 2000 to 2020. Rem Sens 14(19):5034
Hu K, Zhang X, Gouli MR, Liu S, Nie Y (2022b) Retrospective analysis and hazard assessment of Gega glacial lake in the eastern Himalayan syntaxis. Nat Haz Res 2(4):331–342
Islam N, Patel PP (2022) Inventory and GLOF hazard assessment of glacial lakes in the Sikkim Himalayas India. Geocarto Int 37(13):3840–3876
Jiang L, Lin Z, Zhou Z, Luo H, Zheng J, Su D, Song M (2024) Monitoring and disaster assessment of glacier lake outburst in high mountains Asian using multi-satellites and HEC-RAS: a case of Kyagar in 2018. Rem Sens 16(23):4447
Kayastha RB, Maskey S (2024) Glacial lake outburst flood (GLOF) modeling of Tsho Rolpa glacial lake Nepal. Proceed IAHS 387:59–63
Khadka N, Chen X, Nie Y, Thakuri S, Zheng G, Zhang G (2021) Evaluation of glacial lake outburst flood susceptibility using multi-criteria assessment framework in Mahalangur Himalaya. Front Earth Sci 8:601288
Khadka N, Zhang G, Thakuri S (2018) Glacial lakes in the Nepal Himalaya: Inventory and decadal dynamics (1977–2017). Rem Sens 10(12):1913
Khanal NR, Mool PK, Shrestha AB, Rasul G, Ghimire PK, Shrestha RB, Joshi SP (2015) A comprehensive approach and methods for glacial lake outburst flood risk assessment, with examples from Nepal and the transboundary area. Int J Water Resour Dev 31(2):219–237

Khatri B (2025) Impacts of Glacial lake outburst floods on Nepal's national security. Unity J 6(1):63–78
Khullar DR (2006) India: a comprehensive geography. Kalyani Publishers
Komori J, Koike T, Yamanokuchi T, Tshering P (2012) Glacial lake outburst events in the Bhutan Himalayas. Glob Environ Res 16(1):59–70
Kropáček J, Neckel N, Tyrna B, Holzer N, Hovden A, Gourmelen N,... Hochschild V (2015) Repeated glacial lake outburst flood threatening the oldest Buddhist monastery in north-western Nepal. Nat Hazards Earth Syst Sci 15(10):2425–2437
Kumar V, Mehta M, Shukla T (2021) Spatially resolved estimates of glacial retreat and lake changes from Gepang Gath Glacier, Chandra Basin, Western Himalaya, India. J Geol Soc India 97(5):520–526
Kundu S, Khare D, Mondal A (2017a) Landuse change impact on sub-watersheds prioritization by analytical hierarchy process (AHP). Eco Inform 42:100–113
Kundu S, Khare D, Mondal A (2017b) Future changes in rainfall, temperature and reference evapotranspiration in the central India by least square support vector machine. Geosci Front 8(3):583–596
Kundu S, Khare D, Mondal A, Mishra PK (2015) Analysis of spatial and temporal variation in rainfall trend of Madhya Pradesh, India (1901–2011). Environ Earth Sci 73(12):8197–8216
Lala JM, Rounce DR, McKinney DC (2017) Modeling the glacial lake outburst flood process chain in the Nepal Himalaya: reassessing Imja Tsho's hazard
Lamsal D, Sawagaki T, Watanabe T, Byers AC (2016a) Assessment of glacial lake development and prospects of outburst susceptibility: Chamlang South Glacier, eastern Nepal Himalaya. Geomat Nat Haz Risk 7(1):403–423
Lamsal D, Sawagaki T, Watanabe T, Byers AC, McKinney DC (2016b) An assessment of conditions before and after the 1998 Tam Pokhari outburst in the Nepal Himalaya and an evaluation of the future outburst hazard. Hydrol Process 30(5):676–691
Liu JJ, Cheng ZL, Li Y (2014) The 1988 glacial lake outburst flood in Guangxieco Lake, Tibet, China. Nat Hazard 14(11):3065–3075
Mal S, Allen SK, Frey H, Huggel C, Dimri AP (2021) Sectorwise assessment of glacial lake outburst flood danger in the Indian Himalayan region. Mt Res Dev 41(1):R1
Mir RA, Jain SK, Lohani AK, Saraf AK (2018) Glacier recession and glacial lake outburst flood studies in Zanskar basin, western Himalaya. J Hydrol 564:376–396
Mohanty L, Maiti S (2021) Probability of glacial lake outburst flooding in the Himalaya. Resour Environ Sustain 5:100031
Mohanty L, Maiti S, Dixit A (2023a) Spatio-temporal assessment of regional scale evolution and distribution of glacial lakes in Himalaya. Front Earth Sci 10:1038777
Mohanty L, Pateswary V, Maiti S, Mohanty DD (2023b) Future prediction of existing glacial lake's size in the Himalaya by Markov model and glacial surface topography. All Earth 35(1):329–343
Mondal A, Khare D, Kundu S (2016) Change in rainfall erosivity in the past and future due to climate change in the central part of India. Int Soil Water Conserv Res 4(3):186–194
Mondal A, Khare D, Kundu S (2017a) Uncertainty analysis of soil erosion modelling using different resolution of open-source DEMs. Geocarto Int 32(3):334–349
Mondal A, Khare D, Kundu S (2018a) A comparative study of soil erosion modelling by MMF USLE and RUSLE. Geocarto Int 33(1):89–103
Mondal A, Khare D, Kundu S, Mukherjee S, Mukhopadhyay A, Mondal S (2017b) Uncertainty of soil erosion modelling using open-source high resolution and aggregated DEMs. Geosci Front 8(3):425–436
Mondal A, Kundu S, Mukhopadhyay A (2012) Rainfall trend analysis by Mann-Kendall test: a case study of north-eastern part of Cuttack district, Orissa. Int J Geol Earth Environ Sci 2(1):70–78
Mondal A, Lakshmi V, Hashemi H (2018b) Intercomparison of trend analysis of multisatellite monthly precipitation products and gauge measurements for river basins of India. J Hydrol 565:779–790

Mondal A, Le MH, Lakshmi V (2022) Land use, climate, and water change in the Vietnamese Mekong Delta (VMD) using earth observation and hydrological modeling. J Hydrol Reg Stud 42:101132

Mondal A, Pandey A, Mondal S, Kundu S (2025) Impact assessment of glacial lake outburst floods (GLOFs) in the Himalaya using satellite imageries over the last 25 years (2000–2024): a comprehensive review. Society and environment, Remote sensing applications, p 101845

Mondal SK, Patel VD, Bharti R, Singh RP (2023) Causes and effects of Shisper glacial lake outburst flood event in Karakoram in 2022. Geomat Nat Haz Risk 14(1):2264460

Mondal A, Pandey A, Mondal S, Kundu S, Tran TN, Lakshmi V (2026a) Assessing the accuracy of different open-source Digital Elevation Models (DEMs) in morphometric analysis of Kanchi River basin. J Hydrol Reg Stud 65:103429. https://doi.org/10.1016/j.ejrh.2026.103429

Mondal A, Pandey A, Mondal S, Kundu S (2026b) Impact assessment of Glacial Lake Outburst Floods (GLOFs) in the Himalaya using satellite imageries over the last 25 years (2000–2024): a comprehensive review. Remote Sens Appl Soc Environ 41:101845. https://doi.org/10.1016/j.rsase.2025.101845

Mukhopadhyay A, Mondal A, Mukherjee S, Khatua D, Ghosh S, Mitra D, Ghosh T (2014) Forest cover change prediction using hybrid methodology of geoinformatics and Markov chain model: a case study on sub-Himalayan town Gangtok India. J Earth Syst Sci 123(6):1349–1360

Mukhtar MA, Kumar R, Majeed Z (2021) Glacial lake outburst flood (GLOF) hazard assessment of gepang gath pro-glacial lake located in chandra basin, western Himalaya India. Indian J Geosci 75(2):126–136

Nie Y, Liu Q, Liu S (2013) Glacial lake expansion in the Central Himalayas by Landsat images, 1990–2010. PLoS ONE 8(12):e83973

Nie Y, Liu W, Liu Q, Hu X, Westoby MJ (2020) Reconstructing the Chongbaxia Tsho glacial lake outburst flood in the Eastern Himalaya: evolution, process and impacts. Geomorphology 370:107393

Nie Y, Sheng Y, Liu Q, Liu L, Liu S, Zhang Y, Song C (2017) A regional-scale assessment of Himalayan glacial lake changes using satellite observations from 1990 to 2015. Rem Sens Environ 189:1–13

Osti R, Egashira S (2009) Hydrodynamic characteristics of the Tam Pokhari Glacial Lake outburst flood in the Mt. Everest region, Nepal. Hydrol Process Int J 23(20):2943–2955

Osti R, Egashira S, Adikari Y (2013) Prediction and assessment of multiple glacial lake outburst floods scenario in Pho Chu River basin Bhutan. Hydrol Processes 27(2):262–274

Pandey P, Banerjee D, Ali SN, Khan MAR, Chauhan P, Singh S (2022) Simulation and risk assessment of a possible glacial lake outburst flood (GLOF) in the Bhilangna Valley, central Himalaya India. J Earth Syst Sci 131(3):184

Prakash C, Banshtu RS, Sambyal S, Sharma I (2013) Identification and hazard assessment of potentially dangerous glacial lakes in Himachal Pradesh. In: Climate change and himalayan ecosystem-indicator, bio and water resources, pp 12

Qayyum N, Ghuffar S, Ahmad HM, Yousaf A, Shahid I (2020) Glacial lakes mapping using multi satellite PlanetScope imagery and deep learning. ISPRS Int J Geo Inf 9(10):560

Rahman AU (2023) Spatio-temporal appraisal of glacial lakes, susceptibility analysis and potentials of glacial lake outburst floods in the peri-glacial environment of Western Himalayas. Intercont Geoinform Days 6:256–259

Rawat KS, Sahu SR, Singh SK, Mishra AK (2022) Cloudburst analysis in the Nainital district, Himalayan Region, 2021. Discov Water 2(1):12

Rinzin S, Zhang G, Wangchuk S (2021) Glacial lake area change and potential outburst flood hazard assessment in the Bhutan Himalaya. Front Earth Sci 9:775195

Rounce DR, McKinney DC, Lala JM, Byers AC, Watson CS (2016) A new remote hazard and risk assessment framework for glacial lakes in the Nepal Himalaya. Hydrol Earth Syst Sci 20(9):3455–3475

Ruiz-Villanueva V, Allen S, Arora M, Goel NK, Stoffel M (2017) Recent catastrophic landslide lake outburst floods in the Himalayan mountain range. Prog Phys Geogr 41(1):3–28

Saha S, Bera B, Bhattacharjee S, Ghosh D, Tamang L, Shit PK, Sengupta N (2024) Identification of the multiple causes of recent series of landslides and related damage by extreme rainfall and GLOF in Sikkim Himalaya, India, during October 2023. Landslides 21(12):2993–3009

Sattar A, Goswami A, Kulkarni AV (2019) Hydrodynamic moraine-breach modeling and outburst flood routing-a hazard assessment of the South Lhonak lake, Sikkim. Sci Total Environ 668:362–378

Sattar A, Goswami A, Kulkarni AV, Emmer A (2020) Lake evolution, hydrodynamic outburst flood modeling and sensitivity analysis in the Central Himalaya: a case study. Water 12(1):237

Sattar A, Goswami A, Kulkarni AV, Emmer A, Haritashya UK, Allen S, Huggel C (2021a) Future glacial lake outburst flood (GLOF) hazard of the South Lhonak Lake Sikkim Himalaya. Geomorphology 388:107783

Sattar A, Haritashya UK, Kargel JS, Leonard GJ, Shugar DH, Chase DV (2021b) Modeling lake outburst and downstream hazard assessment of the Lower Barun Glacial Lake, Nepal Himalaya. J Hydrol 598:126208

Sattar A, Haritashya UK, Kargel JS, Karki A (2022) Transition of a small Himalayan glacier lake outburst flood to a giant transborder flood and debris flow. Sci Rep 12:12421

Scapozza C, Ambrosi C, Cannata M, Strozzi T (2019) Glacial lake outburst flood hazard assessment by satellite Earth observation in the Himalayas (Chomolhari area, Bhutan). Geographica Helvetica 74(1):125–139

Schmidt S, Nüsser M, Baghel R, Dame J (2020) Cryosphere hazards in Ladakh: the 2014 Gya glacial lake outburst flood and its implications for risk assessment. Nat Hazards 104:2071–2095

Sharma RK, Pradhan P, Sharma NP, Shrestha DG (2018) Remote sensing and in situ-based assessment of rapidly growing South Lhonak glacial lake in eastern Himalaya, India. Nat Hazards 93:393–409

Sherpa SF, Shrestha M, Eakin H, Boone CG (2019) Cryospheric hazards and risk perceptions in the Sagarmatha (Mt. Everest) national park and buffer zone Nepal. Nat Hazards 96:607–626

Shijin W, Shitai J (2015) Evolution and outburst risk analysis of moraine-dammed lakes in the central Chinese Himalaya. J Earth Syst Sci 124:567–576

Shrestha AB, Eriksson M, Mool P, Ghimire P, Mishra B, Khanal NR (2010) Glacial lake outburst flood risk assessment of Sun Koshi basin Nepal. Geomat Nat Haz Risk 1(2):157–169

Shrestha F, Gao X, Khanal NR, Maharjan SB, Shrestha RB, Wu LZ,... Bajracharya SR (2017) Decadal glacial lake changes in the Koshi basin, central Himalaya, from 1977 to 2010, derived from Landsat satellite images. J Mount Sci 14(10):1969–1984

Singhal M, Chauhan V, Singh R, Kumar V, Joshi P, Pachauri RK (2022) Glacial lake outburst floods (GLOFs) and the requisite of wireless sensor based early warning system. In: AIP conference proceedings, vol 2481, no 1. AIP Publishing

Somos-Valenzuela MA, McKinney DC, Byers AC, Rounce DR, Portocarrero C, Lamsal D (2015) Assessing downstream flood impacts due to a potential GLOF from Imja Tsho in Nepal. Hydrol Earth Syst Sci 19(3):1401–1412

Thompson I, Shrestha M, Chhetri N, Agusdinata DB (2020) An institutional analysis of glacial floods and disaster risk management in the Nepal Himalaya. Int J Disast Risk Reduct 47:101567

Uprety Y, Shrestha UB, Rokaya MB, Shrestha S, Chaudhary RP, Thakali A,... Asselin H (2017) Perceptions of climate change by highland communities in the Nepal Himalaya. Clim Develop 9(7):649–661

Veh G (2019) Outburst floods from moraine-dammed lakes in the Himalayas: detection, frequency, and hazard (Doctoral dissertation, Dissertation, Potsdam, Universität Potsdam, 2019)

Veh G, Korup O, von Specht S, Roessner S, Walz A (2019) Unchanged frequency of moraine-dammed glacial lake outburst floods in the Himalaya. Nat Clim Chang 9(5):379–383

Vishwas P, Tiwari KC, Rongali G, Vohra R (2024) Mapping of glacial lakes and glacial lake outburst flood in Kinnaur District, Himachal Pradesh using remote sensing and GIS. In: Advanced geospatial practices in natural environment resource management. IGI Global Scientific Publishing, pp 109–128

Wang W, Xiang Y, Gao Y, Lu A, Yao T (2015) Rapid expansion of glacial lakes caused by climate and glacier retreat in the Central Himalayas. Hydrol Process 29(6):859–874
Wang X, Zhang G, Veh G, Sattar A, Wang W, Allen SK, Xu F (2024) Reconstructing glacial lake outburst floods in the Poiqu River basin, central Himalaya. Geomorphology 449:109063
Wangchuk S, Bolch T, Robson BA (2022) Monitoring glacial lake outburst flood susceptibility using Sentinel-1 SAR data, Google Earth Engine, and persistent scatterer interferometry. Rem Sens Environ 271:112910
Washakh RMA, Chen N, Wang T, Almas S, Ahmad SR, Rahman M (2019) GLOF risk assessment model in the Himalayas: a case study of a hydropower project in the Upper Arun River. Water 11(9):1839
Watanabe T, Lamsal D, Ives JD (2009) Evaluating the growth characteristics of a glacial lake and its degree of danger of outburst flooding: Imja Glacier, Khumbu Himal Nepal. Norsk Geografisk Tidsskrift-Norwegian J Geograph 63(4):255–267
Westoby MJ, Glasser NF, Hambrey MJ, Brasington J, Reynolds JM, Hassan MA (2014) Reconstructing historic Glacial Lake Outburst Floods through numerical modelling and geomorphological assessment: Extreme events in the Himalaya. Earth Surf Proc Land 39(12):1675–1692
Yang L, Lu Z, Ouyang C, Zhao C, Hu X, Zhang Q (2023) Glacial lake outburst flood monitoring and modeling through integrating multiple remote sensing methods and hec-ras. Rem Sens 15(22):5327
Yu Y, Li B, Li Y, Jiang W (2024) Retrospective analysis of glacial lake outburst flood (glof) using ai earth insar and optical images: a case study of south lhonak lake, sikkim. Rem Sens 16(13):2307
Zhang G, Bolch T, Allen S, Linsbauer A, Chen W, Wang W (2019) Glacial lake evolution and glacier–lake interactions in the Poiqu River basin, central Himalaya, 1964–2017. J Glaciol 65(251):347–365
Zheng G, Allen SK, Bao A, Ballesteros-Cánovas JA, Huss M, Zhang G, Stoffel M (2021) Increasing risk of glacial lake outburst floods from future third pole deglaciation. Nat Clim Chang 11(5):411–417

Chapter 3
An Overview of the Hydrological Responses of the Northern Himalayan Basin Under Climate Variability

A. K. Prabhakar

Abstract The significant changes have been observed over the Northern Himalayan (NH) region due to human intervention and climate change. The common issues facing NH regions include winter rainfall, cloudbursts, melting of glaciers, and a rise in the rainfall intensity and temperature. The flood catastrophe, which results in significant losses to property, people, and infrastructure, among other things. Managing the flood is a complex phenomenon under the widely reported global warming in the Himalayan region of India. The current study focuses on the Remote sensing (RS) and Geographical Information System (GIS) applications in hydrological modelling over the NH regions. The article further highlights disturbance in the hydrological cycle due to climate change (viz., precipitation, temperature) and anthropogenic activity, resulting in flood risk, which is a common issue faced in NH regions. Furthermore, the increased anthropogenic activity, deforestation, global warming, and unplanned urban growth have heightened the susceptibility of NH regions to hydrological disasters.

Keywords Hydrological modeling · Climate change · Urban growth

3.1 Introduction

Water is a basic need for all living beings, and today its sustainability is a challenging issue, not only in developing countries like India but all over the world. Climate change (IPCC 1996) has badly affected the availability of water in the northern regions due to significant changes in the hydrological cycle. Global warming significantly affects the occurrence of hydrological processes that changed how available water was distributed over time and space (Chandel and Ghosh 2021). Land use and climate changes could adversely impact the natural water balance components in NH river basins (Rathore et al. 2022). The considerable change is observed due

A. K. Prabhakar (✉)
Department of Civil Engineering, National Institute of Technology, Kurukshetra, India
e-mail: akprabhakar@nitkkr.ac.in; ajayprabhakar2@gmail.com

M. Biswas et al. (eds.), *Himalayan Flood Hazards*, Advances in Geographical and Environmental Sciences, https://doi.org/10.1007/978-981-92-0702-2_3

to ongoing land use changes such as deforestation and urbanization, which disturb the Himalayan region's environment in comparison to earlier times (Sharma et al. 2000; Dimri et al. 2021). The disturbance in the hydrological cycle has led to a more uneven distribution of precipitation, evaporation, and streamflow, resulting in extreme natural hazards like flood and drought conditions.

Himalayan rivers are perennial rivers, fed by both glacial melt and monsoon rainfall, supporting the common water needs of people throughout the year. The atmospheric warming will impact snow cover in winter and heavy rainfall in the monsoon in recent decades across the Indian Himalayas. Hunt et al. (2020) reported that a large proportion of winter precipitation is in the form of rainfall compared to the falling of snow due to the windward side in the northwestern Himalayan regions. Other researchers also noticed that shifting from falling snow to rainfall over the Indian Himalayas (Meher et al. 2018 and Kumar et al. 2022a, b). In recent decades, rainfall patterns have shifted to intense precipitation throughout the monsoon season, causing more intense floods in the NH river basin of India. Bhardwaj et al. (2021) found that the intensity and frequency of rainfall have changed over the Tibetan plateau of the Himalayas regions of India. Most regions of the North-Western Himalayas are highly prone to flash floods due to cloudbursts and heavy rainfall (Kumar et al. 2018). In recent years, these types of extreme events have been occurring over the NH regions, resulting in immense human and economic losses.

Effective management of water resources is essential to conserve water for present and future requirements. The understanding of hydrological processes is a complex phenomenon that can be analyzed through hydrological models. Hydrological modelling is the most effective tool for comprehending the intricate hydrological processes within a watershed in terms of spatial and temporal dimensions (Pandi et al. 2021). The results obtained from the analysis are also associated with uncertainties. Such an assessment is possible only when the variability of rainfall is understood, and modelling of rainfall and runoff can be done accurately. The NH river basin and lower regions are more prone to floods, droughts, and water availability can be analyzed through a hydrological model. More than 80% of the Indian population and the Indian economy depend upon agriculture. Agriculture mainly depends on water availability. The NH region is rich in its natural resources, such as agriculture, forests, and ecosystems (Masoodi and Sundriyal 2020). The major climate of NH regions consists of heavy rainfall, cold waves, and it varies as per the location, elevation, and season. In recent years (i.e., onward from 2010), the heavy rainfall during the monsoon has caused the number of catastrophic floods in various NH regions. The Uttarakhand flood in 2013 was an unexpected flood event; it became a hot topic of research on the study of extreme rainfall.

The rainfall and temperature trend analysis has been conducted to understand the long-term pattern changes compared to the past. Most researchers (Palazzi et al. 2015, Sanjay et al. 2017) have found that anticipated precipitation increases during the monsoon over the various regions of the Indian Himalayas under CMIP5. According to the recent trends under the 13 regional climate models, the southern Himalayas would receive, on average, a 22% rise in monsoon precipitation at the conclusion of the twenty-first century under RCP8.5 scenarios (Sanjay et al. 2017). These types of changes seriously impact the hydrological process, which causes extreme events like floods in most of the NH regions.

This increases the extreme events, such as increasing flood frequency, landslides, biodiversity changes, agricultural impacts, etc., which are clearly visible in NH regions due to continuous development interventions and climate changes (Mir et al. 2015, 2019). The current study is focused on the remote sensing applications in hydrological modelling over the NH regions. The disturbance in the hydrological cycle due to changes in precipitation and temperature, resulting in flood risk, is a common issue faced by the NH-region peoples of India. Hydrological modelling is also being used to study the changes in climate. The understanding of the changed current and the prediction of the hydrological model's major input parameters, viz., precipitation and temperature, by climate models, and their impacts over the NH regions of India. The study also points out the effects of increased anthropogenic activity, deforestation, global warming, and unplanned urban growth, leading to severe flooding and consequences for agriculture and future water availability. The following are a few examples of study areas that have outcomes based on literature reviews presented below in Table 3.1.

The NH regions of India are mostly relying on the principal rivers of the Himalayan region, namely the Indus, Ganges, and Brahmaputra (Fig. 3.1). These rivers sustain the water at downstream regions' distinct needs of agriculture, domestic, hydropower, and other industries during dry periods (Moors et al. 2011; Biemans et al. 2019). The Northern Himalayan (NH) region rivers are a lifeline for millions of people living in states (Jammu & Kashmir, Uttarakhand, Himachal Pradesh, Uttar Pradesh, Bihar, Haryana, and Punjab). The important river basins and sub-basins in the northern region of India are presented in Table 3.2.

3.2 Application of Remote Sensing and GIS

Remote sensing is a powerful technological innovation that can provide much useful information for hydrological modelling. The accuracy of a model depends highly on good data that would be helpful for the development. Remote sensing satellite data have improved the definition of land cover and soil over the watershed, which is an important input for models to determine outputs such as infiltration, evaporation, and runoff coefficients (Ottlé et al. 1989). Conventional methods for classifying land use and land cover are precise but costly, challenging to assess, repetitive, and also time-consuming. The cost for conventional data gathering is about 11 times more as compared to LANDSAT data for a similar watershed area (Glover 1984). Remote sensing can be used to derive the vital input for the hydrological models, viz., satellite imagery.

Geographic Information Systems (GIS) have emerged as a particularly valuable and significant tool to acquire, organize, and display characteristics graphically. GIS is a new technique that provides a common interface for solving complex spatial data to the respective output (McVicar and Jupp 1998). The tight coupling of SWAT

Table 3.1 The study's focus and outcomes are summarised in the literature review

Author (et al., Year)	Study area	Study types & factors	Outcomes
Ahmad et al. (2015)	Northern Himalayan watersheds, Pakistan	SWAT and HydroCAD modeling; land use change scenarios	7.1% increase in runoff, 6.3% decrease in groundwater recharge due to urbanization and deforestation
Ahsan et al. (2023)	Jhelum basin, Himalayas	Multi-model climate ensemble utilizing Bayesian averaging; climate projections (RCP4.5 & RCP8.5)	40 to 46% reduction in river discharge; water yield decline under future scenarios
Jain et al. (2017)	Ganga river basin, India	Hydrologic modeling with SWAT: Snow and rainfed dynamics	Main yield from lateral and groundwater flow; 13–20% runoff from snowmelt
Verma and Dash (2018)	Sirsa river basin, NW Himalaya, India	SWAT modeling using MODIS ET, LULC, and climate variability	12% rise in runoff, decrease in base and lateral flow; 26.5% streamflow reduction in monsoon
Talchabhadel (2022)	Tila river basin, Nepal	CMIP6 climate models; hydrologic modeling; SSP scenarios	Up to 75% variation in water yield; high-altitude regions are highly vulnerable
Khatakho et al. (2021)	Himalayan river basin from the Tibetan Plateau	Evaluation of precipitation products (TRMM, CHIRPS, etc.) with J2000 model	TRMM performed best; MSWEP improved with specific calibration; highlights model sensitivity to precipitation inputs
Aggarwal et al. (2019)	North Western Himalaya, India	Hydrological modelling of sub-basins in Indus and Ganga basins	Glacier and snowmelt critical for lean season flows; significant for hydropower planning
Kumar et al. (2024)	Alaknanda river, Uttarakhand	SWAT model simulation; sensitivity analysis of snowmelt and groundwater parameters	Snowmelt contributes 20–24% to runoff; model highlights need for better hydro-meteorological data
Masiwal et al. (2024)	Nine watersheds in the Northwest Himalayas	Remote sensing and FEWS NET/FLDAS data analysis of hydro-meteorological fluxes	50% decline in snow cover area; basins with high snow cover show lower ET and higher runoff

(continued)

Table 3.1 (continued)

Author (et al., Year)	Study area	Study types & factors	Outcomes
Shea and Immerzeel (2016)	Hindu Kush-Himalaya region	Glacier geometry sensitivity analysis climate scenarios	Projected glacier melts to initially rise, then fall sharply; implications for long-term flow under warming
Li et al. (2017)	Beas basin, Himalayas	GSM-WASMOD glacio-hydrological model under RCP4.5/8.5	Glacier loss up to 80% by 2100; peak runoff reduction projected in August and May
Arora et al. (2008)	Chenab river basin	Snowmelt modeling under temp/rainfall scenarios	T + 3 °C increases snowmelt by 43% and streamflow by 28%; rainfall changes have minor effects
Sharma (1997)	Kosi basin	Land use and climatic analysis; water balance and deterministic modeling	Hydrologic sensitivity is more driven by basin than climate; glacier retreat and deforestation are both impactful
Saran et al. (2010)	Sitla Rao watershed	DEM evaluation for HRU delineation	High-resolution DEMs improved morphometry; limited advantage at the watershed scale
Rautela et al. (2023)	Alaknanda river basin, Uttarakhand	SWAT-based hydrological modeling with snowmelt sensitivity analysis	Snowmelt contributes 20–24%; water yield and ET form 58–63% and 34–39% of precipitation
Aggarwal et al. (2020)	North West Himalaya (Beas, Satluj, Upper Ganga, Jhelum)	Remote sensing & distributed hydrological models, SDGs focus	Snow/glacier/rainfall contributions quantified; future projections show increased floods, decreased snow
Shea and Immerzeel (2016b)	Hindu Kush-Himalaya	Static vs adjusted glacier geometry under warming scenarios	Streamflow first increases (+1 °C), but long-term glacier loss reduces flow; response times are estimated

(continued)

Table 3.1 (continued)

Author (et al., Year)	Study area	Study types & factors	Outcomes
Singh and Bengtsson (2004)	Satluj river basin	Conceptual modelling: temperature and rainfall scenarios	Warming reduces lower basin melt, increases upper melt; summer water availability declines
Kumar et al. (2022a, b)	Beas river basin	HEC-HMS model with CMIP6 SSP scenarios	Extreme floods projected for 2057 (SSP585) and 2068 (SSP245); frequency increases by 2091–2100

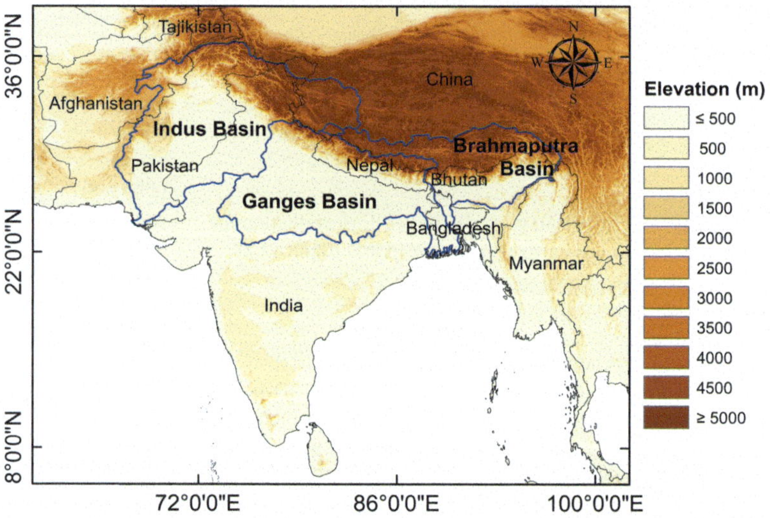

Fig. 3.1 The study area map shows the three major river basins (Ganga, Indus, and Brahmaputra) in the Himalaya, indicated by blue colour. The brown colour scale shows the variation of elevation in the study area

with the GRASS GIS model was used to analyze the water quality at the basin scale (Srinivasan and Arnold 1994). Therefore, the developing countries require cheaper and less time-consuming techniques, such as RS and GIS techniques, for scientific planning and management of watersheds.

Table 3.2 Northern region of India—key river basins and sub-basins

Major basin	Key sub-basins	States covered	Key features
Indus basin	Jhelum, Chenab, Ravi, Beas, Sutlej	Jammu & Kashmir, Himachal Pradesh, Punjab, Haryana	Snow-fed, perennial rivers; shared with Pakistan under the Indus Waters Treaty
Ganga basin	Yamuna, Ramganga, Ghaghara, Gandak, Kosi, Son, Gomti, Tons	Uttarakhand, Uttar Pradesh, Bihar, Jharkhand, West Bengal, Delhi, Haryana, Himachal Pradesh	Largest river basin in India; source of irrigation, domestic water, and religious significance
Brahmaputra basin	Subansiri, Kameng, Manas, Dhansiri, Dibang, Lohit	Arunachal Pradesh, Assam, Meghalaya, Sikkim, parts of West Bengal	Highly braided river, prone to flooding, carries heavy sediment load from Eastern Himalayas

3.3 Hydrological Modelling

Hydrological models can be classified according to function, suitability, application, and advances over the past (Singh 1988). A material, a physical model (Chow et al. 1988), represents a real system, which is similar in properties. A scale model shows similarity on a reduced scale that could be quite similar to the prototype system, viz., laboratory watersheds, lysimeters, and hydraulic models. A mathematical model, symbolic/formal in nature, is a representation of the logical relationship between variables and parameters in the form of equations to find logical outputs. The calibrated and validated hydrological model would be used for a broad spectrum of applications, including forecasting hydrological parameters for water resource management.

In the 1960s, researchers formulated continuous conceptual hydrological simulation models. These models, unlike event models, consider continuous processes and the watershed's soil moisture balance over an extended duration, enabling them to simulate streamflow on a daily, monthly, and seasonal basis (Ponce 1989). Later on, many of these continuous hydrologic models came to light, viz., the Standard Watershed Model (SWM), the SCS curve number method, the rational method, Hydrologic Simulation Package-FORTRAN IV (HSPF), and MIKE Hydro, etc. These model functions are generally controlled by the processes of the drainage system. These models are less useful for ungauged catchments as long-term data are not available (Guo et al. 2021). The hydrological models help to predict continuous streamflow over the catchment and play an essential part in managing the water resources over the region. The hydrological models are employed to forecast future extreme occurrences, such as floods and droughts, over the regions. The lumped model simulates the response from the number of time and space invariant inputs to the output parameter. The distributed models are more reliable for estimating the streamflow from the ungauged catchment. Freeze and Harlan (1969) created the first physically-based distributed model, but the distributed model requires massive amounts of data.

Later on, several hydrological models were developed, such as the distributed Systeme Hydrologique European (SHE) and Semi-distributed models like the Soil and Water Assessment Tool (SWAT), which require less data compared to distributed models. Soft-computing models differ in approach, which set up hydrological models for the prediction of runoff from rainfall on the basis of basic information from the existing data without knowledge of the real hydrological process. The last few decades ANN (Karl and Anil Kumar 2010), Fuzzy rule (Kisi et al. 2006; Lohani et al. 2007), and ANFIS (Chang et al. 2014; Anusree and Varghese 2016) have received more attention in the generation of complex non-linear relationships between the hydrological variables. Consequently, hydrological modelling can be applied to the analysis of any geographic region. The application of hydrological models can effectively achieve a clearer understanding and connection between hydrological parameters in the context of climate change. The range of hydrological models is extensive, and Table 3.3 provides a concise overview of their basic components and applications along with brief descriptions.

3.4 Study of Climate Change Models

The climate models have the potential to understand climate change's effects on the hydrological cycle as a basis for hydrological model development, viz., General Circulation Models (GCMs) and Regional Climate Models (RCMs) (Olsson et al. 2016; Saha and Quinn 2020). The hydrological models used to assess the flood prediction, drought warning, and water quality and availability under the various climatic scenarios (Sood and Smakhtin 2015; Brunner et al. 2021). Under the CMIP5 (GCM) projections, it was found that the more significant rise in temperature and decrease in snow cover occurs over the southwest Himalayan region of the Tibetan plateau during the twenty-first century (Ghatak et al. 2014).

The critical parameters, namely rainfall and temperature, significantly influence future alterations in water resources under different scenarios. The high-resolution climate models come under the RCMs, which are integrated with GCMs and cover a limited area for the study. This method is based on dynamical downscaling (DD), degrading the model output because it consists of systematic biases, resulting in increased uncertainties of predictions. Whereas, Statistical Downscaling (SD) used to relate the atmospheric parameter from GCM to local level observations, biases corrected, gives a more accurate prediction of rainfall and temperature. The procedure for generating future climate scenarios under SD is shown in Fig. 3.2.

The predicted parameters serve as inputs, viz., temperature and rainfall, for executing the hydrological model to forecast various hydrological parameters such as streamflow, evapotranspiration, and existing and future water availability under climate change scenarios from several climate models. The hydrological models are effectively used to anticipate flooding in response to increased anthropogenic activity and climate change.

Table 3.3 The basic components, applications of hydrological modelling, and their brief descriptions

Components	Brief descriptions
Geographic area	Northern Himalayan basin, primarily involving the Indus, Ganga, and Brahmaputra basins
Hydrological challenges	Floods, droughts, uneven water distribution due to climate variability, shifts from snowfall to rainfall
Climatic factors	Global warming, increased frequency and intensity of rainfall, higher temperatures, and changing precipitation patterns
Human & economic impact	Loss of life, property, infrastructure, agricultural impacts, socio-economic disruptions, and biodiversity changes
Hydrological processes	Increased runoff, altered evapotranspiration rates, glacier melt dynamics, and disturbed hydrological cycle
Models & techniques used	Remote sensing, GIS, Hydrological modeling (SWAT, ANN, ANFIS, GCMs, RCMs), Statistical Downscaling (SD)
Observed climate trends	Rise in monsoon precipitation (projected 22% increase under RCP8.5), increased temperature (~0.1 °C per decade), extreme rainfall events becoming more common, increased rainfall intensity, and shifting snow to rain events
Modelling predictions	Reduction in snow cover, increased flooding frequency, drought conditions, variations in future streamflow, projected increase in mean surface runoff (~40%) and evapotranspiration (~10%) under RCP (4.5 and 8.5) scenarios by the century's end
Land use changes	Urbanization, deforestation impacting infiltration, runoff, evaporation; significant environmental disturbances
Impacts on water resources	Short-term increased river discharge (glacier melt), long-term water scarcity risks, variability in water availability impacting agriculture and ecosystem stability
Applications of RS and GIS	Watershed planning, runoff and flood prediction, land use/land cover classification, hydrological model inputs (land cover, soil data, infiltration, evaporation coefficients)
Flood management strategies	Necessity for efficient water management, enhanced hydrological modeling capabilities, accurate rainfall-runoff modeling, adaptive strategies for water retention, and resource allocation
Research gaps & future outlook	Uncertainty in precipitation and temperature modeling, need for higher-resolution climate models, better prediction of extreme events, and further investigation into hydrological modeling accuracy

3.4.1 Precipitation Variability Under Climate Changes

The variability and unpredictability of rainfall patterns may significantly affect agricultural output in India, as the farmers mostly rely on precipitation. Such variability has a detrimental effect on how water supplies are planned and managed to meet the needs of various sectors. Consequently, precipitation is a crucial factor among several hydroclimatic variables, including temperature, precipitation, and humidity

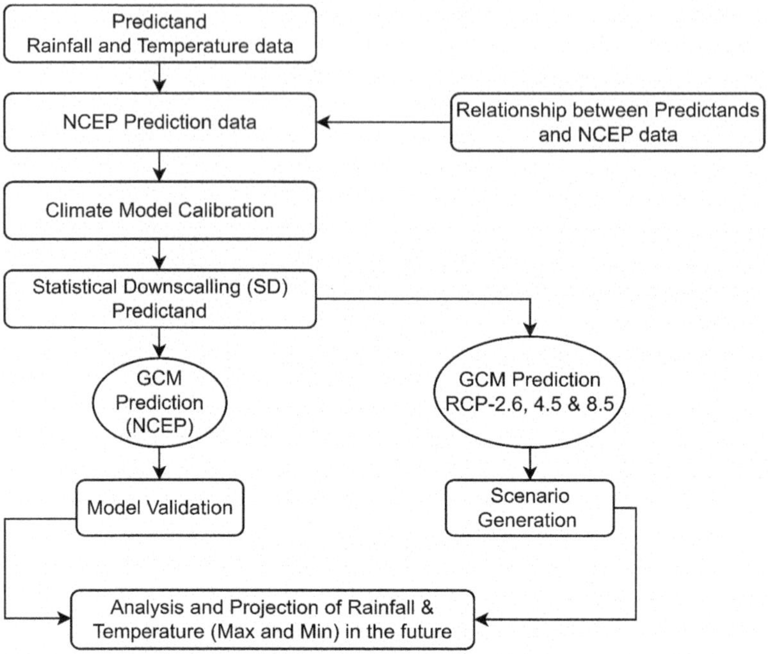

Fig. 3.2 Procedure for generating future climate scenarios

(Meresa et al. 2022). The others researcher's (Giorgi et al. 2019; López et al. 2021) finding that the impacts of global warming strongly indicate significant alterations in rainfall patterns on a worldwide scale. Various regions of Asia had an escalation in severe rainfall occurrences, accompanied by a decline in the total number of rainy days and yearly precipitation. Martel et al. (2021) identified increasing daily trends in both the amplitude and frequency of extreme precipitation events from 1951 to 2000, while also noting a significant decline in the frequency of typical occurrences during the monsoon seasons in central India.

The insignificant decline of annual and monsoon rainfall trends has been noted in Odisha state, alongside an increasing pattern in winter and summer season rainfall (Patra et al. 2012). In the future, due to climate change and population growth, there will be a reduction in freshwater availability, which leads to the study of long-term, both temporal and spatial, variation of rainfall (Lacombe and McCartney 2014) for sustaining the economy of the nation. Liuzzo et al. (2015) investigated annual maximum rainfall series using 65 rain gauges for the period (1950–2008) over the Sicily-Mediterranean area and found an increasing and decreasing trend in rainfall. There are several authors (Das and Meher 2019; Gouda et al. 2023) have found a possible positive trend in extreme rainfall intensity that would cause floods if drainage networks were not properly designed without considering the extreme rainfall events over the Himalayan region of India.

3.4.2 Temperature Variability Under Climate Changes

The rise in temperature of about 0.1 °C is anticipated to last for the next two decades if greenhouse gas and aerosol concentrations are sustained at the levels of the year 2000. In accordance with the Intergovernmental Panel on Climate Change (IPCC) referenced in the Fourth Assessment Report (AR4) (IPCC 2007). The rising trend in annual temperatures is expected to persist at a pace of 0.1–0.5 °C each decade on the high-altitude surfaces of the northern hemisphere (Fan et al. 2020). The average global surface temperature has steadily risen, and numerous GCM models indicate that global warming is a result of heightened greenhouse gas concentrations in the atmosphere over the past century. The estimation of surface temperature variability was better predicted by using long-term time series and found an increasing trend in annual mean temperature (Sanderson et al. 2018). Also, during the twentieth century, the growing trend for mean annual surface air temperature was discovered following the examination of long-term temperature records at roughly 73 stations across India from 1901 to 1982. The yearly average temperature of densely inhabited and heavily industrialised cities, namely Bombay and Calcutta, has revealed an increasing trend of 0.84 °Cand 1.39 °C per 100 years (Hingane 1996), respectively.

The projected increase in daily minimum temperature was determined to exceed that of daily maximum temperature on an annual basis (Wu et al. 2017). The maximum temperature rose by 1.9 °C in November and 1.79 °C in February over a span of 100 years. Mondal et al. (2015) analyzed temperature patterns in India for 107 years (1901–2007) and found wide variations in minimum temperature. A notable rise in mean, minimum, and maximum temperatures was observed, primarily during winter and post-monsoon periods. However, the average annual and trend magnitudes for India were found to be an increase in mean temperature (0.003 °C/year and 1.2%), respectively. Mishra and Lilhare (2016) investigated mean air temperature in Indian sub-continental river basins and found mean air temperature is projected to be increased in Near (0.5–0.8 °C from 2010–2039), Mid (1–2 °C from 2040–2069), and End (1.5–3.5 °C from 2070–2099) term climate change, respectively.

Bhutiyani et al. (2007, 2010) identified an increase in air temperature attributed to a swift escalation in both maximum and minimum temperatures, with the rise in maximum temperature occurring more rapidly in the North West Himalayas (NWH) of India. The most significant increase in maximum temperature during February occurred in the western Himalaya region. A significant rise in temperature was noticed, leading to an excessive melting of glaciers and increased runoff due to the prevalence of a stronger monsoon in the Ladakh region of the Himalayas. The other researchers (Phartiyal et al. 2015); Das and Meher 2019) asserted that elevated temperatures and intensified rainfall due to climate change have impacted the Indian monsoon, potentially resulting in cloudbursts and severe flooding in the Himalayan regions of India.

3.5 Hydrological Modelling Under Climate Change

Global warming occurs due to the emission of greenhouse gases (GHGs) in the atmosphere, which causes significant changes in the global climatic pattern. It also alters the radiative balance of the atmosphere, which would change the climatic variables, viz., the temperature and precipitation (Houghton et al. 1990). The climate variability in precipitation, temperature, and evapotranspiration affects the runoff volume (Waggoner 1990). Climate change causes two extreme events, such as floods and droughts. It is a very pertinent concern for agriculture, the economy, water resources, and the environment. Swain and Patra (2018) reported that climate change decreases annual runoff volume to 21.7% in the 32 catchments spread across India.

Hydrological water balance models are useful in evaluating the effects of climate change on watershed hydrology at the regional level (Arnell 1999). The two most critical hydro-meteorological data, viz., temperature and rainfall, are needed to calibrate and validate hydrological models, which can be obtained from actual data or regional climate models (RCMs). These parameters significantly affect water availability in any region due to climate change. Other meteorological parameters, viz., humidity, radiation, and wind velocity, have also influenced the hydrological models. Reports indicate that the errors in estimating discharge from meteorological inputs are considerably greater than those associated with discharge estimation using the hydrological model (Te Linde et al. (2007). Therefore, the accuracy of meteorological data, viz. observed and RCM data, is a must for enhancing the precision of predicted hydrological output from the hydrological model.

3.5.1 Impacts on the Streamflow

The effects of future climate projections using the PRECIS-RCM (50 km resolution) on the hydrology of the Samat watershed in the Mae Nam Chi basin, Northeast Thailand (Graiprab et al. 2010). Reports indicate that from 2010 to 2050, the regional temperature is projected to rise by approximately 0.80 °C, with rainfall expected to increase by about 4%. Additionally, the average runoff yield is anticipated to rise by 3–5% in comparison to the overall runoff yield of the watershed area. This study examines the effects of climate and land cover change on runoff in the Huifa River Basin, Northeast China, utilising the SWAT model (Zhang et al. 2012). The study indicated that climate change would influence runoff, either increasing or decreasing it, contingent upon variations in precipitation, temperature, radiation, and land cover changes. Furthermore, the results show that annual rainfall variations are − 36.7 mm (1965–1975), − 59.5 mm (1976–1985), + 36.9 mm (1986–1995), and − 45.2 mm (1996–2005) relative to the base period of 1956–1964. Luo et al. (2013) evaluated hydrology at the watershed level, utilising the SWAT model under climate change scenarios A2 and B1 in California. The study observed an increase in precipitation and streamflow during the winter season, alongside a decrease in the summer months.

Mishra and Lilhare (2016) examined changes in hydrological processes within the Indian sub-continental river basin under downscaled and bias-corrected future climate projections utilising the SWAT model. The mean surface runoff and evapotranspiration (ET) are projected to increase by approximately 40% and 10%, respectively, in most river basins under the Representative Concentration Pathways (RCPs) 4.5 and 8.5 scenarios by the end of the twenty-first century. Swain et al. (2022) documented the use of the SWAT hydrological model, employing high-resolution geospatial data to investigate the Tehri reservoir catchment in the lower Himalayan area of India. It was found that high-quality remote sensing data enhances the performance of hydrological models in simulating streamflow over the Himalayan region. According to research by Shukla et al. (2021), which used the hydrological model SWAT to analyse the impact of climate change on streams in the western Himalayan region at Satluj River basin, the melting of snow is a major contributor to the streamflow to the Satluj River at Rampur. The downscaled CORDEX CCSM4 data under RCP4.5 and RCP8.5 forecasts that the future climate in the western Himalayan region will result in increased precipitation and higher temperatures. Precipitation and temperature are projected to rise by around 14% to 21% and 2.18 °C to 5.71 °C, respectively, while future streamflow is expected to increase by 11–19% by the century's end under both scenarios.

3.5.2 Impact on the Frequency and Severity of Floods

The anticipated effects of climate change on Iran's water resources were analysed using the hydrological (SWAT) model, which employed the projected climate scenarios for the periods 2010–2040 and 2070–2100, which required downscaling for 37 climatic sites across the country. Future climate scenarios would result in more rainfall in the country's driest regions and less rainfall in its rainiest ones, according to their findings (Abbaspour et al. 2009). Additionally, the examination of daily rainfall intensities revealed that floods in moist regions were occurring more frequently and with greater intensity. Gosain et al. (2006) evaluated the effects of climate change on 12 river basins in India by utilising the SWAT model. Future scenarios based on HadRM2 (RCM) data indicate that the majority of the country may be affected by the severity of drought and the intensity of floods. In the Cauvery River, a 2.7% increase in rainfall has been noted; however, runoff output has decreased by around 2%, while actual evapotranspiration has risen by about 7.5% as a result of climate change. In the Mahanadi river basin, an increase in precipitation is anticipated, correlating with heightened runoff, water yield, and evapotranspiration in the future.

Flooding in the Northern Himalayan (NH) regions has become increasingly frequent and devastating due to the compounded effects of climate change, glacial melt, and intense precipitation events. These floods, particularly flash floods and glacial lake outburst floods (GLOFs) (Dubey and Goyal 2020; Hazra and Krishna 2022), are often triggered by sudden cloudbursts and excessive monsoon rainfall. The topography of the Himalayas amplifies the intensity of such events, causing rapid runoff, landslides, and severe damage to life and infrastructure. Case studies from Uttarakhand (2013), Leh (2010), and other parts of the region exemplify how localized weather phenomena and glacial dynamics contribute to major flood events (Rawat et al. 2023; Bhambri et al. 2024). The principal factors contributing to the severe flooding in the Kashmir Valley in 2014 include urbanisation, deforestation, encroachment, and climate change. Future flood mitigation measures should focus on regulating encroachment along the banks of the Jhelum River and implementing systematic, planned construction (Chakraborty et al. 2018; Malik 2022). Thus, increased anthropogenic activity, deforestation, and unplanned urban growth have made these areas even more susceptible to future hydrological disasters.

3.5.3 Impacts on Agriculture

The agriculture sector in the NH region is particularly vulnerable to climate-induced hydrological extremes. Floods can result in significant agricultural devastation, loss of arable topsoil, and interruption of planting and harvesting schedules. Since a large portion of the Indian population depends on subsistence and seasonal agriculture, especially in the Himalayan foothills, the socio-economic implications are severe. Flash floods often result in waterlogging and damage to irrigation infrastructure, reducing the productivity of both kharif and rabi crops. Furthermore, soil salinization and deposition of debris from upstream intensify the degradation of arable land. These recurring events undermine food security and place a heavy burden on rural livelihoods, calling for integrated water and land management strategies to enhance agricultural resilience.

3.5.4 Impacts on Future Water Availability

Climate change has significantly altered water availability patterns in the NH river basins. Initially, augmented glacial melt has resulted in elevated streamflow; however, this phenomenon is anticipated to diminish in the long run as glaciers recede. Coupled with erratic rainfall patterns and increased evapotranspiration due to higher temperatures, water scarcity is becoming a critical issue. The shift from snowfall to rainfall, especially during winter, reduces snowpack accumulation, which historically acted as a natural water reservoir. Seasonal variability is intensifying, with surplus water during monsoons leading to floods and acute shortages during dry months. These dynamics pose serious challenges to sustainable water management for agriculture, drinking, hydropower, and ecosystem services. Adaptive strategies, such as efficient irrigation systems, rainwater harvesting, and real-time hydrological monitoring, are vital for mitigating the impact of fluctuating water availability in the Himalayan context.

3.6 Conclusion

The hydrological response of the Himalayan basin to climate change is complex and multifaceted. It involves a mix of increased discharge in rivers due to rapid glacier melt in the short term, followed by long-term decreases in water availability as glaciers retreat over the NH regions of India. The changes in precipitation patterns, erratic rainfall, and an increase in extreme events, such as flash floods and cloudbursts, present additional challenges. The increased anthropogenic activity, deforestation, and unplanned urban growth have heightened the susceptibility of NH regions to future hydrological disasters. Flash floods frequently cause irrigation infrastructure damage and waterlogging, which lowers rabi and kharif agricultural output. These shifts have serious consequences for water security, agriculture, ecosystems, and local communities in the region, highlighting the need for adaptive strategies to mitigate the effects of climate change.

References

Abbaspour KC, Faramarzi M, Ghasemi SS, Yang H (2009) Assessing the impact of climate change on water resources in Iran. Water Resour Res 45(10)

Aggarwal SP, Garg V, Thakur PK, Nikam BR (2019) Hydrological Modelling in North Western Himalaya. In: Sustainable Agriculture Reviews. https://doi.org/10.1007/978-981-13-2128-3_6

Aggarwal SP, Thakur PK, Nikam BR, Garg V (2020) Role of Earth Observation Data and Hydrological Modeling in Supporting UN SDGs in North West Himalaya. ISPRS Archives. https://doi.org/10.5194/ISPRS-ARCHIVES-XLIII-B3-2020-853-2020

Ahmad Z, Ashraf A, Zaheer M, Bashir H (2015) Hydrological response to environment change in Himalayan watersheds: Assessment from integrated modeling approach. J Mt Sci. https://doi.org/10.1007/S11629-013-2892-1

Ahsan S, Bhat MS, Alam A, Farooq H, Shiekh HA (2023) Complementary use of multi-model climate ensemble and Bayesian model averaging for projecting river hydrology in the Himalaya. Environ Sci Pollut Res 30(13):38898–38920

Anusree K, Varghese KO (2016) Streamflow prediction of Karuvannur River basin using ANFIS, ANN and MNLR Models. Procedia Technol 24:101–108

Arnell NW (1999) A simple water balance model for the simulation of streamflow over a large geographic domain. J Hydrol 217(3–4, 314):–335

Arora M, Singh P, Goel NK, Singh RD (2008) Climate variability influences on hydrological responses of a large Himalayan basin. Water Resour Manag 22:1461–1475

Bhambri R, Mehta M, Tiwari SK, Yadav JS, Sain K (2024) High mountain hazards in Uttarakhand. In: Geo-information for Disaster Monitoring and Management. Springer International Publishing, Cham, pp 181–210

Bhardwaj A, Wasson RJ, Chow WT, Ziegler AD (2021) High-intensity monsoon rainfall variability and its attributes: a case study for Upper Ganges Catchment in the Indian Himalaya during 1901–2013. Nat Hazards 105(3):2907–2936

Bhutiyani MR, Kale VS, Pawar NJ (2007) Long-term trends in maximum, minimum and mean annual air temperatures across the Northwestern Himalaya during the twentieth century. Clim Chang 85(1–2):159–177

Bhutiyani MR, Kale VS, Pawar NJ (2010) Climate change and the precipitation variations in the northwestern Himalaya: 1866-2006. Int J Climatol 30(4):535

Biemans H, Siderius C, Lutz AF, Nepal S, Ahmad B, Hassan T et al (2019) Importance of snow and glacier meltwater for agriculture on the Indo-Gangetic Plain. Nat Sustainability 2(7):594–601
Brunner MI, Slater L, Tallaksen LM, Clark M (2021) Challenges in modeling and predicting floods and droughts: A review. Wiley Interdiscip Rev: Water 8(3):e1520
Chakraborty A, Saha S, Sachdeva K, Joshi PK (2018) Vulnerability of forests in the Himalayan region to climate change impacts and anthropogenic disturbances: a systematic review. Reg Environ Chang 18:1783–1799
Chandel VS, Ghosh S (2021) Components of Himalayan river flows in a changing climate. Water Resour Res 57(2):e2020WR027589
Chang FJ, Chiang YM, Tsai MJ, Shieh MC, Hsu KL, Sorooshian S (2014) Watershed rainfall forecasting using neuro-fuzzy networks with the assimilation of multi-sensor information. J Hydrol 508:374–384
Das L, Meher JK (2019) Drivers of climate over the Western Himalayan region of India: A review. Earth Sci Rev 198:102935
Dimri AP, Allen S, Huggel C, Mal S, Ballesteros-Cánovas JA, Rohrer M et al (2021) Climate change, cryosphere and impacts in the Indian Himalayan Region. Curr Sci 120(5):774–790
Dubey S, Goyal MK (2020) Glacial lake outburst flood hazard, downstream impact, and risk over the Indian Himalayas. Water Resour Res 56(4):e2019WR026533
Fan X, Duan Q, Shen C, Wu Y, Xing C (2020) Global surface air temperatures in CMIP6: Historical performance and future changes. Environ Res Lett 15(10):104056
Freeze RA, Harlan RL (1969) Blueprint for a physically based digitally simulated hydrologic response modal. J Hydrol 9:237–258
Ghatak D, Sinsky E, Miller J (2014) Role of snow-albedo feedback in higher elevation warming over the Himalayas, Tibetan Plateau and Central Asia. Environ Res Lett 9(11):114008
Giorgi F, Raffaele F, Coppola E (2019) The response of precipitation characteristics to global warming from climate projections. Earth Syst Dynam 10(1):73–89
Glover J (1984) Landsat feasibility study-Oklawaha basin. Prepared for the St. Johns river water management district, Palatka, Fla.
Gosain AK, Rao S, Basuray D (2006) Climate change impact assessment on hydrology of Indian river basins. Curr Sci:346–353
Gouda KC, Rath SS, Singh N, Ghosh S, Lata R (2023) Extreme rainfall event analysis over the state of Himachal Pradesh in India. Theor Appl Climatol 151(3):1103–1111
Graiprab P, Pongput K, Tangtham N, Gassman PW (2010) Hydrologic evaluation and effect of climate change on the At Samat watershed, Northeastern Region, Thailand. Int Agric Eng J 19(2):12–22
Guo Y, Zhang Y, Zhang L, Wang Z (2021) Regionalization of hydrological modeling for predicting streamflow in ungauged catchments: A comprehensive review. Wiley Interdiscip Rev: Water 8(1):e1487
Hazra P, Krishna AP (2022) Assessment of proglacial lakes in Sikkim Himalaya, India for glacial lake outburst flood (GLOF) risk analysis using HEC-RAS and geospatial techniques. J Geol Soc India 98(3):344–352
Hingane LS (1996) Is a signature of socio-economic impact written on the climate? Clim Chang 32(1):91–102
Houghton JT, Jenkins GJ, Ephraums JJ (1990) Climate Change: The IPCC Scientific Assessment. Cambridge University Press, Cambridge, p 365p
Hunt KM, Turner AG, Shaffrey LC (2020) The impacts of climate change on the winter water cycle of the western Himalaya. Clim Dyn 55(7):2287–2307
IPCC. Second Assessment Report (SAR) (1996) Contribution of working group I to the second assessment report of the intergovernmental panel on climate change. Cambridge University Press, Cambridge, UK. https://www.ipcc-nggip.iges.or.jp/public/gl/invs1.html
IPCC. Climate Change (2007) Climate change impacts, adaptation and vulnerability. Working group II contribution to the intergovernmental panel on climate change—fourth assessment

report summary for policymakers. New York, NY, Cambridge University Press. https://www.ipcc.ch/site/assets/uploads/2018/03/ar4_wg2_full_report.pdf (2007).
Jain SK, Jain S, Jain N, Xu C-Y (2017) Hydrologic modeling of a Himalayan mountain basin by using the SWAT model. Hydrol Earth Syst Sci Discuss. https://doi.org/10.5194/HESS-2017-100
Karl AK, Anil Kumar L (2010) Development of flood forecasting system using statistical and ANN techniques in the downstream catchment of Mahanadi Basin, India. J Water Resour Prot 2010
Khatakho R, Talchabhadel R, Thapa BR (2021) Evaluation of different precipitation inputs on streamflow simulation in Himalayan River basin. J Hydrol. https://doi.org/10.1016/J.JHYDROL.2021.126390
Kisi O, Karahan ME, Şen Z (2006) River suspended sediment modelling using a fuzzy logic approach. Hydrol Process 20(20):4351–4362
Kumar A, Gupta AK, Bhambri R, Verma A, Tiwari SK, Asthana AKL (2018) Assessment and review of hydrometeorological aspects for cloudburst and flash flood events in the third pole region (Indian Himalaya). Pol Sci 18:5–20
Kumar D et al (2022a) Projection of future pluvial flood events over Himalayan river basin under CMIP6 climate data. In: EGU General Assembly Conference. https://doi.org/10.5194/egusphere-egu22-438
Kumar D, Rautela KS, Gandhi RB, Kumar A, Dubey AK (2024) Long-term hydrological simulation of a Himalayan river watershed, Uttarakhand using Soil and Water Assessment Tool. Preprint. https://doi.org/10.22541/au.170668669.92925559/v1
Kumar R, Manzoor S, Vishwakarma DK, Al-Ansari N, Kushwaha NL, Elbeltagi A et al (2022b) Assessment of climate change impact on snowmelt runoff in Himalayan region. Sustainability 14(3):1150
Lacombe G, McCartney M (2014) Uncovering consistencies in Indian rainfall trends observed over the last half century. Clim Chang 123(2):287–299
Li L, Hou Y-K, Xu C-Y, Chen H, Jain SK (2017) Projection of future glacier and runoff change in Himalayan headwater Beas basin by using a coupled glacier and hydrological model. Hydrol Earth Syst Sci Discuss. https://doi.org/10.5194/HESS-2017-525
Liuzzo L, Freni G (2015) Analysis of extreme rainfall trends in Sicily for the evaluation of depth-duration-frequency curves in climate change scenarios. J Hydrol Eng 20(12):04015036
Lohani AK, Goel NK, Bhatia KS (2007) Deriving stage–discharge–sediment concentration relationships using fuzzy logic. Hydrol Sci J 52(4):793–807
López-Moreno JI, Pomeroy JW, Morán-Tejeda E, Revuelto J, Navarro-Serrano FM, Vidaller I, Alonso-González E (2021) Changes in the frequency of global high mountain rain-on-snow events due to climate warming. Environ Res Lett 16(9):094021
Luo Y, Ficklin DL, Liu X, Zhang M (2013) Assessment of climate change impacts on hydrology and water quality with a watershed modeling approach. Sci Total Environ 450:72–82
Malik IH (2022) Anthropogenic causes of recent floods in Kashmir Valley: A study of 2014 flood. SN Soc Sci 2(8):162
Martel JL, Brissette FP, Lucas-Picher P, Troin M, Arsenault R (2021) Climate change and rainfall intensity–duration–frequency curves: Overview of science and guidelines for adaptation. J Hydrol Eng 26(10):03121001
Masiwal A, Singla S, Singhal G, Thakur PK, Garg V (2024) Variability of Hydro-Meteorological Fluxes in North West Himalayan Basins for Hydrological and Sustainability Studies. ISPRS Archives. https://doi.org/10.5194/isprs-archives-xlviii-3-2024-427-2024
Masoodi HUR, Sundriyal RC (2020) Richness of non-timber forest products in Himalayan communities—diversity, distribution, use pattern and conservation status. J Ethnobiol Ethnomed 16:1–15
McVicar TR, Jupp DLB (1998) The current and potential operational uses of remote sensing to aid decisions on drought exceptional circumstances in Australia: A review. Agric Syst 57(3):399–468

Meher JK, Das L, Benestad RE, Mezghani A (2018) Analysis of winter rainfall change statistics over the Western Himalaya: The influence of internal variability and topography. Int J Climatol 38(S1):e475–e496

Meresa H, Tischbein B, Mekonnen T (2022) Climate change impact on extreme precipitation and peak flood magnitude and frequency: observations from CMIP6 and hydrological models. Nat Hazards 111(3):2649–2679

Mir BH, Akhoon A, Rather NA (2015) Assessing impact of climate change on surface water resources in Dal Catchment. Elixir Civil Engg 83:33247–33254

Mir BH, Kumar R, Rather NA (2019) Water resources scenario of Himalayan region, applied agricultural practices for mitigating climate change. vol 2, Ch 18

Mishra V, Lilhare R (2016) Hydrologic sensitivity of Indian sub-continental river basins to climate change. Glob Planet Chang 139:78–96

Mondal A, Khare D, Kundu S (2015) Spatial and temporal analysis of rainfall and temperature trend of India. Theor Appl Climatol 122(1–2):143–158

Moors EJ, Groot A, Biemans H, van Scheltinga CT, Siderius C, Stoffel M et al (2011) Adaptation to changing water resources in the Ganges basin, northern India. Environ Sci Pol 14(7):758–769

Olsson J, Arheimer B, Borris M, Donnelly C, Foster K, Nikulin G et al (2016) Hydrological climate change impact assessment at small and large scales: key messages from recent progress in Sweden. Climate 4(3):39

Ottlé C, Vidal-Madjar D, Girard G (1989) Remote sensing applications to hydrological modeling. J Hydrol 105(3–4):369–384

Palazzi E, von Hardenberg J, Terzago S, Provenzale A (2015) Precipitation in the Karakoram-Himalaya: a CMIP5 view. Clim Dyn 45:21–45

Pandi D, Kothandaraman S, Kuppusamy M (2021) Hydrological models: a review. Int J Hydrol Sci Technol 12(3):223–242

Patra JP, Mishra A, Singh R, Raghuwanshi NS (2012) Detecting rainfall trends in twentieth century (1871–2006) over Orissa State, India. India Clim Change 111(3–4):801–817

Phartiyal B, Singh R, Kothyari GC (2015) Late-quaternary geomorphic scenario due to changing depositional regimes in the Tangtse valley, Trans-Himalaya, NW India. Palaeogeogr Palaeoclimatol Palaeoecol 422:11–24

Ponce VM (1989) Engineering hydrology, Principles and practices, Prentice-Hall, Englewood Chiffs, New Jersey. Available http://ponce.sdsu.edu/330textbookhydrology_chapters.html. Accessed Dec 2010

Rathore P, Roy A, Karnatak H (2022) Predicting the future of species assemblages under climate and land use land cover changes in Himalaya: A geospatial modelling approach. Clim Chang Ecol 3:100048

Rautela KS, Kumar D, Gandhi RB, Kumar A, Dubey A (2023) Long-term hydrological simulation for the estimation of snowmelt contribution of Alaknanda River Basin. Aqua, Uttarakhand using SWAT. https://doi.org/10.2166/aqua.2023.176

Rawat M, Ahmed R, Jain SK, Lohani AK, Rongali G, Tiwari KC (2023) Glacier–glacial lake changes and modeling glacial lake outburst flood in Upper Ganga Basin, India. Model Earth Syst Environ 9(1):507–526

Saha GC, Quinn M (2020) Integrated surface water and groundwater analysis under the effects of climate change, hydraulic fracturing and its associated activities: A case study from Northwestern Alberta, Canada. Hydrology 7(4):70

Sanderson BM, Oleson KW, Strand WG, Lehner F, O'Neill BC (2018) A new ensemble of GCM simulations to assess avoided impacts in a climate mitigation scenario. Clim Chang 146:303–318

Sanjay J, Krishnan R, Shrestha AB, Rajbhandari R, Ren GY (2017) Downscaled climate change projections for the Hindu Kush Himalayan region using CORDEX South Asia regional climate models. Adv Clim Chang Res 8(3):185–198

Saran S, Sterk G, Peters P, Dadhwal VK (2010) Evaluation of digital elevation models for delineation of hydrological response units in a Himalayan watershed. Geocarto Int. https://doi.org/10.1080/10106040903051967

Sharma KP (1997) Impact of land-use and climatic changes on hydrology of the Himalayan Basin: A case study of the Kosi Basin. *Dissertation*

Sharma KP, Vorosmarty CJ, Moore B (2000) Sensitivity of the Himalayan hydrology to land-use and climatic changes. Clim Chang 47:117–139

Shea JM, Immerzeel WW (2016) Glaciological and hydrological sensitivities in the Hindu Kush—Himalaya. In: EGU General Assembly Conference

Shea JM, Immerzeel WW (2016b) An assessment of basin-scale glaciological and hydrological sensitivities in the Hindu Kush–Himalaya. Ann Glaciol. https://doi.org/10.3189/2016AOG71A073

Shukla S, Jain SK, Kansal ML (2021) Hydrological modelling of a snow/glacier-fed western Himalayan basin to simulate the current and future streamflows under changing climate scenarios. Sci Total Environ 795:148871

Singh P, Bengtsson L (2004) Hydrological sensitivity of a large Himalayan basin to climate change. Hydrol Process. https://doi.org/10.1002/HYP.1468

Singh VP (1988) Hydrologic Systems, vol 1. Rainfall-runoff modeling, Prentice Hall, New Jersey

Sood A, Smakhtin V (2015) Global hydrological models: a review. Hydrol Sci J 60(4):549–565

Srinivasan R, Arnold JG (1994) Integration of a basin-scale water quality model with GIS. JAWRA J Am Water Resour Assoc 30(3):453–462

Swain JB, Patra KC (2018) Impact assessment of land use/land cover and climate change on streamflow regionalization in an ungauged catchment. J Water Clim Chang jwc2018161

Swain S, Mishra SK, Pandey A, Pandey AC, Jain A, Chauhan SK, Badoni AK (2022) Hydrological modelling through SWAT over a Himalayan catchment using high-resolution geospatial inputs. Environ Challenges 8:100579

Talchabhadel R (2022) Hydrologic response to climate change: A case from a high-mountain river basin. In: EGU General Assembly Conference. https://doi.org/10.5194/egusphere-egu22-503

Te Chow V, Maidment DR, Mays LW (1988) Applied Hydrology. McGraw Hill, Book Company, New York, pp 114–142

Te Linde AH, Aerts JCJH, Hurkmans RTWL, Eberle M (2007) Comparing model performance of two rainfall-runoff models in the Rhine basin using different atmospheric forcing data sets. Hydrol Earth Syst Sci Discuss 4(6):4325–4360

Verma N, Dash P (2018) Hydrologic Response to Land Use Change and Climate Variability in an Ungauged Basin, North-Western Himalaya, India. Sci Geogr J. https://doi.org/10.15551/SCIGEO.V63I1.377

Waggoner PE (1990) Climate change and US water resources. John Wiley, New York

Wu J, Xu Y, Gao XJ (2017) Projected changes in mean and extreme climates over Hindu Kush Himalayan region by 21 CMIP5 models. Adv Clim Chang Res 8(3):176–184

Zhang A, Zhang C, Fu G, Wang B, Bao Z, Zheng H (2012) Assessments of impacts of climate change and human activities on runoff with SWAT for the Huifa River Basin, Northeast China. Water Resour Manag 26(8):2199–2217

Chapter 4
Assessment of Heavy Precipitation-Induced Flood Studies on the Western Himalayas from 2002 to 2024: A Review

Anupam Pandey, Arun Mondal, Partha Paul, Souvik Mondal, and Surendra Kumar Chandniha

Abstract The Western Himalayas are one of the important flood hotspot zones, which impact on the entire ecosystem. To assess the current flood status of Western Himalayas triggered by heavy precipitation, authors have conducted a detailed review of past flood studies on the Western Himalayas published during 2002–2024 and observed that India is facing more frequent flood hazards than any other Western Himalayan extended countries (Pakistan and China) especially during summer monsoon season (June–September). We found that rate of publication was maximum (47.83%) during 2012–2018. Authors and organization around the world (India, Pakistan, China, USA, UK, Italy, Germany, Singapore, and Japan, etc.) have contributed to the Western Himalayan flood studies but the Indian researchers and organizations contributed more (53.35 and 37.57%) in flood studies than other countries researchers and organization (China (27.72 and 26.46%), and Pakistan (14.61, 12.70%)). The hydro-meteorological data and satellite imagery were mainly used by the authors to assess the spatio-temporal distribution of flood, their impact, and associated vulnerabilities with the assistance of different models (WRF, HEC-RAS, COSMO, MM5, NETRA, LIS, etc.) and different statistical methods (Mann Kendall test, correlation, regression, RMSE, and NASH). We also found that the Western Himalayan region faced significant damage (loss of life and property) due to the floods. It may be expected that research on the Western Himalayan floods will be increasing in the future, as flood events are expected to increase under a warming climate.

Keywords Western Himalayas · Heavy precipitation · Flood · Scopus

A. Pandey · A. Mondal (✉) · P. Paul · S. Mondal
Department of Geography, University of Allahabad, Prayagraj, India
e-mail: amondal@allduniv.ac.in

S. K. Chandniha
BRSM College of Agricultural Engineering and Technology & Research Station, IGKV, Mungeli, India

M. Biswas et al. (eds.), *Himalayan Flood Hazards*, Advances in Geographical and Environmental Sciences, https://doi.org/10.1007/978-981-92-0702-2_4

4.1 Introduction

Globally, floods are one of the costliest natural disasters, which affect the agricultural productivity, freshwater availability, and infrastructure (Hallegatte et al. 2013; Schewe et al. 2014; Lesk et al. 2016; Winsemius et al. 2016; Kundu et al. 2017a, b; Kumar and Mishra 2020; Swarnkar et al. 2021). The frequency and magnitude of floods have increased considerably in the recent decades around the globe under global warming (Westra et al. 2014; Lehmann et al. 2015; Mondal et al. 2018a, b; Papalexiou and Montanari 2019; Chinita et al. 2021). Floods have caused a loss of 1 trillion USD and affected 1.65 billion people in the last 40 years (UNDRR 2020). It is estimated that floods will occur more frequently in the near future due to climate change (Peijun 2016). Climate change causes extreme precipitation events (Mondal et al. 2017a, b; Pendergrass 2018; Zhang et al. 2021; Zhou et al. 2021) by modifying the general circulation pattern of temperature and precipitation (Hock et al. 2019). The extreme precipitation led flash floods cause widespread damage to the mountainous regions of the world, including the Himalaya (Raghuvanshi and Agarwal 2024).

The Himalayas are a global hotspot of catastrophic floods. In the recent years, Himalayas have experienced some cascading flood events that have impacted the people in the Himalayas in a multifaceted way. The floods in this region mainly occurred due to heavy rainfall and cloudburst during the summer monsoon season (Nandargi and Dhar 2011; Mukhopadhyay et al. 2014; Mondal et al. 2016; Mahto et al. 2023). Some notable flood events are the Dhauli Ganga valley flood (2021), Uttarakhand flood (2022), Indian Himalaya flood (2022), Pakistan flood (2022), Himachal Pradesh flood (2023), North India flood (2023) caused huge damage in this region (Mondal et al. 2022; Sain et al. 2022; Samantray and Gouda 2024; Bazai et al. 2024; Gupta et al. 2024; Kushwaha et al. 2024). The increased frequency of flood events in the Himalayas has posed a risk to water and food security in South Asian countries (IPCC 2022). Floods have affected more than 1 billion people in South Asia in last two decades (Nanditha et al. 2023). The increasing frequency of floods in the Himalayas is a consequence of observable hydro-meteorological changes under warming climate (Bookhagen 2010; Nandargi and Dhar 2011).

Marked increases in temperature and precipitation have been observed in the Western Himalayan region in the last few decades (Namgyal et al. 2025). It has experienced an increase in extreme flows during 1980–2003 (Kundu et al. 2015). This region has witnessed some high magnitude flood events is the last decade, such as 2010 flood in Khyber Pakhtunkhwa, 2010 flood in Ladakh, 2014 flood in Kashmir, 2018 flood in Nubra valley, 2022 flood in Pakistan; 2023 flood in Himachal Pradesh, (Rahman and Khan 2013; Kar and Tiwari 2016; Banerjee and Dimri 2019; Ahmad et al. 2020; Thayyen et al. 2020; Shekhar et al. 2024; Wang et al. 2024a, b). The precipitation in Western Himalayas is mainly governed by southwest monsoonal winds in summer and subtropical westerly jet winds during winters (Simpson 1921; Pisharoty and Desai 1956; Webster et al. 1998; Dimri et al. 2015).

There have been multiple studies that examine the extreme precipitation led flood events in the Western Himalayas (Martius et al. 2013; Kar and Tiwari 2016; Romshoo et al. 2018; Narasimha Rao et al. 2021; Gupta et al. 2022; Sain et al. 2022; Raghuvanshi and Agarwal 2024; Samantray and Gouda 2024; Gupta et al. 2024; Kushwaha et al. 2024; Shekhar et al. 2024; Bazai et al. 2024). Martius et al. (2013) utilized ERA-Interim data, Tropical Rainfall Measuring Mission (TRMM) data, and Rain gauge data to identify the factors that led to the Pakistan flood, 2010. Kar and Tiwari (2016) used India Meteorological Department (IMD) rainfall data, Tropical Rainfall Measuring Mission Multi-Satellite Precipitation Analysis (TMPA), and MODIS snow cover data and configured the Weather Research and Forecasting (WRF) model to simulate the heavy rainfall event that occurred in Jammu and Kashmir, 2014. Romshoo et al. (2018) utilized rainfall data from the Indian Meteorological Department (IMD), Integrated Multi-Satellite Retrievals for GPM (IMERG), and INSAT-3D, along with discharge data and field survey to assess the extent and the causes of Jhelum basin flood, 2014. Rao et al. used Indian Meteorological Department (IMD) rainfall data, ERA5 precipitation data at 0.25° resolution, INSAT-3D satellite data and also configured the Weather Research and Forecasting (WRF) model to simulate the rainfall event that occurred in Yamunanagar, 2016. Sain et al. (2022) utilized rainfall data and Google Earth images to identify the spatio-temporal distribution of rainfall and assess the damage caused by the Amarnath flood, 2022. Gupta et al. (2024) used ERA5 hourly precipitation data at $0.25° \times 0.25°$ resolution, State Data Centre (SDC) precipitation and discharge data, and Sentinel-1 SAR imagery to analyze flood over Himachal Pradesh, 2023. Bazai et al. (2024) used Pakistan Meteorological Department (PMD) rainfall data, flood discharge data to analyze the extent of the flood. Alongside this, authors also conducted field survey to assess the damage caused by the Swat River flood, 2022. Notably, most of the studies mainly focused on analyzing the processes that caused these type of extreme event, flood extent and their impact on the environment and socio-cultural setting. However, none of the studies have given a complete picture of the floods of the western Himalayas. Hence, the present study is undertaken to provide readers a complete information about the flood event that occurred in the Western Himalaya in last 20 years.

The overall objective of this paper is to identify the areas in the western Himalayas that were more affected by floods and the time of their occurrence (summer, monsoon, autumn, winter, or spring). The authors also want to analyze the research contribution by the researchers in flood studies. The present study also wants to find the types of data (primary or secondary) and methods (primary survey, statistical analysis, model run, satellite image analysis, etc.) used in analyzing flood events to provide detailed insight into the popular tools and techniques that have been in use in flood studies. Lastly, this study will assess the extent of damage caused by the floods in the western Himalayas.

4.2 Study Area

The Himalayas are a crescent-shaped mountain range extending for over 2500 km from the south of the Indus Valley beyond Nanga Parbat in the west to Namcha Barwa in the east and spanning over 6 South Asian countries (India, China, Nepal, Bhutan, Pakistan, and Myanmar). The range width varies from 350 km in the west to 150 km in the east (Sathyakumar et al. 2020). The entire Himalayas have been categorized into three major regions: (i) Western Himalayan region, (ii) Central Himalayan region, and (iii) Eastern Himalayan region. The Western Himalayan region is one of the largest regions of the Himalayas (Das and Meher 2019). The Western Himalayan region extends from 30°29′4″ N to 37°34′30″ N and from 72°41′30″ E to 80°39′47″ E and covers a 292,114 km^2 area. It spans across countries like India, Pakistan, and China. In India, it encompasses the states of Jammu and Kashmir, Ladakh, Himachal Pradesh, and parts of Uttarakhand. The Western Himalayas consist of four parallel mountain ranges (Zaskar, Pir Panjal, Dhaola Dhar, and Sivalik). The highest point of the study area is Nanga Parbat (8126 m). The study region is a source of major rivers of South Asia like the Beas, Chenab, Ravi, and Jhelum. The climate of the Western Himalayas varies with altitude from subtropical to wet-temperate and moist temperate to cold arid. The mean annual rainfall in the western Himalayas is around 75–150 mm. The forest types in the Western Himalayas also change with changing elevation. The tropical Sal forests are found in the foothill region. On the other hand, broadleaved forests are found at an elevation of 1000–2000 m. Coniferous forests are observed at an elevation of 2000–3000 m, and sub-alpine forests are found at >3000 m. The alpine meadows and shrubs are observed at an elevation of >4000 m (Fig. 4.1).

4.3 Database and Methodology

For the present research work, authors thoroughly searched western Himalayan flood-related articles (2002 to 2024) in the Scopus database with the Keywords "**Flood**" AND "**Western Himalayas**" AND "**Heavy precipitation**" AND "**Cloudburst**". Authors included original research papers and book chapters all written in English, in the present study. Scopus database provides wide range of papers from every discipline and cited data (Hasana et al. 2022). All selected articles (69) on the Western Himalayan flood were retrieved from the Scopus database in CSV format. After that, a thorough reading of the articles was done, and various information related to the West Himalayan flood (flood occurrence place, year of flood event, season of flood, flood cause, methods used in flood studies, and damage caused by flood) was tabulated. Based on these data, different graphs were prepared to show the year-wise article publication, journal, and organization contributions. Various maps were also prepared based on tabulated data and extracted bibliometric information to show the location of the flood area, satellite imagery-based studies, contribution of author and

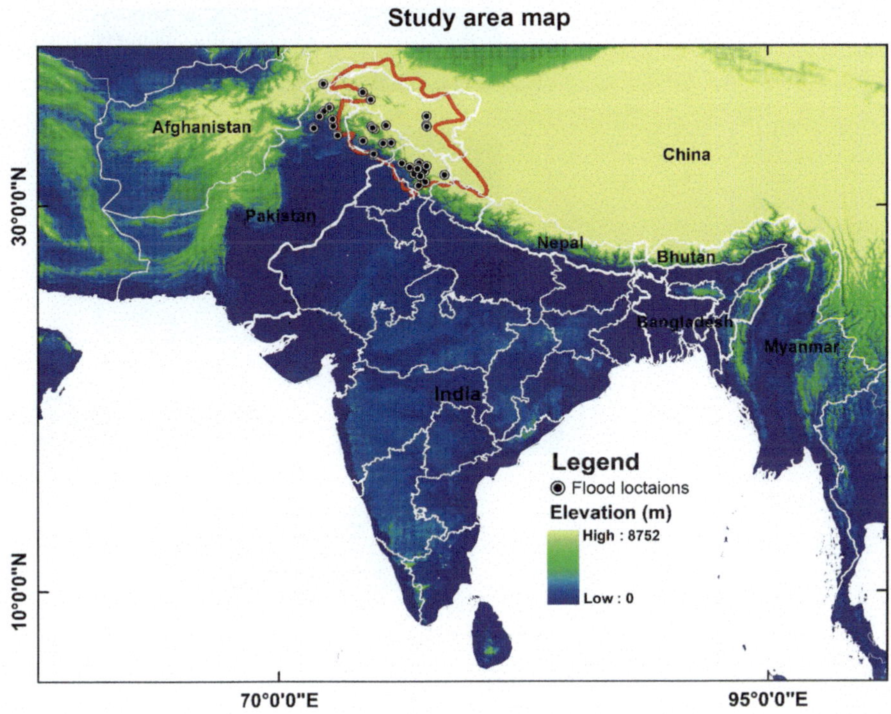

Fig. 4.1 Location of the Western Himalayas (red polygon) in South Asia and the distribution of flood locations across the Western Himalayas. The color scale shows the elevation (m) variation in South Asia and Western Himalayas

organizations in the western Himalayas flood studies, and their collaboration. The methodological flow chart is depicted in Fig. 4.2.

4.4 Results

4.4.1 Flood Occurrence Place

4.4.1.1 Location

Figure 4.1 shows the location of all the flood events that occurred in the Western Himalayas in the last two decades. It is visible that more of the floods occurred in the south-western part of the Western Himalayas. On the other hand, the northern part of the Western Himalayas experienced least flood events. The flood events in this part of Himalayas mainly occurred due to the dual effects of monsoonal winds and western disturbances.

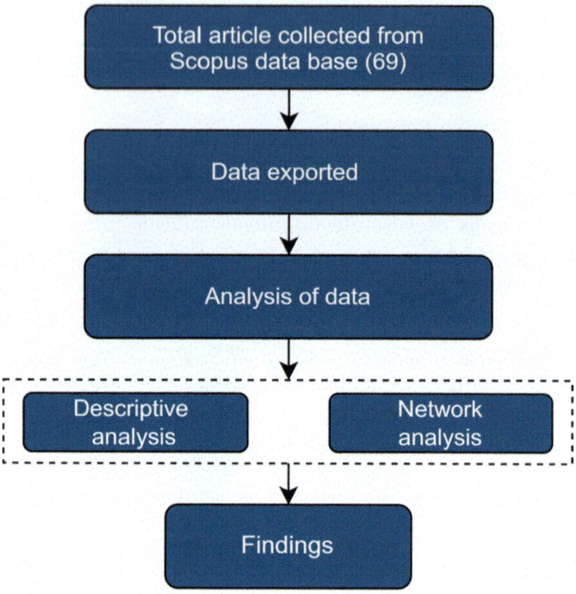

Fig. 4.2 Methodological flow chart of the entire study

4.4.1.2 Study Area-Wise Publication

Figure 4.3 shows the country-wise flood events. Among the western Himalayan countries (India, Pakistan, and China), India was experienced more flood events (53.62%) followed by Pakistan (42.03%). There were few flood events that were experienced by both India and Pakistan (4.35%). In India, Jammu and Kashmir was the most flooded area (36.15%) followed by Himachal Pradesh (24.32%), and Ladakh (24.32%) (Fig. 4.4).

4.4.2 Flood Occurrence Seasons

Floods over the western Himalayas mainly occurred in the summer monsoon season (June to September) caused by heavy rainfall and cloudbursts. During this season, moisture-laden monsoonal winds are obstructed by the mighty Himalayas. As a result, they move upwards and cause heavy rainfall in the Himalayas and their surrounding regions. Meanwhile, there was one flood event that occurred in the western Himalayas other than the monsoon season. The Pakistan, 2022 flood occurred in the summer season (March to June) due to the extreme rainfall (Otto et al. 2023; Cui et al. 2024).

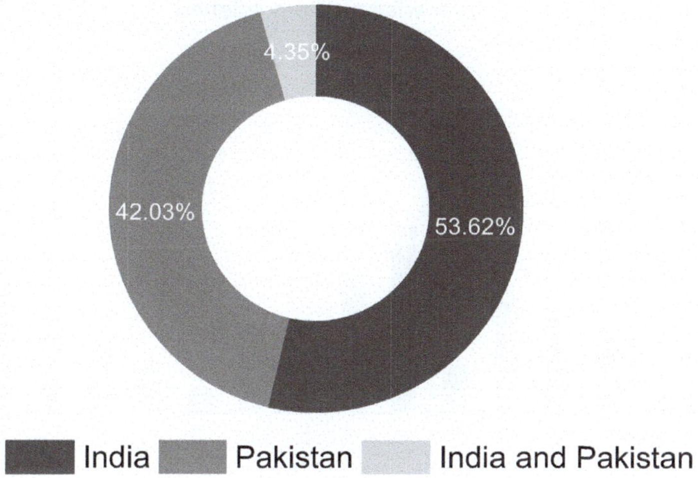

Fig. 4.3 Study area wise article publication on Western Himalayan flood studies. India with 53.62% experienced most flood events

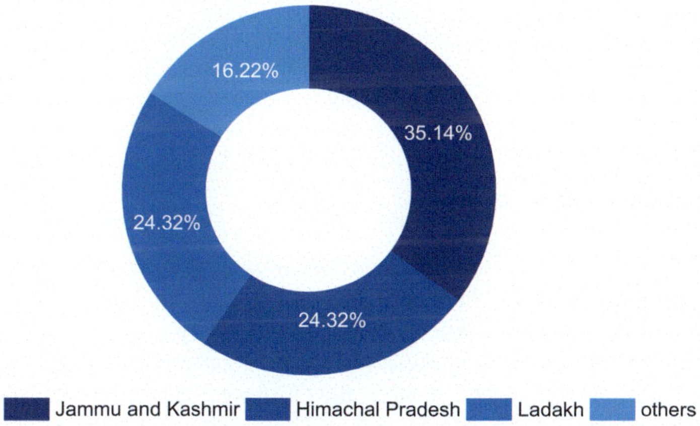

Fig. 4.4 Study area wise publication on flood studies in Indian states of western Himalayas. Jammu and Kashmir with 34.14% experienced most flood events

4.4.3 Publication Result

4.4.3.1 Publication Trend

Figure 4.5 shows that before 2012, article publications were very less ($\leq$3/year). Meanwhile, a rise in article publications on western Himalayan flood studies was observed from 2012 onwards. Since 2012, the year-wise article publications were $\geq$ 4/ year. The only exceptions to this were 2014, 2019, 2020, and 2021 ($\leq$ 3/year). From

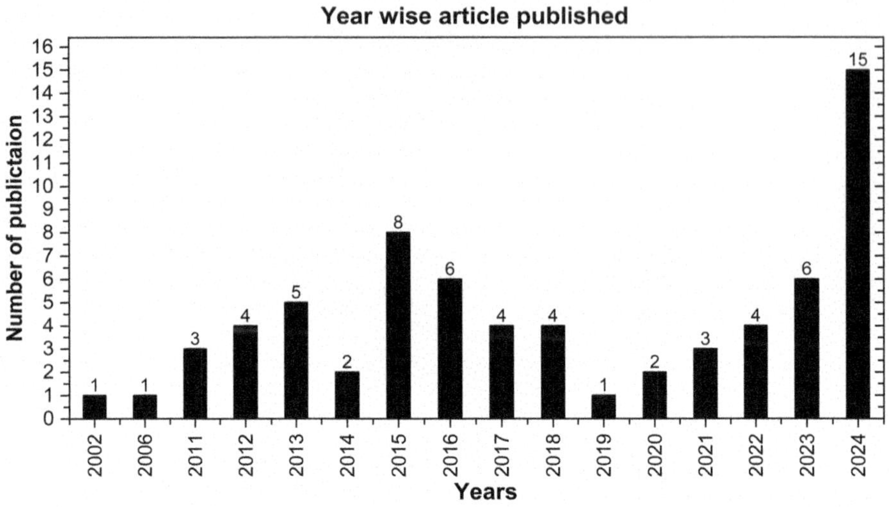

Fig. 4.5 Bar graph shows the year-wise article publications on Western Himalayan flood with higher fluctuation from 2012 onwards

2012 to 2018, article publications were maximum (n = 33). Among the observed years (2002 to 2024), the maximum number of articles on flood studies was published in 2024 (n = 15).

4.4.3.2 Author Contribution

Figure 4.6 shows the collaboration among prominent authors in flood studies. A total of 267 authors came together to form 45 clusters and 766 links, with a total link strength of 786. Authors of the articles were represented by a separate node and with a unique color. The links among these nodes show authors collaboration in flood studies (Zhao 2017). Among the clusters, the cluster with red color was the biggest (n = 21 authors). Among the authors (n = 267), 97.76% were associated with only 1 article (n = 245), and only 0.75% were associated with 3 articles (n = 2), and the rest 7.49% were associated with 2 articles (n = 20). S.C. Bhan and Amit Kumar were the most prominent authors, with 3 articles. Among the authors, Peng Cui was the most connected with a total link strength of 16. On the other hand, A.P. Dimri received the highest citation (n = 150) compared to the more prominent authors like S.C. Bhan and Amit Kumar.

4.4.3.3 Organization Contribution

Figure 4.7 shows the association between various institutions and their shared influence in the western Himalayan flood studies. A total of 189 institutions formed 50

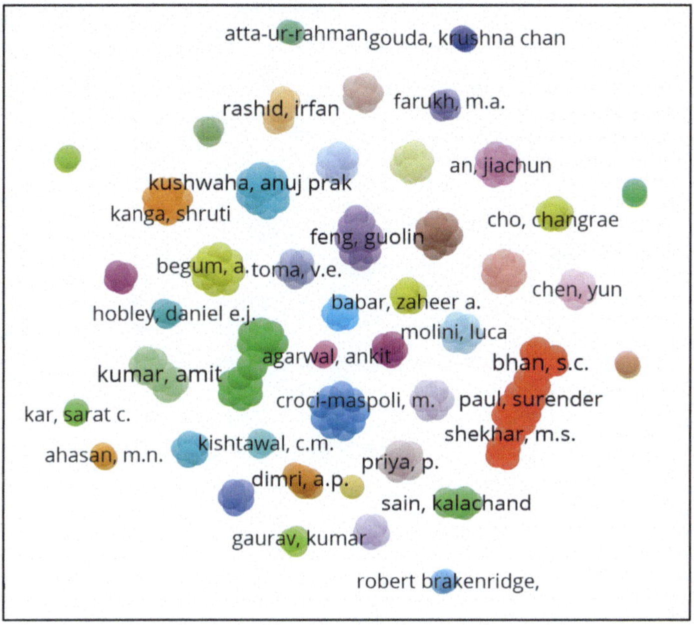

Fig. 4.6 Author connectivity map shows the contribution of researchers to flood studies over the Western Himalayas. A total of 267 authors formed 45 clusters. Each cluster is represented with unique color

clusters and 505 links with a total link strength of 506. Among the clusters, the most prominent cluster was 1 (19 authors) represented with red color. The Chinese Academy of Sciences was the most influential (5 articles) among other institutions. Other institutions like the University of Washington, Indian Meteorological Department, Pacific North-west National Laboratory, and Royal Netherlands Meteorological Institute also published quality articles on the Himalayan flood (2 articles, respectively).

4.4.3.4 Journal Contribution

A total of 69 articles were published in 49 journals. Among the journals, 83.67% of the journals published only 1 flood article, 10.20% published 2 articles, and only 6.12% published 3 or more articles. Among the top 8 journals, Natural Hazards was the most prominent journal and published the maximum number of articles (n = 12). All the top journals had a good impact factor and were indexed in quality indices (Fig. 4.8).

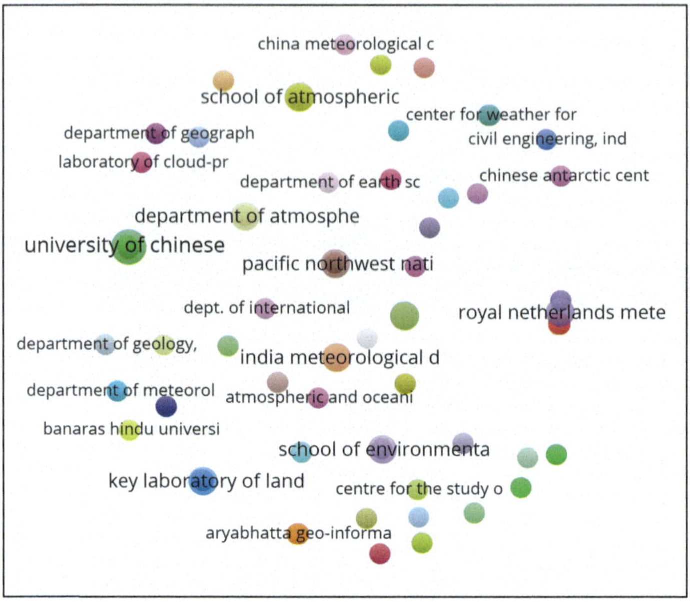

Fig. 4.7 Organization connectivity map shows the contributions of academic institutions to flood studies over the Western Himalayas. A total of 189 institutions formed 50 clusters. Each cluster is represented with unique color

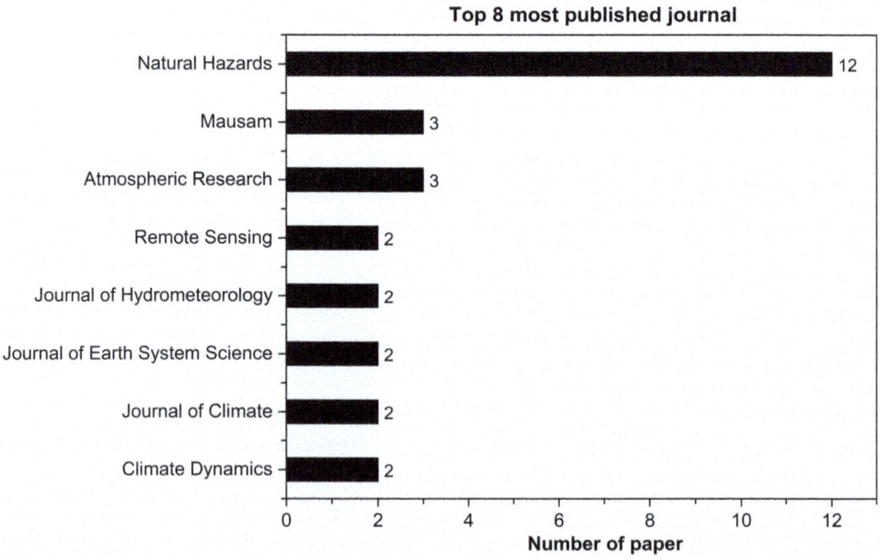

Fig. 4.8 Bar graph shows the top 8 Journals publishing quality articles on Western Himalayan floods from 2002 to 2024. Natural Hazards is the most prominent journal with highest number of articles

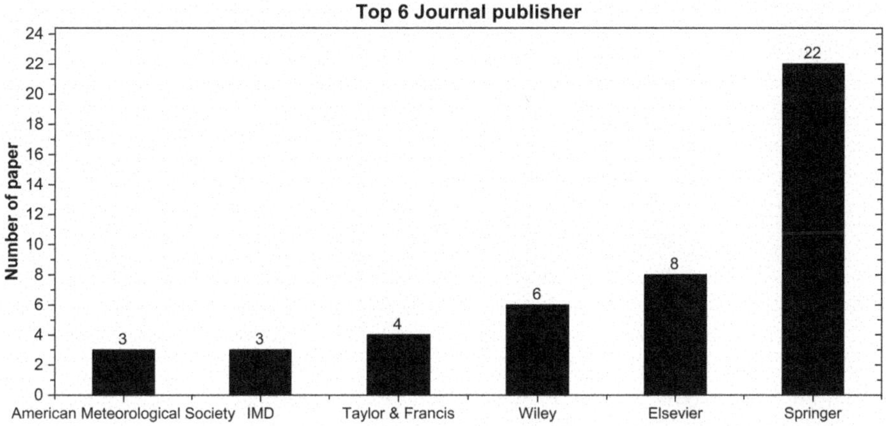

Fig. 4.9 Bar graph shows the top 6 journal publisher publishing quality articles on Western Himalayan floods from 2002 to 2024. Springer is the most prominent publisher with highest number of publications

4.4.3.5 Publisher Contribution

A total of 69 articles were published by 24 publishers from 2002 to 2024. Among the publishers, 4 publishers (Springer, Elsevier, Wiley, and Taylor and Francis) published 57.97% (n = 40) articles during 2002 to 2024. These 4 publishers represent only 16.67% of the total publishers. Meanwhile, 42.03% of the articles (n = 29) were published by rest 83.33% of the publishers (n = 20). Among the top 6 journals, American Meteorological Society and IMD were the least productive journals (n = 3) (Fig. 4.9).

4.4.4 Data Used

The authors of the flood studies used both primary and secondary data to evaluate flood events in the western Himalayas. Authors mainly collected primary data by field surveys, conducting interviews of flood-affected people, and real-time observation, etc. Authors (Das et al. 2006; Arya 2011; Thayyen et al. 2013; Hobley et al. 2012; Rahman and Khan 2013; Bahadar et al. 2015; Sangode et al. 2017; Gupta et al. 2022; Malik et al. 2022; Chand et al. 2024; Bazai et al. 2024) used primary data like thickness of the deposited debris, water level during and aftermath of the flood, channel width, extent of flood inundated area, rainfall duration, information related to time of occurrence of flood, and damage to assess the various flood events (Himalaya, 2003; Leh, India, 2010; Leh, India, 2010; Ladakh, India, 2010; Ladakh, India, 2010; Pakistan, 2010; Ladakh India, 2010; Himachal Pradesh, India, 2021; Kashmir valley,

India, 2014; Kullu valley, Himachal Pradesh, India, 2018; Swat river basin, Pakistan, 2022).

Secondary data like hydro-meteorological and satellite data are very crucial for assessing various aspects of floods, such as rainfall amount during floods, peak discharge, flood extent, and vulnerability. The authors used various kinds of secondary data (hydro-meteorological and satellite data) to assess the flood events and their associated effects in the western Himalayas. Hydrological data like river discharge and water level data were used by the authors (Mondal et al. 2012; Hobley et al. 2012; Thayyen et al. 2013; Martius et al. 2013; Syvitski and Brakenridge 2013; Bahadar et al. 2015; Romshoo et al. 2018; Wang et al. 2024a, b) to assess the water level and water discharge during flood, to run HEC-RAS model, and to know the atmospheric process associated with flood event in their research work (Ladak, India, 2010; Leh, India, 2010; Martius et al. 2013; Syvitski and Brakenridge 2013; Bahadar et al. 2015; Kashmir, 2014; Pakistan, 2022).

Authors also used different kinds of meteorological data (rainfall data and air temperature data) in their flood studies. Indian Meteorological Department (IMD) rainfall data were used by various authors (Mondal et al. 2025; Sain et al. 2022; Gupta et al. 2022; Chand et al. 2024; Kanwar et al. 2024; Shekhar et al. 2024; Khan et al. 2024) to assess the rainfall amount, and rainfall variability during flood, flood vulnerability, etc., of flood events (Amaranth holy cave, Jammu and Kashmir, India, 2022; Himachal Pradesh, India, 2021; Himachal Pradesh, India, 2018; Himachal Pradesh, India; Himachal Pradesh, India, 2023; Jammu and Kashmir, India, 2021). Meanwhile, satellite-based rainfall data like Tropical Rainfall Measuring Mission (TRMM) also used by the authors (Thayyen et al. 2013; Ray et al. 2015; Bhatt et al. 2017; Kumar et al. 2017; Sangode et al. 2017) to find out the spatial, temporal, and seasonal rainfall variation over flood-affected areas (Leh, India, 2010; Jammu and Kashmir, India, 2014; Kashmir, India, 2014; Jammu and Kashmir, India, 2014; Ladakh, India, 2010).

Some of the authors (Das et al. 2006; Kumar et al. 2014; Ullah and Shouting 2013; Ahasan and Khan 2013; Martius et al. 2013; Kumar et al. 2014; Bhan et al. 2015; Muhammad Tahir et al. 2015; Priya et al. 2015; Rasmussen et al. 2015; Yadav et al. 2015; Cho et al. 2016; Viterbo et al. 2016; Kar and Tiwari 2016; Priya et al. 2017; Shukla et al. 2017; Rashid et al. 2019; Van der Schrier et al. 2018; Banerjee and Dimri 2019; Di Capua et al. 2021; Narasimha Rao et al. 2021; Nanditha et al. 2023; Ma et al. 2023a, b; Otto et al. 2023; Raghuvanshi and Agarwal 2024; Wang et al. 2024a, b; Gupta et al. 2024; Kushwaha et al. 2024; Kanwar et al. 2024; Cui et al. 2024) used various precipitation data set of different sources (MORPH, CMORPH, JAXA, IMERG, GCM, PMD, CHIRPS, REANALYSIS data ERA5, NOAA) to analyze the trend in precipitation, spatio-temporal distribution of precipitation, etc., in their respective flood studies (Himachal Pradesh, India, 2003; Leh, India, 2010; Pakistan, 2010; North-west Pakistan, 2010; Pakistan, 2022; Leh, India, 2010; Leh, India, 2010; Pakistan, 2010; North-west India and Pakistan, 2010; Pakistan, 2010; Western Himalayan, 2013; North India, 2013; Pakistan, 2010; Jammu and Kashmir, India, 2014; Western Himalayas, 2013; Western Himalaya, 2013; Kashmir, India, 2014; Pakistan, 2010; Leh, India, 2010; Pakistan, 2010; Yamunanagar, India, 2016;

Pakistan, 2022; Pakistan, 2022; Pakistan, 2022; Himachal Pradesh, India, 2023; Pakistan, 2022; Pakistan, 2022; Himachal Pradesh, India, 2023; North India, 2023; Himachal Pradesh, India; Pakistan, 2022).

Figure 4.10 shows the year-wise satellite imagery used by the authors (Gaurav et al. 2011; Kumar et al. 2014; Ahasan and Khan 2013; Bhan et al. 2015; Yadav et al. 2015; Kumar and Acharya 2016; Bhatt et al. 2017; Viterbo et al. 2016; Kumar et al. 2017; Romshoo et al. 2018; Bisht et al. 2018; Khalid et al. 2018; Banerjee and Dimri 2019; Ahmad et al. 2020; Thayyen et al. 2020; Narasimha Rao et al. 2021; Nanditha et al. 2023; Chen et al. 2024; Wang et al. 2024a, b; Bazai et al. 2024; Gupta et al. 2024; Kushwaha et al. 2024) in flood studies (Pakistan, 2010; Leh, India, 2010; North-western Pakistan, 2010; Leh, India, 2010; Western Himalayas, 2013; Kashmir, India, 2014; Jammu and Kashmir, 2014; Pakistan, 2010; Jammu and Kashmir, India, 2014; Kashmir, India, 2014; Himachal Pradesh, India; Pakistan, 2010; Leh, India, 2010; Kashmir, India, 2014; Ladakh, India, 2018; Yamunanagar, India, 2016; Pakistan, 2022; Pakistan, 2022; Pakistan, 2022; Pakistan, 2022; Himachal Pradesh, India, 2023; North India, 2023). Landsat imageries (Landsat 4, Landsat TM, Landsat MSS, Landsat ETM, Landsat 7, Landsat 8) were used in more number of articles ($n = 16$) followed by Sentinel imageries ($n = 11$). Among Landsat imagery, Landsat 8 was used in more number of articles ($n = 7$). Meanwhile, in Sentinel Imageries, Sentinel-2 was used more by the authors ($n = 6$). INSAT, Meteosat-7, Meteosat-5, Radarsat-2, and MODIS were the least used by the authors.

4.4.5 Methods Used

The authors utilized various methods to assess the western Himalayan flood. Based on the methods, this paper is divided into four major aspects, i.e., primary survey, model, statistical analysis, and satellite image analysis. The primary survey method was utilized by various authors (Das et al. 2006; Arya 2011; Thayyen et al. 2013; Hobley et al. 2012; Rahman and Khan 2013; Kumar et al. 2017; Romshoo et al. 2018; Sangode et al. 2017; Bisht et al. 2018; Malik et al. 2022; Gupta et al. 2022; Bazai et al. 2024; Mondal et al. 2026; Chand et al. 2024) to assess the flood extent, water level after flood, flood impact, etc., in their studies (Himachal Pradesh, India, 2003; Leh, India, 2010; Leh, India, 2010; Ladakh, India, 2010; Ladakh, India, 2010; Jammu and Kashmir, India, 2014; Kashmir, India, 2014; Ladakh, India, 2010; Himachal Pradesh, India; Kashmir, India, 2014; Himachal Pradesh, India, 2021; Pakistan, 2022; Himachal Pradesh, India, 2018). On the other hand, water system-based reputed models (WRF, HEC-RAS, COSMO, MM5, NETRA, and LIS etc.) were used by various authors (Das et al. 2006; Kumar et al. 2014; Thayyen et al. 2013; Ullah and Shouting 2013; Ahasan and Khan 2013; Martius et al. 2013; Kumar et al. 2014; Muhammad Tahir et al. 2015; Bahadar et al. 2015; Cho et al. 2016; Viterbo et al. 2016; Kar and Tiwari 2016; Shukla et al. 2017; Sangode et al. 2017; Van der Schrier et al. 2018; Mehmood et al. 2021; Narasimha Rao et al. 2021; Otto et al. 2023; Nanditha et al. 2023) to predict the intensity of flood and its future directions, and to map

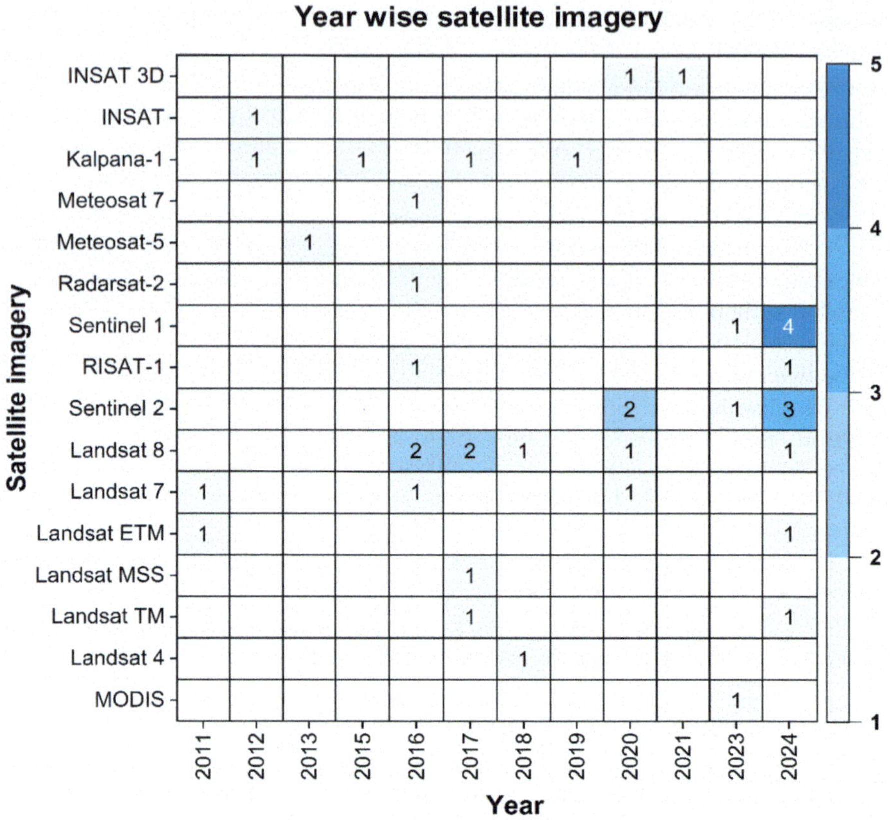

Fig. 4.10 Heat map shows the year-wise distribution of different satellite imagery used in flood studies. The year 2024 recorded highest number of satellite imagery-based flood studies

flood plain areas in their flood studies (Himachal Pradesh, India, 2003 Leh, India, 2010; Leh, India, 2010; Pakistan, 2010; North-Western Pakistan, 2010; Pakistan, 2010; Jammu and Kashmir, India, 2010; Pakistan, 2010; Pakistan, 2010; North India, 2013; Pakistan, 2010; Jammu and Kashmir, India, 2014; Western Himalayas, 2013; Ladakh, India, 2010; Pakistan, 2010; Pakistan, 2010; Yamunanagar, India, 2016; Pakistan, 2022; Pakistan, 2022).

Authors (Rahman and Khan 2013; Hartmann and Andresky 2013; Hartmann and Buchanan 2014; Feige et al. 2014; Yamada et al. 2016; Viterbo et al. 2016; Van der Schrier et al. 2018; Rashid et al. 2019; Khalid et al. 2018; Banerjee and Dimri 2019; Mehmood et al. 2021; Di Capua et al. 2021; Ullah et al. 2021; Sain et al. 2022; Ahmad et al. 2022; Ahsan et al. 2023; Ma et al. 2023a, b; Samantray and Gouda 2024; Wang et al. 2024a, b; Gupta et al. 2024; Kanwar et al. 2024; Cui et al. 2024; Shekhar et al. 2024; Raghuvanshi and Agarwal 2024) also utilized various statistical analysis (Mann Kendall test, correlation, regression, RMSE, and NASH) to assess the trends in precipitation, annual and seasonal precipitation change in their respective studies (Ladakh, India, 2010; Indus River basin, 1951–2010; Pakistan, 1979–2011; Pakistan, 2010; Pakistan and North west India; 2010; Pakistan, 2010; Pakistan, 2010; Kashmir, India, 2014; Pakistan, 2010; Leh, India, 2010; Pakistan, 2010; Pakistan, 2010; Pakistan, 1981–2018; Amarnath holy cave, Jammu and Kashmir, India 2022; Kashmir, India, 2014; North-west Himalayas, 1980–2016; Pakistan, 2022; Indian Himalayas, 2022; Pakistan, 2022; Himachal Pradesh, India, 2023; Himachal Pradesh, India; Palistan, 2022; Himachal Pradesh, 2023; Himachal Pradesh, 2023).

Satellite imagery (Landsat 4, TM, MSS, ETM, 7, 8, MODIS, Sentinel−1, Sentinel−2, INSAT, Kalpana-1, etc.) processing was by various authors (Gaurav et al. 2011; Kumar et al. 2014; Ahasan and Khan 2013; Bhan et al. 2015; Yadav et al. 2015; Kumar and Acharya 2016; Bhatt et al. 2017; Viterbo et al. 2016; Kumar et al. 2017; Romshoo et al. 2018; Bisht et al. 2018; Khalid et al. 2018; Banerjee and Dimri 2019; Ahmad et al. 2020; Thayyen et al. 2020; Narasimha Rao et al. 2021; Nanditha et al. 2023; Chen et al. 2024; Wang et al. 2024a, b; Bazai et al. 2024; Gupta et al. 2024; Kushwaha et al. 2024) to analyze the flood extent and depth, flood inundation and vulnerable areas in their flood studies (Pakistan, 2010; Leh, India, 2010; North-western Pakistan, 2010; Leh, India, 2010; Western Himalayas, 2013; Kashmir, India, 2014; Jammu and Kashmir, 2014; Pakistan, 2010; Jammu and Kashmir, India, 2014; Kashmir, India, 2014; Himachal Pradesh, India; Pakistan, 2010; Leh, India, 2010; Kashmir, India, 2014; Ladakh, India, 2018; Yamunanagar, India, 2016; Pakistan, 2022; Pakistan, 2022; Pakistan, 2022; Pakistan, 2022; Himachal Pradesh, India, 2023; North India, 2023).

In addition to these methods, Khalid et al. (2018) and Gupta et al. (2022) used morphometric analysis of the upper Indus basin and Bhagsu Nala to assess the geomorphological and hydrological characteristics in their respective flood studies (Pakistan, 2010; Himachal Pradesh, India, 2021).

4.5 Damage

Floods in the western Himalayas caused heavy casualties and huge damage to public and private property. Massive economic loss (public infrastructure and house damage, etc.) reported by various authors (Gardner 2002; Ahmad et al. 2011; Arya 2011; Gaurav et al. 2011; Kumar et al. 2014; Hobley et al. 2012; Hashmi et al. 2012; Thayyen et al. 2013; Syvitski and Brakenridge 2013; Hartmann and Andresky 2013; Feige et al. 2014; Kumar et al. 2014; Muhammad Tahir et al. 2015; Rasmussen et al. 2015; Bahadar et al. 2015; Mishra 2015; Ray et al. 2015; Kumar and Acharya 2016; Kumar et al. 2017; Romshoo et al. 2018; Sangode et al. 2017; Van der Schrier et al. 2018; Bisht et al. 2018; Rashid et al. 2019; Ahmad et al. 2020; Thayyen et al. 2020; Sain et al. 2022; Malik et al. 2022; Gupta et al. 2022; Gouda et al. 2023; Ma et al. 2023a, b; Otto et al. 2023; Sawa 2023; Nanditha et al. 2023; Wang et al. 2024a, b; Bazai et al. 2024; Gupta et al. 2024; Kushwaha et al. 2024; Cui et al. 2024; Shekhar et al. 2024; Mahmood et al. 2024; Khan et al. 2024; Chand et al. 2024; Raghuvanshi and Agarwal 2024) in their flood studies. Agricultural loss (damaged to cropped area and water logging) was reported by the authors (Ahmad et al. 2011; Hashmi et al. 2012; Hartmann and Andresky 2013; Syvitski and Brakenridge 2013; Feige et al. 2014; Bahadar et al. 2015; Mishra 2015; Kumar and Acharya 2016; Gouda et al. 2023; Malik et al. 2022; Nanditha et al. 2023; Ma et al. 2023a, b; Otto et al. 2023; Sawa 2023; Chand et al. 2024; Wang et al. 2024a, b; Bazai et al. 2024; Gupta et al. 2024; Kushwaha et al. 2024; Mahmood et al. 2024; Khan et al. 2024) and heavy casualties (human and animal live loss) were also reported by authors (Ahmad et al. 2011; Gaurav et al. 2011; Hashmi et al. 2012; Thayyen et al. 2013; Kumar et al. 2014; Hobley et al. 2012; Hartmann and Andresky 2013; Syvitski and Brakenridge 2013; Feige et al. 2014; Kumar et al. 2014; Ray et al. 2015; Bhan et al. 2015; Muhammad Tahir et al. 2015; Rasmussen et al. 2015; Bahadar et al. 2015; Mishra 2015; Kumar et al. 2017; Van der Schrier et al. 2018; Ahmad et al. 2020; Gupta et al. 2022; Gouda et al. 2023; Sain et al. 2022; Malik et al. 2022; Nanditha et al. 2023; Ma et al. 2023a, b; Otto et al. 2023; Sawa 2023; Chand et al. 2024; Wang et al. 2024a, b; Bazai et al. 2024; Gupta et al. 2024; Kushwaha et al. 2024; Cui et al. 2024; Shekhar et al. 2024; Mahmood et al. 2024; Khan et al. 2024) in their respective studies. All of the reported floods have adversely impacted in one way or another, but some caused more devastation in terms of life loss, infrastructure damage, and crop damage. For instance, **Ladakh Himalaya, India, 2010** flood caused devastation and damaged near about 71 towns and villages. At least 255 people died due to heavy overnight rainfall triggered flash floods (Sangode et al. 2017). **Leh, India, 2010 flood** caused huge loss of life and property. It claimed lives of 200 people and injured 400 people (Bhan et al. 2015). **Pakistan, 2010 flood** claimed 1000 people's lives. This flood affected 507,000 ha of standing crops, 52,000 cattle, 900 educational institutions, and damaged 200 thousand houses (Rahman and Khan 2013). **Jammu and Kashmir, India, 2014** flood caused havoc and claimed 300 people's lives. Preliminary damage over 3.50 lakh structures including 2.50 lakh residential houses (Ray et al. 2015). **Himachal Pradesh, India, 2023** flood caused huge economic losses of 10.59 crores. It claimed 72 lives and injured 92 people (Raghuvanshi and Agarwal 2024).

4.6 Conclusion

The present study provides a detailed description of floods caused by heavy precipitation in the Western Himalayas. Based on the objectives, the authors have made the following observation. Authors have found that the South-western part of the western Himalayas experienced more flood events (60.87%) than other parts of the Western Himalayas due to the dual effects of monsoonal winds and western disturbance. Notably, the majority of these floods occurred in the summer monsoon season (June–September). It is observed that among the Western Himalayas countries (India, Pakistan, and China), India (53.63%) was affected more by extreme floods in the last two decades. It may be due to the extent of the Western Himalayas in India (965.65 Km^2) and nature of the Indian climate influenced by monsoonal winds. Between 2002 and 2024, more articles (47.83%) were published from 2012 to 2018. It shows that during this period, Western Himalayas experienced major floods. S.C. Bhan and Amit Kumar contributed more to flood studies and collaborated more with other authors. Among the institutions, the Chinese Academy of Sciences contributed more to evaluating the flood events. Different institutions all around the world have participated in Western Himalayan flood studies. Natural Hazard was the more influential journal with the highest number of article (n = 12), and among the publishers, Springer, Elsevier, Wiley, and Taylor and Francis were the leading publishers in publishing quality articles on the Western Himalayan floods. We observed that, authors of the articles utilized various kinds of data (primary and secondary). Secondary data was used in more studies, especially hydro-meteorological and satellite data. In hydro-meteorological data, rainfall data (IMD, MORPH, CMORPH, JAXA, CHIRPS, TRMM, etc.) and in satellite imagery Landsat images (Landsat 4, Landsat MSS, Landsat TM, Landsat ETM, Landsat 7, and Landsat 8) were used more by the authors. Based on the availability of data, authors selected appropriate methods (primary survey, water system-based model, statistical analysis, and satellite image analysis) to accurately predict the flood event. Floods have wreaked havoc in the Himalayas in the last two decades. Huge economic loss and loss of life, damage to houses, roads, highways, and agricultural lands were reported from the flood studies.

In last, it can be said that this study will be helpful for the future researchers to understand the flood hotspot in Western Himalayas and to select appropriate data and method for analyzing flood events, their extent, and associated vulnerabilities. Additionally, such studies will also help the decision maker to develop a proper flood risk mitigation strategies for the West Himalayan region (Kashyap et al. 2025).

References

Ahasan MN, Khan AQ (2013) Simulation of a flood producing rainfall event of 29 July 2010 over north-West Pakistan using WRF-ARW model. Nat Hazards 69(1):351–363

Ahmad F, Kazmi SF, Pervez T (2011) Human response to hydro-meteorological disasters: a case study of the 2010 flash floods in Pakistan. J Geogr Reg Plan 4(9):518

Ahmad T, Pandey AC, Kumar A (2020) Impact of 2014 Kashmir flood on land use/land cover transformation in Dal Lake and its surroundings, Kashmir valley. SN Appl Sci 2(4):681

Ahmad T, Pandey AC, Kumar A (2022) Long-term precipitation monitoring and its linkage with flood scenario in changing climate conditions in Kashmir valley. Geocarto Int 37(19):5497–5522

Ahsan S, Bhat MS, Alam A, Sheikh HA, Farooq H (2023) Hydrological extremes and climatic controls on streamflow in Jhelum basin, NW Himalaya. Theor Appl Climatol 151(3):1729–1752

Arya R (2011) Leh floods 2010: an extreme geological event. Disaster Dev 5:103–129

Bahadar I, Shafique M, Khan T, Tabassum I, Ali MZ (2015) Flood hazard assessment using hydro-dynamic model and GIS/RS tools: a case study of Babuzai-Kabal tehsil Swat Basin, Pakistan. J Himal Earth Sci 48(2):129

Banerjee A, Dimri AP (2019) Comparative analysis of two rainfall retrieval algorithms during extreme rainfall event: a case study on cloudburst, 2010 over Ladakh (Leh), Jammu and Kashmir. Nat Hazards 97:1357–1374

Bazai NA, Alam M, Cui P, Hao W, Khan AP, Waseem M, Ahmed T (2024) Dynamics and impacts of monsoon-induced geological hazards: a 2022 flood study along the Swat River in Pakistan. Nat Hazards Earth Syst Sci Discuss 2024:1–27

Bhan SC, Devrani AK, Sinha V (2015) An analysis of monthly rainfall and the meteorological conditions associated with cloudburst over the dry region of Leh (Ladakh), India. Mausam 66(1):107–122

Bhatt CM, Rao GS, Farooq M, Manjusree P, Shukla A, Sharma SVSP, Dadhwal VK (2017) Satellite-based assessment of the catastrophic Jhelum floods of September 2014, Jammu & Kashmir, India. Geomat Nat Haz Risk 8(2):309–327

Bisht S, Chaudhry S, Sharma S, Soni S (2018) Assessment of flash flood vulnerability zonation through geospatial technique in high altitude Himalayan watershed, Himachal Pradesh India. Remote Sens Appl Soc Environ 12:35–47

Bookhagen B (2010) Appearance of extreme monsoonal rainfall events and their impact on erosion in the Himalaya. Geomat Nat Haz Risk 1(1):37–50. https://doi.org/10.1080/19475701003625737

Chand K, Jamwal AK, Meraj G, Thakur T, Farooq M, Kumar P, Debnath J (2024) Integrating geoenvironmental and socioenvironmental analyses for flood vulnerability assessment in the Kullu Valley, Himachal Pradesh, India. Int J Disaster Risk Reduct 108:104494

Chen F, Zhang M, Zhao H, Guan W, Yang A (2024) Pakistan's 2022 floods: spatial distribution, causes and future trends from Sentinel-1 SAR observations. Remote Sens Environ 304:114055

Chinita MJ, Richardson M, Teixeira J, Miranda PM (2021) Global mean frequency increases of daily and sub-daily heavy precipitation in ERA5. Environ Res Lett 16(7):074035

Cho C, Li R, Wang SY, Yoon JH, Gillies RR (2016) Anthropogenic footprint of climate change in the June 2013 Northern India flood. Clim Dyn 46:797–805

Cui L, Meng J, Li Y, An J, Zou Z, Zhong L, Wu G (2024) Spatiotemporal evolution characteristics of 2022 Pakistan severe flood event based on multi-source satellite gravity observations. Remote Sens 16(9):1601

Das L, Meher JK (2019) Drivers of climate over the Western Himalayan region of India: a review. Earth Sci Rev 198:102935

Das S, Ashrit R, Moncrieff MW (2006) Simulation of a Himalayan cloudburst event. J Earth Syst Sci 115:299–313

Di Capua G, Sparrow S, Kornhuber K, Rousi E, Osprey S, Wallom D, Coumou D (2021) Drivers behind the summer 2010 wave train leading to Russian heatwave and Pakistan flooding. NPJ Clim Atmos Sci 4(1):55

Dimri AP, Niyogi D, Barros AP, Ridley J, Mohanty UC, Yasunari T, Sikka DR (2015) Western disturbances: a review. Rev Geophys 53(2):225–246

Feige F, Ahmadbabar Z, Liguo S, Feizhi X, Yunchen Y, Weitang W (2014) Heavy rainfall in Pakistan during 27–29 July 2010: role of atmospheric energy conversion characteristics. Mausam 65(1):57–66

Gardner JS (2002) Natural hazards risk in the Kullu district, Himachal Pradesh, India. Geogr Rev 92(2):282–306

Gaurav K, Sinha R, Panda PK (2011) The Indus flood of 2010 in Pakistan: a perspective analysis using remote sensing data. Nat Hazards 59:1815–1826

Gouda KC, Rath SS, Singh N, Ghosh S, Lata R (2023) Extreme rainfall event analysis over the state of Himachal Pradesh in India. Theor Appl Climatol 151(3):1103–1111

Gupta V, Ram BK, Kumar S, Sain K (2022) A case study of the 12 July 2021 Bhagsunath (McLeod Ganj) flash flood in Dharamshala, Himachal Pradesh: a warning against constricting natural drainage. J Geol Soc India 98(5):607–610

Gupta V, Syed B, Pathania A, Raaj S, Nanda A, Awasthi S, Shukla DP (2024) Hydrometeorological analysis of July-2023 floods in Himachal Pradesh, India. Nat Hazards 120(8):7549–7574

Hallegatte S, Green C, Nicholls RJ, Corfee-Morlot J (2013) Future flood losses in major coastal cities. Nat Clim Chang 3(9):802–806

Hartmann H, Andresky L (2013) Flooding in the Indus River basin—a spatiotemporal analysis of precipitation records. Glob Planet Chang 107:25–35

Hartmann H, Buchanan H (2014) Trends in extreme precipitation events in the Indus River Basin and flooding in Pakistan. Atmos Ocean 52(1):77–91

Hasana U, Swain SK, George B (2022) A bibliometric analysis of ecotourism: a safeguard strategy in protected areas. Reg Sustainability 3(1):27–40

Hashmi HN, Siddiqui QTM, Ghumman AR, Kamal MA, Mughal HUR (2012) A critical analysis of 2010 floods in Pakistan. Afr J Agric Res 7(7):1054–1067

Hobley DE, Sinclair HD, Mudd SM (2012) Reconstruction of a major storm event from its geomorphic signature: the Ladakh floods, 6 August 2010. Geology 40(6):483–486

Hock R, Rasul G, Adler C, Cáceres B, Gruber S, Hirabayashi Y, Zhang Y (2019) High mountain areas. In: Pörtner H-O, Roberts DC, Masson-Delmotte V, Zhai P, Tignor M, Poloczanska E, Mintenbeck K, Alegría A, Nicolai M, Okem A, Petzold J, Rama B, Weyer NM (eds) IPCC special report on the ocean and cryosphere in a changing climate, pp 131–202

IPCC (2022) Climate change 2022: impacts, adaptation, and vulnerability. In: Pörtner H-O, Roberts DC, Tignor M, Poloczanska ES, Mintenbeck K et al (eds) Contribution of working group II to the sixth assessment report of the intergovernmental panel on climate change. Cambridge University Press, p 3056. https://doi.org/10.1017/9781009325844

Kanwar N, Kuniyal JC, Rautela KS, Singh L, Pandey DC (2024) Longitudinal assessment of extreme climate events in Kinnaur district, Himachal Pradesh, North-Western Himalaya, India. Environ Monit Assess 196(6):557

Kar SC, Tiwari S (2016) Model simulations of heavy precipitation in Kashmir, India, in September 2014. Nat Hazards 81:167–188

Kashyap A, Cook K, Behera MD (2024) Geomorphic imprint of high mountain floods: insight from the 2022 hydrological extreme across the upper Indus terrain in NW Himalayas. EGUsphere 2024:1–36

Kashyap A, Cook KL, Behera MD (2025) Geomorphic imprint of high-mountain floods: insights from the 2022 hydrological extreme across the upper Indus River catchment in the northwestern Himalayas. Earth Surf Dyn 13(1):147–166. https://doi.org/10.5194/esurf-13-147-2025

Khalid B, Cholaw B, Alvim DS, Javeed S, Khan JA, Javed MA, Khan AH (2018) Riverine flood assessment in Jhang district in connection with ENSO and summer monsoon rainfall over upper Indus Basin for 2010. Nat Hazards 92:971–993

Khan AN (2013) Analysis of 2010-flood causes, nature and magnitude in the Khyber Pakhtunkhwa. Pak Nat Hazards 66(2):887–904

Khan I, Kumar P, Bahuguna H (2024) Cloudburst catastrophe: impact of cloudburst triggered Debris Laden flash flood at Honzar Village, Kishtwar District, Jammu and Kashmir. Disaster 25:25–39

Kumar A, Asthana AKL, Priyanka RS, Jayangondaperumal R, Gupta AK, Bhakuni SS (2017) Assessment of landslide hazards induced by extreme rainfall event in Jammu and Kashmir Himalaya, Northwest India. Geomorphology 284:72–87

Kumar A, Houze RA Jr, Rasmussen KL, Peters-Lidard C (2014) Simulation of a flash flooding storm at the steep edge of the Himalayas. J Hydrometeorol 15(1):212–228

Kumar R, Acharya P (2016) Flood hazard and risk assessment of 2014 floods in Kashmir Valley: a space-based multisensor approach. Nat Hazards 84:437–464

Kumar R, Mishra V (2020) Increase in population exposure due to dry and wet extremes in India under a warming climate. Earth's Future 8(12):e2020EF001731

Kundu S, Khare D, Mondal A (2017a) Landuse change impact on sub-watersheds prioritization by analytical hierarchy process (AHP). Eco Inform 42:100–113

Kundu S, Khare D, Mondal A (2017b) Future changes in rainfall, temperature and reference evapotranspiration in the Central India by least square support vector machine. Geosci Front 8(3):583–596

Kundu S, Khare D, Mondal A, Mishra PK (2015) Analysis of spatial and temporal variation in rainfall trend of Madhya Pradesh, India (1901–2011). Environ Earth Sci 73(12):8197–8216

Kushwaha AP, Solanki H, Vegad U, Mahto SS, Mishra V (2024) Land and atmospheric drivers of the 2023 flood in India. Earth Space Sci 11(10):e2024EA003750

Lehmann J, Coumou D, Frieler K (2015) Increased record-breaking precipitation events under global warming. Clim Chang 132:501–515

Lesk C, Rowhani P, Ramankutty N (2016) Influence of extreme weather disasters on global crop production. Nature 529(7584):84–87
Ma Q, Lei H, Feng T, Hu R, Niu M, Hu Z, Feng G (2023a) Impact of spring Tibetan Plateau snow cover on extreme precipitation in Pakistan in July and August 2022. Atmos Res 295:107007
Ma Y, Hu X, Chen Y, Hu Z, Feng T, Feng G (2023b) Different characteristics and drivers of the extraordinary Pakistan rainfall in July and August 2022. Remote Sens 15(9):2311
Mahmood S, Abbas F, Hussain S (2024) Cause and damages assessment of 2022-flood in Khyber Pakhtunkhwa, Pakistan. Int J Innov Sci Technol 6:637–651
Mahto SS, Nayak MA, Lettenmaier DP, Mishra V (2023) Atmospheric rivers that make landfall in India are associated with flooding. Commun Earth Environ 4(1):120
Malik IH, Hashmi NI, S. (2022) Ethnographic account of flooding in North-Western Himalayas: a study of Kashmir Valley. GeoJournal 87(2):1265–1283
Martius O, Sodemann H, Joos H, Pfahl S, Winschall A, Croci-Maspoli M, Wernli H (2013) The role of upper-level dynamics and surface processes for the Pakistan flood of July 2010. Q J R Meteorol Soc 139(676):1780–1797
Mehmood A, Jia S, Lv A, Zhu W, Mahmood R, Saifullah M, Adnan RM (2021) Detection of spatial shift in flood regime of the Kabul river basin in Pakistan, causes, challenges, and opportunities. Water 13(9):1276
Mishra AK (2015) A study on the occurrence of flood events over Jammu and Kashmir during September 2014 using satellite remote sensing. Nat Hazards 78(2):1463–1467
Mondal A, Khare D, Kundu S (2016) Change in rainfall erosivity in the past and future due to climate change in the central part of India. Int Soil Water Conserv Res 4(3):186–194
Mondal A, Khare D, Kundu S (2017b) Uncertainty analysis of soil erosion modelling using different resolution of open-source DEMs. Geocarto Int 32(3):334–349
Mondal A, Khare D, Kundu S (2018b) A comparative study of soil erosion modelling by MMF, USLE and RUSLE. Geocarto Int 33(1):89–103
Mondal A, Khare D, Kundu S, Mukherjee S, Mukhopadhyay A, Mondal S (2017a) Uncertainty of soil erosion modelling using open-source high resolution and aggregated DEMs. Geosci Front 8(3):425–436
Mondal A, Kundu S, Mukhopadhyay A (2012) Rainfall trend analysis by Mann-Kendall test: a case study of north-eastern part of Cuttack district, Orissa. Int J Geol Earth Environ Sci 2(1):70–78
Mondal A, Lakshmi V, Hashemi H (2018a) Intercomparison of trend analysis of multisatellite monthly precipitation products and gauge measurements for river basins of India. J Hydrol 565:779–790
Mondal A, Le MH, Lakshmi V (2022) Land use, climate, and water change in the Vietnamese Mekong Delta (VMD) using earth observation and hydrological modeling. J Hydrol Reg Stud 42:101132
Mondal A, Pandey A, Mondal S, Kundu S (2025) Impact assessment of Glacial Lake outburst floods (GLOFs) in the Himalaya using satellite imageries over the last 25 years (2000–2024): a comprehensive review. In: Remote sensing applications: society and environment, p 101845
Mondal A, Pandey A, Mondal S, Kundu S, Tran TND, Lakshmi V (2026) Assessing the accuracy of different open-source Digital Elevation Models (DEMs) in morphometric analysis of Kanchi River basin. J Hydrol: Reg Stud 65:103429. https://doi.org/10.1016/j.ejrh.2026.103429
Muhammad Tahir K, Yin Y, Wang Y, Babar ZA, Yan D (2015) Impact assessment of orography on the extreme precipitation event of July 2010 over Pakistan: a numerical study. Adv Meteorol 2015(1):510417
Mukhopadhyay A, Mondal A, Mukherjee S, Khatua D, Ghosh S, Mitra D, Ghosh T (2014) Forest cover change prediction using hybrid methodology of geoinformatics and Markov chain model: a case study on sub-Himalayan town Gangtok, India. J Earth Syst Sci 123(6):1349–1360
Namgyal P, Sarkar S, Kumar R (2025) Vulnerability assessment of rural households to climate change using livelihood vulnerability framework approach in the trans-Himalayan region of Ladakh, India. Anthropocene 49:100467

Nandargi S, Dhar ON (2011) Extreme rainfall events over the Himalayas between 1871 and 2007. Hydrol Sci J 56(6):930–945

Nanditha JS, Kushwaha AP, Singh R, Malik I, Solanki H, Chuphal DS, Mishra V (2023) The Pakistan flood of August 2022: causes and implications. Earth's Future 11(3):e2022EF003230

Narasimha Rao N, Paul S, Skekhar MS, Singh GP, Mitra AK, Bhan SC (2021) Unprecedented heavy rainfall event over Yamunanagar, India during 14 July 2016: an observational and modelling study. Meteorol Appl 28(6):e2039

Nepal S, Shrestha AB (2015) Impact of climate change on the hydrological regime of the Indus, Ganges and Brahmaputra river basins: a review of the literature. Int J Water Resour Dev 31(2):201–218

Otto FE, Zachariah M, Saeed F, Siddiqi A, Kamil S, Mushtaq H, Clarke B (2023) Climate change increased extreme monsoon rainfall, flooding highly vulnerable communities in Pakistan. Environ Res Clim 2(2):025001

Papalexiou SM, Montanari A (2019) Global and regional increase of precipitation extremes under global warming. Water Resour Res 55(6):4901–4914

Peijun S (2016) Atlas of global change risk of population and economic systems. In: IHDP/future earth-integrated risk governance project series. ISBN 978–981–16-6690-2 ISBN 978–981–16-6691-9 (eBook)

Pendergrass AG (2018) What precipitation is extreme? Science 360(6393):1072–1073

Pisharoty PR, Desai BN (1956) Western disturbances and Indian weather. Mausam 7(4):333–338

Priya P, Krishnan R, Mujumdar M, Houze RA (2017) Changing monsoon and midlatitude circulation interactions over the Western Himalayas and possible links to occurrences of extreme precipitation. Clim Dyn 49:2351–2364

Priya P, Mujumdar M, Sabin TP, Terray P, Krishnan R (2015) Impacts of Indo-Pacific Sea surface temperature anomalies on the summer monsoon circulation and heavy precipitation over Northwest India–Pakistan region during 2010. J Clim 28(9):3714–3730

Raghuvanshi AS, Agarwal A (2024) Multiscale dynamics of transient merging between western disturbances and monsoonal lows: connections to the July 2023 flood in Himachal Pradesh. Atmos Res 304:107401

Rashid I, Parray AA, Romshoo SA (2019) Evaluating the performance of remotely sensed precipitation estimates against in-situ observations during the September 2014 mega-flood in the Kashmir Valley. Asia Pac J Atmos Sci 55:209–219

Rasmussen KL, Hill AJ, Toma VE, Zuluaga MD, Webster PJ, Houze RA Jr (2015) Multiscale analysis of three consecutive years of anomalous flooding in Pakistan. Q J R Meteorol Soc 141(689):1259–1276

Ray K, Bhan SC, Bandopadhyay BK (2015) The catastrophe over Jammu and Kashmir in September 2014: a meteorological observational analysis. Curr Sci 109:580–591

Romshoo SA, Altaf S, Rashid I, Dar RA (2018) Climatic, geomorphic and anthropogenic drivers of the 2014 extreme flooding in the Jhelum basin of Kashmir, India. Geomat Nat Haz Risk 9(1):224–248

Sain, K., Mehta, M., & Kumar, V. (2022). Heavy rainfall-triggered flash floods around the Amarnath holy cave

Samantray P, Gouda KC (2024) A comprehensive study of atmospheric dynamics associated with cloudburst events in 2022 over Indian Himalayan Region. J Earth Syst Sci 133(3):151

Sangode SJ, Meshram DC, Rawat S, Kulkarni Y, Chate DM, Gudadhe SS (2017) Sedimentary and geomorphic signatures of a cloud burst and triggered flash floods in the Indus valley of Ladakh Himalaya. Himal Geol 38(1):12–29

Sathyakumar, S., Mungee, M., & Pal, R. (2020). Biogeography of the mountain ranges of South Asia

Sawa JN (2023) New waves of flash floods and risks perception of climate change: An overview of the 2022 flood disaster in Pakistan. Pak J Humlties Soc Sci Res 6(2):72–82

Schewe J, Heinke J, Gerten D, Haddeland I, Arnell NW, Clark DB, Kabat P (2014) Multimodel assessment of water scarcity under climate change. Proc Natl Acad Sci 111(9):3245–3250

Scoccimarro E, Gualdi S, Bellucci A, Zampieri M, Navarra A (2013) Heavy precipitation events in a warmer climate: results from CMIP5 models. J Clim 26(20):7902–7911

Shekhar MS, Gupta P, Sharma V, Nag K, Paul S, Saklani DR, Singh A (2024) Study of natural disaster in Manali Valley (Himachal Pradesh), India on 09 July 2023. Mausam 75(4):1111–1116

Shukla BP, Kishtawal CM, Pal PK (2017) Satellite-based nowcasting of extreme rainfall events over Western Himalayan region. IEEE J Sel Top Appl Earth Obs Remote Sens 10(5):1681–1686

Simpson GC (1921) The south-west monsoon. Q J R Meteorol Soc 47(199):151–171

Saravana Kumar M, Shekhar MS, Rama Krishna SSVS, Bhutiyani MR, Ganju A (2012) Numerical simulation of cloud burst event on August 05, 2010, over Leh using WRF mesoscale model. Nat Hazards 62:1261–1271

Swarnkar S, Mujumdar P, Sinha R (2021) Modified hydrologic regime of upper Ganga basin induced by natural and anthropogenic stressors. Sci Rep 11(1):19491

Syvitski JP, Brakenridge GR (2013) Causation and avoidance of catastrophic flooding along the Indus River, Pakistan. GSA Today 23(1):4–10

Thayyen RJ, Dimri AP, Kumar P, Agnihotri G (2013) Study of cloudburst and flash floods around Leh, India, during August 4–6, 2010. Nat Hazards 65:2175–2204

Thayyen RJ, Singh MK, Dimri AP (2020) Visual evidence of constrained area of a cloudburst, 12 June 2018, Tirisha village, Nubra valley, Ladakh, India. J Clim Chang 6(2):47–57

Ullah K, Shouting G (2013) A diagnostic study of convective environment leading to heavy rainfall during the summer monsoon 2010 over Pakistan. Atmos Res 120:226–239

Ullah W, Wang G, Lou D, Ullah S, Bhatti AS, Ullah S, Ali G (2021) Large-scale atmospheric circulation patterns associated with extreme monsoon precipitation in Pakistan during 1981–2018. Atmos Res 253:105489

UNDRR (2020) Human cost of disasters: An overview of the last 20 years 2000–2019. United Nations for Disaster Risk Reduction (UNDRR), Geneva, Switzerland

Van der Schrier G, Rasmijn LM, Barkmeijer J, Sterl A, Hazeleger W (2018) The 2010 Pakistan floods in a future climate. Clim Chang 148:205–218

Vellore RK, Krishnan R, Pendharkar J, Choudhury AD, Sabin TP (2014) On the anomalous precipitation enhancement over the Himalayan foothills during monsoon breaks. Clim Dyn 43:2009–2031

Viterbo F, von Hardenberg J, Provenzale A, Molini L, Parodi A, Sy OO, Tanelli S (2016) High-resolution simulations of the 2010 Pakistan flood event: sensitivity to parameterizations and initialization time. J Hydrometeorol 17(4):1147–1167

Wang J, Li K, Hao L, Xu C, Liu J, Qu Z, Sun Y (2024a) Disaster mapping and assessment of Pakistan's 2022 mega-flood based on multi-source data-driven approach. Nat Hazards 120(4):3447–3466

Wang Y, Cui P, Zhang CD, Zhang GT, Hao JS, Yuan X, Wang L (2024b) Antecedent snowmelt and orographic precipitation contributions to water supply of Pakistan disastrous floods, 2022. Adv Clim Chang Res 15(3):419–430

Webster PJ, Magana VO, Palmer TN, Shukla J, Tomas RA, Yanai MU, Yasunari T (1998) Monsoons: processes, predictability, and the prospects for prediction. J Geophys Res Oceans 103(C7):14451–14510

Westra S, Fowler HJ, Evans JP, Alexander LV, Berg P, Johnson F, Roberts N (2014) Future changes to the intensity and frequency of short-duration extreme rainfall. Rev Geophys 52(3):522–555

Winsemius HC, Aerts JC, Van Beek LP, Bierkens MF, Bouwman A, Jongman B, Ward PJ (2016) Global drivers of future river flood risk. Nat Clim Chang 6(4):381–385

Yadav BP, Kumar N, Tomar CS (2015) Analysis of heavy precipitation events over western Himalayan region. Int J Earth Atmos Sci 2:90–96

Yamada TJ, Takeuchi D, Farukh MA, Kitano Y (2016) Climatological characteristics of heavy rainfall in northern Pakistan and atmospheric blocking over western Russia. J Clim 29(21):7743–7754

Zhang W, Furtado K, Wu P, Zhou T, Chadwick R, Marzin C, Sexton D (2021) Increasing precipitation variability on daily-to-multiyear time scales in a warmer world. Sci Adv 7(31):eabf8021

Zhao X (2017) A scientometric review of global BIM research: analysis and visualization. Autom Constr 80:37–47. https://doi.org/10.1016/j.autcon.2017.04.002
Zhou ZQ, Xie SP, Zhang R (2021) Historic Yangtze flooding of 2020 tied to extreme Indian Ocean conditions. Proc Natl Acad Sci 118(12):e2022255118

Chapter 5
At-Site Flood Quantile Estimation Using Multiparameter Distributions in the Gandak River, Himalayas

Robindro Singh Khwairakpam, Sananda Kundu, and Sukanta Singh Moirangthem

Abstract Modelling peak flood discharges and estimating flood frequency on a river provides essential information for designing and managing hydraulic structures. Bihar is one of the most flood-prone Indian states, with numerous Himalayan-originated rivers in Northern Bihar experiencing frequent flooding. The study aimed to analyse the at-site flood frequency to estimate peak discharge in the Gandak River basin, India. For the analysis, annual peak discharges from the Triveni gauging site, with a 46-year data record, were used. Using L-moments, the maximum likelihood estimate, and the method of moments, the parameters of ten advanced multiparameter distribution models were estimated. Quantiles of different return periods were estimated using the parameter estimation methods. Three types of goodness-of-fit tests, Kolmogorov–Smirnov, Anderson–Darling, and the Chi-squared were performed to identify the best-fit distribution among the models. The results showed that the Generalised Extreme Value (GEV) and Log-Pearson 3 distributions are the most suitable in the study area. The near-perfect positive relationship between the observed and simulated data confirmed the selection of the most suitable models. The study recommended adopting a return period of more than 500 years for the design of hydraulic structures in the Gandak River basin.

Keywords At-site flood frequency · Gandak river · Multiparameter distributions · Quantile estimation

5.1 Introduction

Flood, a widespread global disaster (Liu et al. 2023), is caused by high stream flow that overtops natural or artificial banks and submerges the surrounding land (Saravanan and Abijith 2022). The impacts of floods are more significant than those

R. S. Khwairakpam · S. Kundu (✉) · S. S. Moirangthem
Department of Geography, Manipur University, Imphal, Manipur, India
e-mail: skundu@manipuruniv.ac.in

M. Biswas et al. (eds.), *Himalayan Flood Hazards*, Advances in Geographical and Environmental Sciences, https://doi.org/10.1007/978-981-92-0702-2_5

of any other disaster, in terms of loss of property, livestock, and human life, as well as damage to the surrounding environment. From 1990 to 2023, more than 5900 flood events of various types occurred worldwide, resulting in the loss of approximately seven million lives, affecting about 400 million people, and causing economic losses of US$1.7 trillion (CRED 2024). Worldwide, floods account for 22% of all disasters, 21% of deaths, 23% of economic losses, and 46% of the population affected by all kinds of disasters (CRED 2024). In the rivers of North India, the magnitude and intensity of floods are elevated by climate change and river channel siltation (Sahay and Srivastava 2014). Numerous Himalayan rivers drain into the plains of Bihar, making it the most flood-prone state in India, with 76% of its population residing in areas vulnerable to flooding and 73% of its territory affected by flooding (Acharya and Prakash 2019). Northern Bihar is one of the regions in India that is affected by floods almost every year (Tripathi et al. 2019). The remarkable floods in North Bihar occurred in 1987, 1995, 1998, 2000, 2001, 2003, 2004, 2008, 2010, 2013, 2016, 2017, 2018 and 2019 (Tripathi et al. 2020a). The Gandak catchment in India and Nepal has experienced significant flooding over the past two decades, resulting in property damage and millions of deaths (Chaubey et al. 2021). The leading causes of floods in the lower Gandak basin are heavy rainfall throughout the monsoon months in the upper catchment and rapid snowmelt at higher altitudes in the Himalayan region, driven by high temperatures (Prasad and Rishideo 2018).

Analysing the frequency and probability estimation of flood occurrence is fundamental to minimise property damage and flood-related fatalities. Flood frequency analysis involves two approaches: at-site and regional analyses. At-site analysis requires extensive streamflow data, whereas regional analysis provides reliable estimates of flood occurrence for ungauged catchments (Mangukiya and Sharma 2024). At-site analysis offers accurate information for hydraulic structure design. At-site flood frequency analysis, which depends on a relatively long duration of observed flow data at a particular location, is the most straightforward approach to designing flood estimation (Rahman et al. 2013). Analysis of at-site flood frequency primarily relies on annual flood series, either (i) Annual Maximum Series (AMS) or (ii) Partial Duration Series (PDS). The AMS calculates annual peak flow, while the PDS considers all flood peaks above a predetermined flood base. PDS offers extensive data and improves frequency estimates, but selecting a flood threshold is challenging without prior knowledge (Karim et al. 2017). Due to limited observed data, at-site FFA faces challenges, leading to the inappropriate selection of extreme value probability distributions and improper parameter estimation (Hu et al. 2020). A statistical distribution with a predetermined exceedance probability is assigned to the data employing a specific estimation approach, presuming independence and consistency to estimate a flood event (Saghafian et al. 2014). Choosing a suitable probability distribution and related parameter estimation process is crucial in at-site flood frequency analysis. The accuracy of other flood estimation techniques, such as regional flood estimation and rainfall-runoff modelling, is assessed using at-site flood frequency analysis (Rahman et al. 2013). The choice of a probability distribution is essential in flood frequency analysis, as incorrect choices can lead to under- or overestimation of flood estimates, especially at higher return periods (Khan et al. 2023). Since at-site flood frequency is the most direct method in a gauge area, it is more widely used worldwide (Haddad and Rahman 2008; Ul Hassan et al. 2019; Valentini et al. 2024). Cunnane (1989) explored worldwide assessments and recommendations for multiple distribution models and parameter estimation techniques. Kousar et al. (2020) highlighted the various distribution models recommended in different countries. Several authors adopted at-site FFA in different catchments of India, such as the Ghagra River (Verma and Rai 2023), the Mahanadi River (Samantaray et al. 2021; Sahoo and Ghose 2021), the Jhelum River (Bhat et al. 2019), and the Narmada River (Ray and Goel 2019).

Distribution models applied in at-site frequency analysis can be classified into two types: (i) two-parameter models, such as Gumbel, Pareto, Exponential, Beta, and Normal (Ilinca and Anghel 2023; Valentini et al. 2024; Bhuyan et al. 2016); (ii) three or more parameter (multiparameter) models, such as Generalised Extreme Value (GEV), Generalised Pareto, Log-Pearson 3, Wakeby, and Generalised Logistic (Heidarpour et al. 2017; Das 2020; Hansen 2015). Some models have two versions that differ in the number of parameters, such as the Weibull, Pearson Type 5, Lognormal, and Log-logistics distributions (Mamoon and Rahman 2017; Ilinca et al. 2023). Two-parameter (2P) distributions exhibit some biased estimation, while three-parameter (3P) distributions consistently estimate (Khan et al., 2023). According to Bobee et al. (1993), flood modelling distributions should have at least three parameters to handle a wide range of flood statistics with flexibility. 2-parameter distributions yield satisfactory results, but multiparameter distributions provide a better fit between observed and predicted data (Cassalho et al. 2017). Therefore, the present study attempts to apply multiparameter probability distribution models.

Previous studies in the Gandak mainly focus on mapping flood-inundation areas using satellite imagery (Tripathi et al. 2020b, 2022; Chakravorty and Jha 2022). Sahay and Sehgal (2014) developed Wavelet-ANFIS models to forecast daily monsoon flows in the Gandak River in Bihar, India. Chaubey et al. (2021) employed an integrated approach, combining assessments of extreme weather events and geomorphology to identify potential flooding along the Gandak River. Prasad and Rishideo (2018) attempted to analyse peak discharge at the lower Gandak River using a shorter data duration (33 years). Kumar et al. (2016) evaluated the effectiveness of the TRMM and CMORPH satellite precipitation products for analysing extreme events in the southern Himalayan Gandak River basin. Limited studies in the region focus on estimating peak flood events (Singla et al. 2014; Sahay and Srivastava 2014). Floods are a prevalent hazard along the Gandak River and its neighbouring basins, including Burhi Gandak and Bagmati. The Gandak River basin exhibits unique rainfall variability, underscoring the need to monitor flood events and trends across its sub-watersheds (Chaubey et al. 2021). Estimating peak flood events is critical for designing hydraulic structures, such as bridges, culverts, weirs, and flood control structures, in the lower Gandak region. Therefore, the present study aims to estimate peak discharge events in the Gandak River using advanced probability distribution models and to select the most suitable distributions for the study area.

5.2 Materials and Methods

5.2.1 Study Area

The Gandak basin is a transboundary basin situated in the central Himalayas between 85°22′-88°21′ E and 26°47′-29°12′ N (Rai et al. 2018). The Gandak River, originating from the Nhubine Himal Glacier in Nepal, flows southeast and connects with the Ganges in India. It comprises six main sub-watersheds: Kali Gandak, Budhi Gadak, Marshandi, Rapti, Seti, and Trisuli. The Gandak River has a unique topography, characterised by glaciers, snowfields, and mountainous terrain in its upper catchment (Cui et al. 2022). The Kali Gandaki flows through the deepest Gorge in the world. The Pachnad, Sonha, and Gandak rivers converge at Triveni, forming the international boundary between Nepal and India. After that, the river flows through the alluvial and flood-prone plain of Bihar, India, before joining the Ganga on the other side of Patna (Fig. 5.1). The entire catchment area is 46600 km^2, of which the lower Gandak plain constitutes 16%, and the upper catchment, including Tibet and Nepal, constitutes 83% (Prasad and Rishideo 2021). The lower Gandak river basin receives monsoon rainfall and snowmelt from the massive Himalayan catchment upstream, rendering it susceptible to natural disasters like landslides, droughts, avalanches, and glacial lake outburst floods (Kumar et al. 2022). Due to the narrow, low-lying basin in India, during high discharge events, vast volumes of water and sediment must pass through it, causing flooding of the plain.

5.2.2 Data

The Central Water Commission provides daily water flow data for the Gandak River through the India Water Resource Information System (India-WRIS) portal. Due to limited gauge availability, Triveni, a gauged site on the Gandak River, was selected for analysis. The annual maximum series of observed flow data for 46 years from 1970 to 2018 for the gauged site was collected. However, due to data unavailability, the analysis did not consider flow data from 2001, 2002 and 2003. A minimum of 30 years of continuous discharge data is necessary to derive reliable and accurate discharges for different return periods (Pawar et al. 2023). Therefore, the record length of the data used in the present study is adequate for performing flood hydrological frequency analysis. Table 5.1 summarises the mean, skewness, standard deviation, and kurtosis of the observed data at the gauged site. The peak discharge of each year, its exceedance probability, and the return period is also given in Table 5.1.

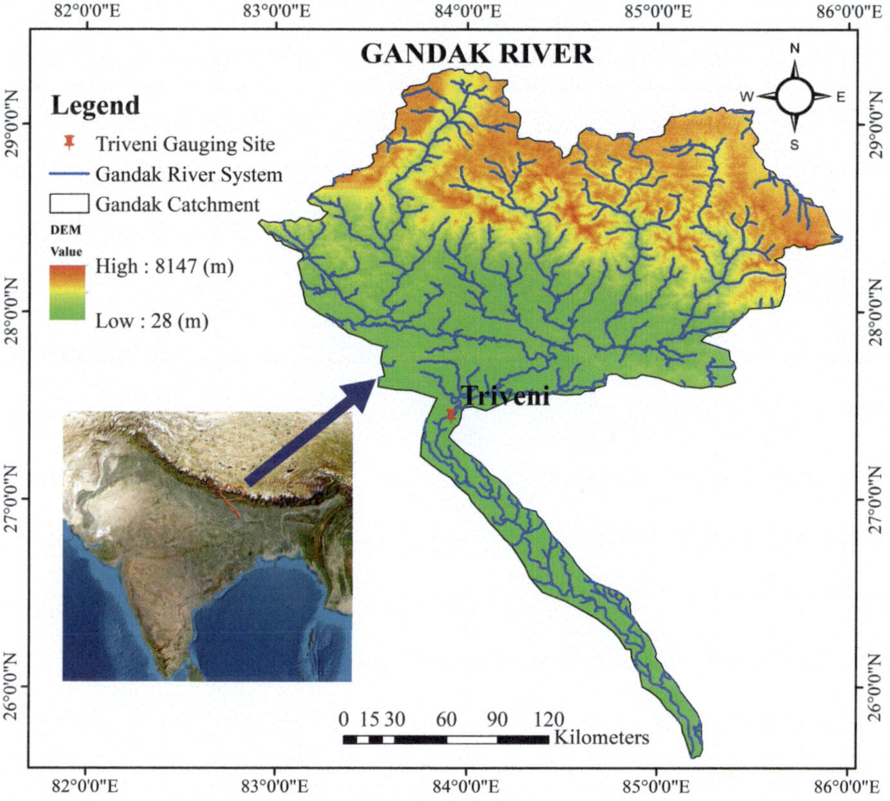

Fig. 5.1 Location of the Gandak River and the study area

5.2.3 Methodology

5.2.3.1 Probability Distribution Model

Choosing a suitable probability distribution is crucial for describing the at-site flood frequency analysis at a specific location. Ten distribution models are selected from the available models based on their greater frequency of use across studies and recommendations in different countries. These include (i) Gamma (3P), (ii) Generalised Extreme Value (GEV), (iii) Generalised Gamma (4P), (iv) Generalised Logistics, (v) Generalised Pareto, (vi) Log-Pearson 3, (vii) Lognormal (3P), (viii) Log-Logistics (3P), (ix) Wakeby, and (x) Weibull (3P). GEV, Generalised Logistics, Generalised Pareto, Log-Pearson 3, and Wakeby are advanced distribution models, and all have three parameters, except the Wakeby distribution, which has five (Anghel et al. 2023). Gamma (3P), Generalised Gamma (4P), Lognormal (3P), Log-Logistics (3P), and Weibull (3P) are non-negative distribution models (Loveridge et al. 2013).

Table 5.1 Descriptive statistics of discharge data at Triveni gauge

Year	Annual maximum series (AMS)	Descending order (M)	Corresponding year	Rank	Exceedance probability (P)	Return period (T)
1970	21086.85	21086.85	1970	1	2.13	47.00
1971	6413.3	15650	2010	2	4.26	23.50
1972	12633.31	13902.04	1974	3	6.38	15.67
1973	13402.64	13402.64	1973	4	8.51	11.75
1974	13902.04	13250	2007	5	10.64	9.40
1975	12423.9	12700	1986	6	12.77	7.83
1976	10322.8	12700	1999	7	14.89	6.71
1977	10715.6	12633.31	1972	8	17.02	5.88
1978	10885.03	12423.9	1975	9	19.15	5.22
1979	6795	12200	1987	10	21.28	4.70
1980	8700	11785	2017	11	23.40	4.27
1981	9824.52	11080	1984	12	25.53	3.92
1982	7820	10885.03	1978	13	27.66	3.62
1983	7322.32	10715.6	1977	14	29.79	3.36
1984	11080	10500	1989	15	31.91	3.13
1985	7197.35	10454	2014	16	34.04	2.94
1986	12700	10322.8	1976	17	36.17	2.76
1987	12200	10220	1998	18	38.30	2.61
1988	10200	10200	1988	19	40.43	2.47
1989	10500	9915	2011	20	42.55	2.35
1990	9400.3	9824.52	1981	21	44.68	2.24
1991	4050	9739	2013	22	46.81	2.14
1992	6950	9400.3	1990	23	48.94	2.04
1993	7900	9290	2000	24	51.06	1.96
1994	5913	9200	1996	25	53.19	1.88
1995	7750	8997	2012	26	55.32	1.81
1996	9200	8895.79	2016	27	57.45	1.74
1997	6052	8700	1980	28	59.57	1.68
1998	10220	8650	2009	29	61.70	1.62
1999	12700	7915	2005	30	63.83	1.57
2000	9290	7900	1993	31	65.96	1.52
2004	7858	7858	2004	32	68.09	1.47
2005	7915	7820	1982	33	70.21	1.42
2006	6830	7750	1995	34	72.34	1.38
2007	13250	7322.32	1983	35	74.47	1.34

(continued)

Table 5.1 (continued)

Year	Annual maximum series (AMS)	Descending order (M)	Corresponding year	Rank	Exceedance probability (P)	Return period (T)
2008	6710	7197.35	1985	36	76.60	1.31
2009	8650	6950	1992	37	78.72	1.27
2010	15650	6830	2006	38	80.85	1.24
2011	9915	6795	1979	39	82.98	1.21
2012	8997	6710	2008	40	85.11	1.18
2013	9739	6554.39	2018	41	87.23	1.15
2014	10454	6413.3	1971	42	89.36	1.12
2015	5764.87	6052	1997	43	91.49	1.09
2016	8895.79	5913	1994	44	93.62	1.07
2017	11785	5764.87	2015	45	95.74	1.04
2018	6554.39	4050	1991	46	97.87	1.02
Maximum discharge (Q_{max})			21086.85			
Minimum discharge			4050			
Mean discharge (Q_{mean})			9663.44			
Standard deviation (Q_{std})			3061.84			
Coefficient of variation (%)			32%			
Q_{max}/Q_{mean} ratio			2.18			
$Q_{if} = (Q_{std} + Q_{mean})$			12725.28			

5.2.3.2 Parameter Estimation Methods

Three approaches to estimating distribution parameters are examined in this study: the L-moment method, the Maximum Likelihood Estimate, and the Method of Moments. The L-Moment is a straightforward, reliable, and objective tool for flood modelling. The probability-weighted moments (PWMs) of data arranged in ascending order form the foundation for L-Moments (Jaiswal et al. 2021). Hosking and Wallis (1997) described the steps of parameter estimation by L-Moments and quantile functions for the various distribution models. The Method of Moments relies on statistical properties such as variance, skewness, and mean. However, short data series and high-order moments frequently yield incorrect values (Anghel and Ilinca 2023). The Maximum Likelihood Estimate determines parameters by optimising the log-likelihood function of a probability distribution (Langat et al. 2019). The parameter types and parameter estimation methods for the selected distribution models are given in Table 5.2.

Table 5.2 Parameter types and parameter estimation methods

Sl. no.	Distribution models	Parameters	Parameter estimation methods (Choi et al. 2011)
1	Gamma (3P)	Shape (α), scale (β), location (γ)	Maximum likelihood estimate
2	Generalised extreme value (GEV)	Shape (k), scale (σ), location (μ)	Method of L-moments
3	Generalised gamma (4P)	Shape (k), shape (α), Scale (β), location (γ)	Maximum likelihood estimate
4	Generalised pareto	Shape (k), scale (σ), location (μ)	Method of L-moments
5	Generalised logistics	Shape (k), scale (σ), location (μ)	Method of L-moments
6	Log-logistics (3P)	Shape (α), scale (β), location (γ)	Maximum likelihood estimate
7	Lognormal (3P)	Scale (σ), location (μ), location (γ)	Maximum likelihood estimate
8	Log-pearson 3	Shape (α), scale (β), location (γ)	Method of moments
9	Wakeby	Shape (α), scale (β), location (γ), location (ξ), δ	Method of L-moments
10	Weibull (3P)	Shape (α), scale (β), location (γ)	Maximum likelihood estimate

5.2.3.3 Quantile Estimation

A *flood quantile* is defined as a flood event with a specific exceedance probability statistically determined using extreme value theory (Mudersbach et al. 2017). It represents a specific value of flood discharge associated with a given probability of occurrence within a specific period. The average interval between flood events is called the return period or recurrence interval. The probability of occurrence, or probability of exceedance, is the chance that a flood will attain or exceed a particular magnitude during a specific period (Ul Hassan et al. 2019). The probability that the exceedance for a given flood (Q) with a return period (T) may be exceeded once in return period (T) years is expressed as follows:

$$P(Q_T \leq q) = \frac{1}{T} \tag{5.1}$$

where Q_T is the estimated peak flood of a return period (T).

The present study followed the return periods used by Umar et al. (2021) to estimate peak discharge. The quantile values of the return periods are 2, 5, 10, 25, 50, 100, 200 and 500 years and are estimated using the Inverse Quantile Function of the Cumulative Distribution Function (CDF) (Cassalho et al. 2018). The CDF is the probability that the variable has a value less than or equal to 'x'. The CDF shows the probability of non-exceedance for a given magnitude. Details of the quantile estimation for the different distributions are described in Fawad et al. (2022) and Khan et al. (2023).

5.2.3.4 Goodness-of-Fit Test

The goodness-of-fit test of a statistical model indicates how well it fits a given data set. The difference between the observed and predicted values by the model under consideration is usually summarised using measures of goodness-of-fit (Guru and Jha 2015). Three types of goodness-of-fit test: Kolmogorov–Smirnov (K-S) test, Anderson–Darling (A-D) test, and the Chi-squared test (X^2) are performed to select the most suitable distribution model for estimation of flood frequency in the study area. The K-S statistic is a nonparametric test to ascertain whether a sample is representative of an assumed continuous distribution (Kumar 2019). If the value of the test statistic exceeds the critical value derived from a table at a level of significance (α), the null hypothesis (H0: data follows specified distribution) is rejected. The A-D test is a version of the K-S test that prioritises the tails over the distribution's centre (Farooq et al. 2018). If the test statistic of the A-D test is greater than the critical value at a significance level, the null hypothesis that the distribution is normal is rejected. The X^2 test establishes whether a sample is representative of a population with a particular distribution. Data binning affects the test statistic's value (Ke et al. 2018). The significance level for the three types of tests is set at $\alpha = 0.05$ (95% confidence level).

5.3 Results and Discussion

The annual peak discharge at the Triveni gauging site of the Gandak River on the international boundary between India and Nepal ranges from approximately 4000 cumecs to 21000 cumecs. The highest peak discharge at the gauging site is about 21100 cumecs, recorded in 1970. The discharge volumes of the northern Indian rivers are relatively more significant than those of the Peninsular rivers (Gupta et al. 2022). The present analysis did not consider flow data from 2001, 2002 and 2003 due to unavailability. The statistical characteristics of the observed data at the Triveni gauging site are summarised in Table 5.1. The $Q_{\mathrm{max}}/Q_{\mathrm{mean}}$ ratio confirmed that the most significant recorded discharge during the study period at the site is approximately two times higher than the mean annual discharge. The exceedance probability and return period of the highest discharge are 2.13 and 47 years, respectively. The mean discharge and standard deviation of the data are approximately 9660 cumecs and 3100 cumecs, respectively. The coefficient of variation is 32%, depicting moderate interannual variability. However, some studies considered a coefficient of variation below 70% to indicate low interannual variability (Umar et al. 2021; Pawar et al. 2020). The annual maximum series for the Triveni gauging site is shown in Fig. 5.2. The AMS graph shows higher peak discharges in 1970, 1973, 1974, 2010 and 2007. Three methods are used to estimate the model parameters.

Maximum Likelihood Estimate measures the parameters of Gamma (3P), Generalised Gamma (4P), Log-logistics (3P), Lognormal (3P) and Weibull (3P) distributions. The parameters of the GEV, Generalised Pareto, Generalised Logistics and Wakeby models are estimated using L-moments. The Method of Moments calculates the parameters of the Log-Pearson 3 distribution. The parameter values for each probability distribution model, estimated using the parameter estimation methods, are given in Table 5.3.

The PDF graph in Fig. 5.3 shows that the discharge data is positively skewed. The data skew at this location indicates the need to develop alternative distributions that adequately model the data. Positive skewness indicates a few extreme flood events with large values, whereas most peak flows are smaller and more common. This condition indicates a high likelihood of minor to moderate flooding, yet severe flooding remains a significant threat in the area.

One of the main aims of Flood frequency estimation is to determine the extreme quantiles of the distribution at the gauging site. Therefore, the quantile estimates for return periods 2, 5, 10, 25, 50, 100, 200 and 500 years are computed by applying quantile functions and parameter values of distributions. Table 5.4 presents quantile estimates with a 95% confidence level for various return periods. Generalised Pareto, which provided better results (Panigrahi et al. 2020), is estimated to be higher than other distributions in the lower return period but much less for the higher return periods.

Log-logistic (3P) estimates quantile values more accurately than any other distributional model at higher return periods. The predicted extreme values estimated

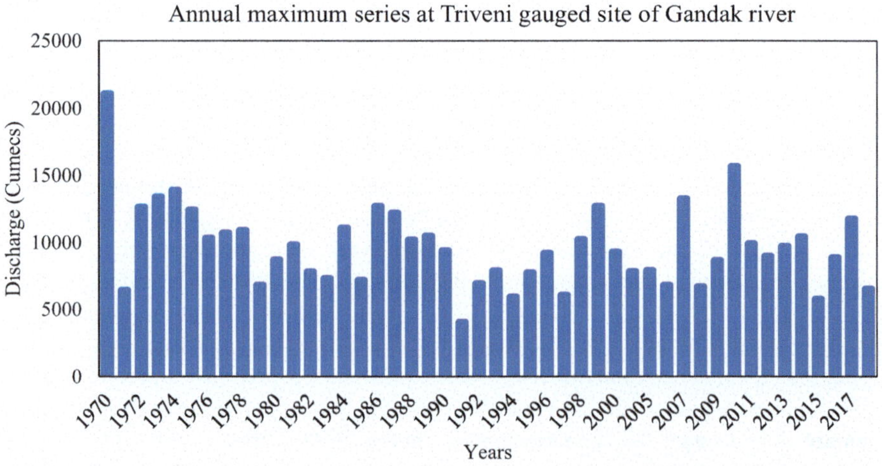

Fig. 5.2 Annual Maximum Series (AMS) of observed discharge data of the triveni gauging site

Table 5.3 Estimated parameter values of different distributions

	Distributions	Parameter values
1	Gamma (3P)	$\alpha = 6.4749$ $\beta = 1165.0755$ $\gamma = 2119.7149$
2	Generalised extreme value	$k = -0.0347$ $\sigma = 2470.9863$ $\mu = 8319.4186$
3	Generalised gamma (4P)	$k = 0.7636$ $\alpha = 12.8442$ $\beta = 282.2730$ $\gamma = 1557.8558$
4	Generalised logistic	$k = 0.1478$ $\sigma = 1601.3153$ $\mu = 9364.1065$
5	Generalised pareto	$k = -0.4849$ $\sigma = 6126.5711$ $\mu = 5537.6120$
6	Log-logistic (3P)	$\alpha = 5.6128$ $\beta = 8940.8930$ $\gamma = 295.7049$
7	Log-pearson 3	$\alpha = 3432.4192$ $\beta = -0.0016$ $\gamma = 66.4810$
8	Lognormal (3P)	$\sigma = 0.2913$ $\mu = 9.1710$ $\gamma = -369.5181$
9	Wakeby	$\alpha = 10963.5647$ $\beta = 3.0699$ $\gamma = 1940.4031$ $\delta = 0.0969$ $\xi = 4821.0367$
10	Weibull (3P)	$\alpha = 2.0654$ $\beta = 6774.6371$ $\gamma = 3655.7493$

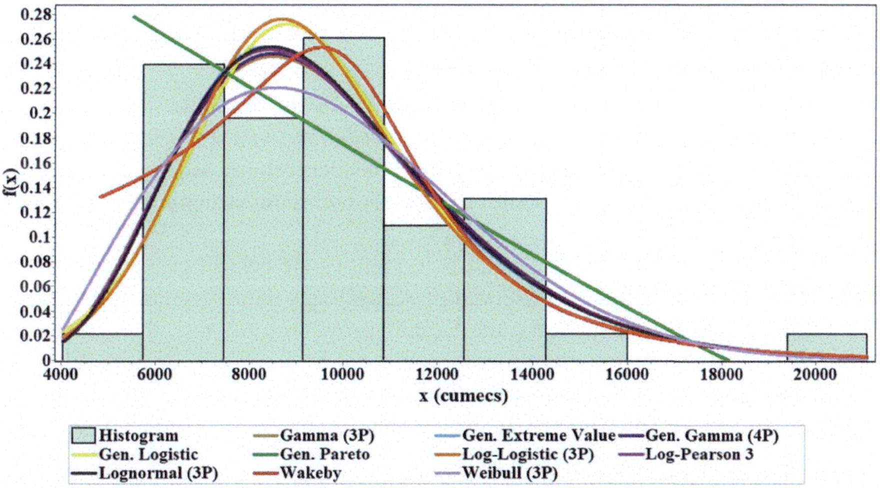

Fig. 5.3 Probability Density Function (PDF) of observed data at Triveni gauging site

by the selected distribution models at the gauging site are displayed in Fig. 5.4. The predicted discharge for the 2-year return period is much closer to the observed mean. It indicates that the Gandak River at the Triveni gauging site experiences relatively frequent, regular low-flow events rather than being dominated by infrequent extreme floods. The 100-year flood level is considered the minimum criterion for the design of hydraulic structures in various basins and countries (Sathe et al. 2012; Ren et al. 2017). However, in the present study, no distribution model can estimate a discharge with a 100-year return period that is much higher than the recorded

Table 5.4 Quantile estimation for different return periods by the selected distribution models

Distributions	Return periods (years)							
	2	5	10	25	50	100	200	500
Gamma (3P)	9278.83	11979.8	13624.1	15561.5	16916.6	18206.4	19446.5	21026.8
GEV	9219.33	11930.9	13668.3	15799.8	17336.3	18824.8	20272.4	22129.4
Gen. gamma (4P)	9278.68	11970.5	13625.5	15594.3	16984.3	18317.5	19608.7	21267.7
Gen. logistics	9264.11	11727.8	13421.1	15759.7	17687.8	19797.4	22120.1	25567.9
Gen. pareto	9144.19	12382.8	14035.3	15519.1	16276.3	16817.3	17203.9	17551
Log-logistics (3P)	9236.6	11741.5	13520.6	16045.9	18181.7	20569.3	23254.7	27340.6
Log-pearson 3	9232.74	11948.7	13668.6	15772.6	17298.9	18795.8	20277.7	22228.8
Lognormal (3P)	9244.41	11915.1	13594.5	15639.1	17116.4	18561.3	19988.2	21862.1
Wakeby	9358.2	11746.4	13394.9	15721.9	17622.4	19654.9	21828.6	24935.5
Weibull (3P)	9328.84	12185.8	13800.9	15587.3	16768.9	17846.5	18843.3	20062.7

maximum discharge. Gamma (3P), Generalised Pareto and Weibull (3P) cannot estimate much higher than the maximum observed discharge, even at the 500-year return period. These conditions confirm the need to consider a flood with a return period of more than 500 years for the design of various hydraulic structures in the lower Gandak basin.

Another main objective of this paper is to identify the best-fit probability distribution model by processing the observed peak discharge data. To achieve this goal, three types of goodness-of-fit tests: the Kolmogorov–Smirnov test, the Anderson–Darling test and the Chi-squared test are conducted by taking the critical values at a 95% confidence level. The critical values at $\alpha = 0.05$ significance level in the K-S are the same (0.1963) in all ten distribution models. Similar conditions are also found in the A-D test. However, the critical values in the X^2 test are not the same, as the number of trials and the degrees of freedom determine them. The statistical results of the tests are given in Table 5.6, with an allotting ranking of the models. The ranking is based on the difference between the critical value and the test statistic; the smaller the difference, the better the fit. Rank 1 denotes the most suitable model to model the discharge data, and rank 2 becomes the second-best-fit distribution model and so on. In the K-S test, not all the distribution models are rejected. However, the Generalised Pareto and Wakeby models are rejected in the A-D region, as their test statistics exceed the given critical values. These two models are also not applicable to the Chi-squared test.

In the present analysis, a single distribution does not consistently provide the best-fit for flood estimation at the Triveni gauging site. Therefore, a particular distribution cannot be selected as the most suitable model. Similar challenges are also observed in

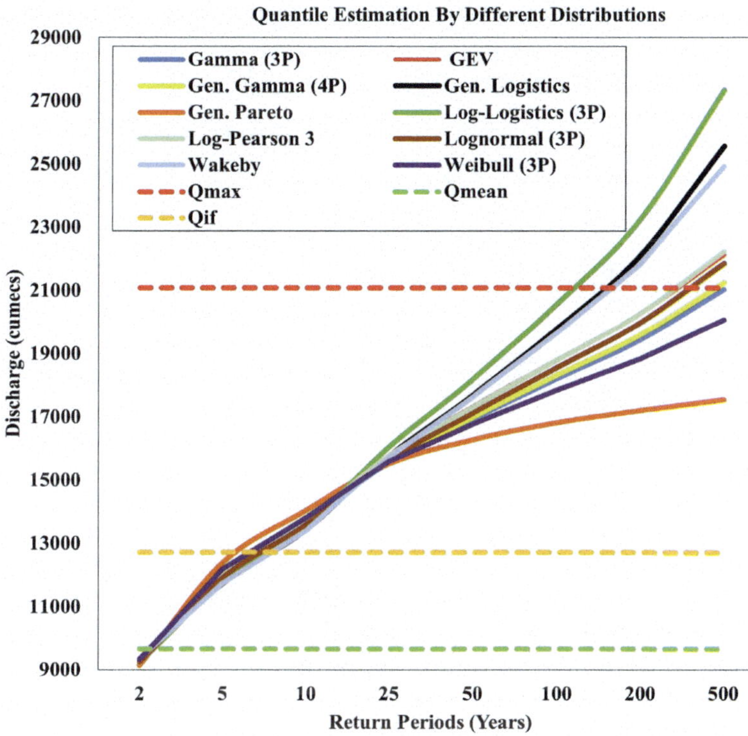

Fig. 5.4 Quantile estimation by the different distribution models at the Triveni gauging site

studies by Drissia (2019) across various basins in Kerala, Ul Hassan et al. (2019) in the Torne River of Sweden and Finland and Rahman et al. (2013) in the 127 catchments in Australia. Although Wakeby and Generalised Pareto have more parameters than the other selected models, they do not emerge as the best-fit model in three types of tests.

A relative scoring method developed by Mamoon and Rahman (2017) is applied in the present study to select a single most suitable distribution model for the site. For each of the three tests, the ten probability distributions are ranked independently on a scale of 1 to 10, with rank 1 denoting the best-fit distribution, rank 2 the second-best and so on. The first-rank probability distribution received a maximum score of 10, the second-rank distribution received a score of 9 and so on. The model with the highest cumulative score across all three goodness-of-fit tests is deemed the most suitable distribution (Tables 5.5 and 5.6).

The results of the relative coring method indicate that GEV and Log-Pearson 3 are the two highest-performing models. The peak discharge values estimated from these two distributions for different return periods are almost identical (Fig. 5.5). Therefore, these two models are the most suitable for the Triveni gauging site on the Gandak River. Similar results are also found in some Himalaya-originated rivers,

Table 5.5 Goodness-of-fit test results of the distribution models

	Kolmogorov–smirnov (k-s) test				Anderson–darling (a-d) test				Chi-squared (x2) test			
Distributions	Statistic	Critical value at 0.05	Reject	Rank	Statistic	Critical value at 0.05	Reject	Rank	Statistic	Critical value at 0.05	Reject	Rank
Gamma (3P)	0.064	0.1963	No	3	0.2074	2.5018	No	5	2.9683	11.0705	No	7
GEV	0.0614	0.1963	No	1	0.2042	2.5018	No	4	2.9351	11.0705	No	3
Gen. gamma (4P)	0.0658	0.1963	No	6	0.2035	2.5018	No	3	2.9631	11.0705	No	6
Gen. logistic	0.0805	0.1963	No	10	0.226	2.5018	No	6	2.2872	11.0705	No	1
Gen. pareto	0.0723	0.1963	No	8	4.1899	2.5018	Yes	10				
Log-logistic (3P)	0.0801	0.1963	No	9	0.2354	2.5018	No	7	2.3352	11.0705	No	2
Log-pearson 3	0.0615	0.1963	No	2	0.2028	2.5018	No	2	2.9462	11.0705	No	4
Lognormal (3P)	0.0649	0.1963	No	5	0.1979	2.5018	No	1	2.9521	9.4877	No	5
Wakeby	0.0679	0.1963	No	7	4.0607	2.5018	Yes	9				
Weibull (3P)	0.6415	0.1963	No	4	0.3423	2.5018	No	8	3.3746	11.0705	No	8

Table 5.6 Relative score of the distributions in three types of tests

Distributions	Kolmogorov–smirnov		Anderson–darling		Chi-squared		Total score
	Rank	Score	Rank	Score	Rank	Score	
Gamma (3p)	3rd	8	5th	6	7th	4	18
GEV	1st	10	4th	7	3rd	8	25
Gen. gamma (4p)	6th	5	3rd	9	6th	5	19
Gen. logistics	10th	1	6th	5	1st	10	16
Gen. pareto	8th	8	10th	0		0	8
Log-logistics (3p)	9th	2	7th	4	2nd	9	15
Log-pearson 3	2nd	9	2nd	9	4th	7	25
Lognormal (3p)	5th	6	1st	10	5th	6	22
Wakeby	7th	4	9th	0		0	4
Weibull (3p)	4th	7	8th	3	8th	3	13

such as the Jhelum River basin (Umar et al. 2021) and the Swat River (Farooq et al. 2018). Figure 5.5 shows that the predicted discharge of the 200-year return period cannot exceed the maximum observed discharge value (Q_{max}). The Q_{max} is lower than the expected value of a 500-year flood. Therefore, the present analysis indicates that a flood return period exceeding 500 years should be considered when designing various hydraulic structures in the lower Gandak basin.

An attempt is made to determine the relationship between the observed and predicted AMS using the Log-Pearson 3 and GEV models. Since the PDFs of these models are nearly identical, their distributions are also nearly identical (Fig. 5.6). The obtained value of R^2 in the Log-Pearson 3 is 0.9607, while that of GEV is 0.9612. These values indicate a strong positive correlation between the observed and predicted peak discharges at the gauging site. Therefore, GEV and Log-Pearson distribution models are considered the most suitable probability distribution models at the Triveni gauging site of the Gandak River.

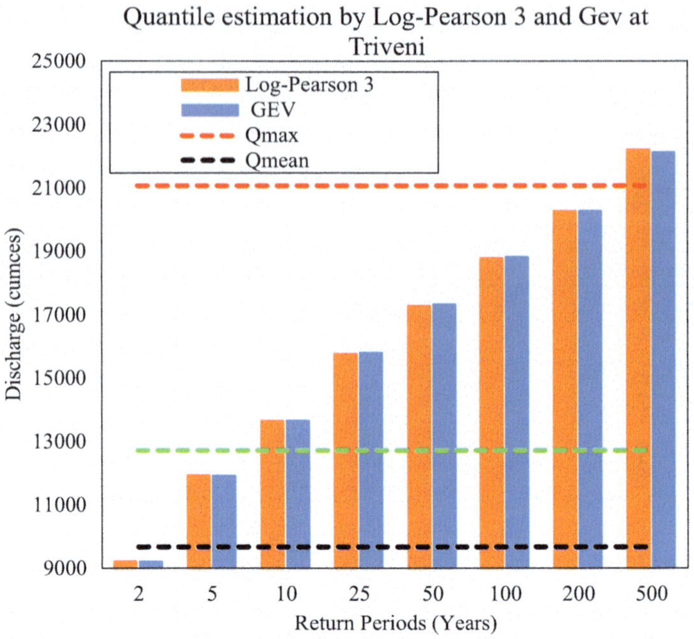

Fig. 5.5 Quantile estimation for the return periods based on log-pearson and GEV models

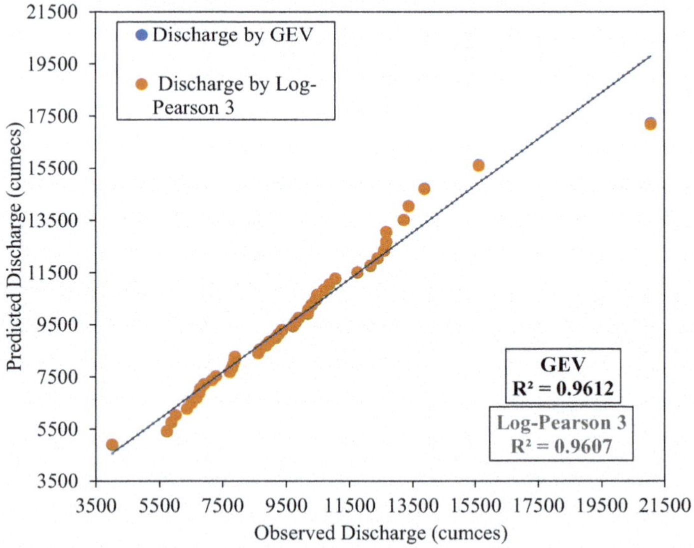

Fig. 5.6 Relation between observed and predicted maximum series based on the GEV model

5.4 Conclusion

The at-site flood frequency analysis for the Triveni point on the Gandak River is conducted using peak discharge data spanning approximately 46 years. The elevated, rugged topography of the Himalayas dominates the upstream reaches of the Gandak River before the Triveni gauging site. After the Triveni gauging site, the Gandak River enters the plain region of Bihar state, India. Bihar is one of the regions most affected by floods. Therefore, frequent monitoring of water discharge at the Triveni gauging site is essential for early warning for the lower Gandak region. The discharge recorded at the gauging site is substantial, indicating a considerable discharge in the lower Gandak region of Bihar. The present estimation adopted advanced and multiparameter probability distribution models, which consistently estimate flood statistics and provide a better fit between observed and predicted data. The application of many goodness-of-fit tests makes it difficult to choose the most suitable distribution that consistently fits the observed data. The study concludes that the Log-Pearson 3 and GEV distributions are the most suitable for the Gandak River basin, based on a relative scoring method. The analysis showed that the highest recorded discharge at the Triveni point is almost equal to the design flood for the 500-year return period. Therefore, constructing hydraulic structures in the lower Gandak requires adopting a design flood with a return period greater than 500 years. The study offers insights into estimating design floods for various return periods in the basin, enabling the selection of a more suitable probability distribution model for peak flood prediction. The limited number of gauging sites in the Gandak River is a significant constraint for the analysis. The future exploration of the present study can be extended by adding two-parameter models to verify whether the number of parameters influences the selection of the most suitable distribution. The results of the present study can be used to prepare flood hazard maps and to design hydraulic structures in the study area.

Acknowledgement The authors thank the Central Water Commission for providing the necessary data through the India Water Resources Information System (India-WRIS https://indiawris.gov.in/wris/#/) used in the present study. The authors are also thankful to the Centre for Research on the Epidemiology of Disasters (CRED), University of Louvain, Belgium, for providing information about flood events in its EM-DAT: Emergency Events Database (https://public.emdat.be/).

References

Acharya A, Prakash A (2019) When the river talks to its people: local knowledge-based flood forecasting in the Gandak River basin, India. Environ Dev 31:55–67

Anghel CG, Ilinca C (2023) Evaluation of various generalized Pareto probability distributions for flood frequency analysis. Water 15(8):1557

Anghel CG, Stanca SC, Ilinca C (2023) Extreme events analysis using LH-moments method and quantile function family. Hydrology 10(8):159

Bhat MS, Alam A, Ahmad B, Kotlia BS, Farooq H, Taloor AK, Ahmad S (2019) Flood frequency analysis of river Jhelum in Kashmir basin. Quat Int 507:288–294

Bhuyan MK, Jena J, Bhunya PK (2016) At-site flood analysis using exponential and generalized logistic models in partial duration series (PDS). Int J Eng Technol 8(1):501–514

Bobee B, Cavadias G, Ashkar F, Bernier J, Rasmussen P (1993) Towards a systematic approach to comparing distributions used in flood frequency analysis. J Hydrol 142(1–4):121–136

Cassalho F, Beskow S, de Mello CR, de Moura MM, Kerstner L, Ávila LF (2018) At-site flood frequency analysis coupled with multiparameter probability distributions. Water Resour Manag 32:285–300

Cassalho F, Beskow S, Vargas MM, Moura MMD, Ávila LF, Mello CRD (2017) Hydrological regionalization of maximum stream flows using an approach based on L-moments. RBRH 22(00):e27

Chakravorty B, Jha R (2022) Assessment of Spatio-temporal variation of waterlogged areas lying between lower Gandak and Burhi Gandak basins, India. J Indian Soc Remote Sens 50(3):583–596

Chaubey PK, Srivastava PK, Gupta A, Mall RK (2021) Integrated assessment of extreme events and hydrological responses of indo-Nepal Gandak River basin. Environ Dev Sustain 23(6):8643–8668

Choi JS, Hong S, Chi SB, Lee HB, Park CK, Kim HW et al (2011) Probability distribution for the shear strength of seafloor sediment in the KR5 area for the development of manganese nodule miner. Ocean Eng 38(17–18):2033–2041

CRED: Centre for Research on the Epidemiology of Disasters. (2024). Emergency Events Database (EM-DAT). https://public.emdat.be/data. Accessed on 5 May, 2024

Cui B, Zhang Y, Wang Z, Gu C, Liu L, Wei B et al (2022) Ecological risk assessment of trans-boundary region based on land-cover change: a case study of Gandaki river basin. Himalayas Land 11(5):638

Cunnane C (1989) Statistical distribution for flood frequency analysis. WMO Oper Hydrol Rep No 33 WMO-No 718 Geneva Switz

Das S (2020) Assessing the regional concept with sub-sampling approach to identify probability distribution for at-site hydrological frequency analysis. Water Resour Manag 34:803–817

Drissia TK, Jothiprakash V, Anitha AB (2019) Flood frequency analysis using L moments: a comparison between at-site and regional approach. Water Resour Manag 33:1013–1037

Farooq M, Shafique M, Khattak MS (2018) Flood frequency analysis of river swat using log Pearson type 3, generalized extreme value, Normal, and Gumbel max distribution methods. Arab J Geosci 11:1–10

Fawad M, Cassalho F, Ren J, Chen L, Yan T (2022) State-of-the-art statistical approaches for estimating flood events. Entropy 24(7):898

Gupta H, Reddy KK, Gandla V, Paridula L, Chiluka M, Vashisth B (2022) Freshwater discharge from the large and coastal peninsular rivers of India: a reassessment for sustainable water management. Environ Sci Pollut Res:1–18

Guru N, Jha R (2015) Flood frequency analysis of Tel Basin of Mahanadi River system, India using annual maximum and POT flood data. Aquat Procedia 4:427–434

Haddad K, Rahman A (2008) Investigation on at-site flood frequency analysis in south-East Australia. IEM J, J Inst Eng, Malays 69(3):59–64

Hansen CR (2015) Comparison of regional and at-site frequency analysis methods for the estimation of southern Alberta extreme rainfall. Can Water Resour J/Rev Can RessourS HydrS 40(4):325–342

Heidarpour B, Saghafian B, Yazdi J, Azamathulla HM (2017) Effect of extraordinary large floods on at-site flood frequency. Water Resour Manag 31:4187–4205

Hosking JRM, Wallis JR (1997) Regional frequency analysis: an approach based on L-moments. Cambridge University Press, New York

Hu L, Nikolopoulos EI, Marra F, Anagnostou EN (2020) Sensitivity of flood frequency analysis to data record, statistical model, and parameter estimation methods: an evaluation over the contiguous United States. J Flood Risk Manag 13(1):e12580

Ilinca C, Anghel CG (2023) Frequency analysis of extreme events using the univariate Beta family probability distributions. Appl Sci 13(7):4640

Ilinca C, Stanca SC, Anghel CG (2023) Assessing flood risk: LH-moments method and univariate probability distributions in flood frequency analysis. Water 15(19):3510

Jaiswal RK, Nayak TR, Lohani AK, Galkate RV (2021) Regional flood frequency modeling for a large basin in India. Nat Hazards:1–17

Karim F, Hasan M, Marvanek S (2017) Evaluating annual maximum and partial duration series for estimating frequency of small magnitude floods. Water 9(7):481

Ke Q, Jonkman SN, Van Gelder PH, Bricker JD (2018) Frequency analysis of storm-surge-induced flooding for the Huangpu river in shanghai, China. J Mar Sci Eng 6(2):70

Khan Z, Rahman A, Karim F (2023) An assessment of uncertainties in flood frequency estimation using bootstrapping and Monte Carlo simulation. Hydrology 10(1):18

Kousar S, Khan AR, Ul Hassan M, Noreen Z, Bhatti SH (2020) Some best-fit probability distributions for at-site flood frequency analysis of the Ume River. J Flood Risk Manag 13(3):e12640

Kumar B, Patra KC, Lakshmi V (2016) Daily rainfall statistics of TRMM and CMORPH: a case for trans-boundary Gandak River basin. J Earth Syst Sci 125(5):919–934

Kumar B, Roy D, Lakshmi V (2022) Impact of temperature and precipitation lapse rate on hydrological modelling over Himalayan Gandak River basin. J Mt Sci 19(12):3487–3502

Kumar R (2019) Flood frequency analysis of the Rapti river basin using log Pearson type-III and Gumbel extreme Value-1 methods. J Geol Soc India 94(5):480–484

Langat PK, Kumar L, Koech R (2019) Identification of the most suitable probability distribution models for maximum, minimum, and mean streamflow. Water 11(4):734

Liu Q, Yuan J, Yan W, Liang W, Liu M, Liu J (2023) Association of natural flood disasters with infectious diseases in 168 countries and territories from 1990 to 2019: a worldwide observational study. Glob TransitS 5:149–159

Loveridge M, Rahman A, Hill P, Babister M (2013) Investigation into probabilistic losses for design flood estimation: a case study for the Orara River catchment. NSW Australas J Water Resour 17(1):13–24

Mamoon AA, Rahman A (2017) Selection of the best fit probability distribution in rainfall frequency analysis for Qatar. Nat Hazards 86(1):281–296

Mangukiya NK, Sharma A (2024) Alternate pathway for regional flood frequency analysis in data-sparse region. J Hydrol 629:130635

Mudersbach C, Bender J, Netzel F (2017) An analysis of changes in flood quantiles at the gauge Neu Darchau (Elbe River) from 1875 to 2013. Stoch Env Res Risk A 31:145–157

Panigrahi B, Paramjita D, Giri M, Paul JC (2020) Frequency analysis for prediction of maximum flood discharge in Mahanadi River basin. Int J Curr Microbiol App Sci 9(8):3626–3639

Pawar UV, Hire PS, Gunjal RP, Patil AD (2020) Modeling of magnitude and frequency of floods on the Narmada River: India. Model Earth Syst Environ 6(4):2505–2516

Pawar U, Try S, Muttil N, Rathnayake U, Suppawimut W (2023) Frequency and trend analyses of annual peak discharges in the lower Mekong Basin. Heliyon 9(9)

Prasad R, Rishideo J (2021) Soil characteristics and land use in flood prone lower Gandak plain. Int J Appl Res 7(12):148–158

Prasad R, Rishideo J (2018) Peak discharge analyses for flood Management in Lower Gandak Basin. In: Mal S, Singh R, Huggel C (eds) Climate change, extreme events and disaster risk reduction. Sustainable Development Goals Series, Springer, Cham

Rahman AS, Rahman A, Zaman MA, Haddad K, Ahsan A, Imteaz M (2013) A study on selection of probability distributions for at-site flood frequency analysis in Australia. Nat Hazards 69:1803–1813

Rai R, Zhang Y, Paudel B, Acharya BK, Basnet L (2018) Land use and land cover dynamics and assessing the ecosystem service values in the trans-boundary Gandaki River basin, Central Himalayas. Sustainability 10(9):3052

Ray LK, Goel NK (2019) Flood frequency analysis of Narmada River basin in India under nonstationary condition. J Hydrol Eng 24(8):05019018

Ren M, He X, Kan G, Wang F, Zhang H, Li H et al (2017) A comparison of flood control standards for reservoir engineering for different countries. Water 9(3):152

Saghafian B, Golian S, Ghasemi A (2014) Flood frequency analysis based on simulated peak discharges. Nat Hazards 71:403–417

Sahay RR, Srivastava A (2014) Predicting monsoon floods in rivers embedding wavelet transform, genetic algorithm and neural network. Water Resour Manag 28:301–317

Sahoo A, Ghose DK (2021) Flood frequency analysis for menace gauging station of Mahanadi River, India. J Inst Eng (India) Ser A 102(3):737–748

Samantaray S, Sahoo A, Agnihotri A (2021) Assessment of flood frequency using statistical and hybrid neural network method: Mahanadi River basin, India. J Geol Soc India 97:867–880

Saravanan S, Abijith D (2022) Flood susceptibility mapping of northeast coastal districts of Tamil Nadu India using multi-source geospatial data and machine learning techniques. Geocarto Int 37(27):15252–15281

Sathe BK, Khir MV, Sankhua RN (2012) Rainfall analysis and design flood estimation for upper Krishna River basin catchment in India. Int J Sci Eng Reasearch 3(8)

Singla S, Haldar R, Khosa R, Singla R, Rajeev R (2014) Frequency analysis of annual one day to five consecutive days maximum rainfall for Gandak river basin. Int J Eng Technol 3(2):93

Tripathi G, Pandey AC, Parida BR, Kumar A (2020a) Flood inundation mapping and impact assessment using multi-temporal optical and SAR satellite data: a case study of 2017 flood in Darbhanga district, Bihar, India. Water Resour Manag 34:1871–1892

Tripathi G, Pandey AC, Parida BR, Shakya A (2020b) Comparative flood inundation mapping utilizing multi-temporal optical and SAR satellite data over North Bihar region: a case study of 2019 flooding event over North Bihar. In: Spatial information science for natural resource management. IGI Global, pp 149–168

Tripathi G, Parida BR, Pandey AC (2019) Spatio-temporal rainfall variability and flood prognosis analysis using satellite data over North Bihar during the august 2017 flood event. Hydrology 6(2):38

Tripathi G, Phulwari BS, Parida BR, Pandey AC, Behera MD (2022) Flood inundation and flood-water depth mapping using synthetic aperture radar data in the Gandak River basin. In: Handbook of Himalayan ecosystems and sustainability, vol 2. CRC Press, pp 89–107

Ul Hassan M, Hayat O, Noreen Z (2019) Selecting the best probability distribution for at-site flood frequency analysis; a study of Torne River. SN Appl Sci 1(12):1629

Umar S, Lone MA, Goel NK (2021) Modeling of peak discharges and frequency analysis of floods on the Jhelum river, North Western Himalayas. Model Earth Syst Environ 7:1991–2003

Valentini MHK, Beskow S, Beskow TLC, de Mello CR, Cassalho F, da Silva MES (2024) At-site flood frequency analysis in Brazil. Nat Hazards 120(1):601–618

Verma R, Rai SC (2023) Evaluation of spatio-temporal variability of a flood using the hydrological process of flood frequency in ghagra river basin, India. J Geol Soc India 99(12):1671–1682

Chapter 6
Assessment of Himalayan Floods Using HEC-RAS 2D Flood Modeling: A Case Study from a Part of Beas River

Ansha Mumtaz

Abstract Flood modeling has become a central component of water resource planning and disaster risk reduction, particularly for those regions prone to extreme rainfall, high-intensity runoff, and flash flooding such as the Indian Himalayas. This detailed spatial understanding is crucial for targeted interventions. In conclusion, this research provides vital, actionable information for district-level disaster planning, land-use management, and the design of climate-resilient infrastructure within the Beas River basin. The generated flood hazard maps and LULC impact analyses for both current (2022) and projected (20% increase, 50% increase, and flood 2025) scenarios serve as essential tools for policymakers and emergency managers to enhance preparedness, mitigate potential damages, and ultimately build greater resilience against future flood events in the face of changing climate patterns.

Keywords Flood modeling · Comparative study · LULC analysis

6.1 Introduction

Floods are one of the most frequent and costly natural disasters in the world, which occur when water overflows onto land. Floods are produced by a range of hydrometeorological phenomena such as heavy rainfall, snowmelt, glacial lake outburst, tropical cyclones, and riverbank failure. Floods globally result in tremendous loss of life, infrastructure loss, displacement of populations, and long-term economic and environmental consequences (Kundzewicz et al., 2014; Zhang et al., 2012). Flood frequency and intensity doubled in the past decades, which are primarily attributed to climate change causing abnormal and augmented precipitation. In the developing world, especially in Asia, Africa, and South America, flood risk is augmented by deteriorated infrastructure, land-use planning, deforestation, and uncontrolled urbanization. These anthropogenic processes exacerbate the natural flood risks by hindering

A. Mumtaz (✉)
Symbiosis Institute of Geoinformatics, Symbiosis International (Deemed University), Pune, India
e-mail: mumtazansha@gmail.com

M. Biswas et al. (eds.), *Himalayan Flood Hazards*, Advances in Geographical and Environmental Sciences, https://doi.org/10.1007/978-981-92-0702-2_6

natural drainage, augmenting the surface runoff, and disturbing sensitive ecosystems. South Asia is especially susceptible to river and flash flooding as the region's topography is mountainous, the climate is monsoonal, and the population is high. India, Bangladesh, Nepal, and Pakistan are recurrently hit by disastrous flood disasters during summer monsoons, with far too often disastrous socioeconomic consequences. In India, the Himalayan region encounters a unique set of challenges in flood management. The western Himalayas, characterized by their steep slopes, fragile geological formations, and rapidly changing land-use patterns, are especially vulnerable to flash floods and landslides (Pandey et al., 2013; Sanyal & Lu, 2004). The interplay of extreme weather conditions and human activities, including infrastructure development and deforestation, has significantly increased the flood risk in these mountainous regions. A pertinent example is the Beas River basin in Himachal Pradesh, which is an important tributary of the Indus River system. The Beas River, which starts at the Rohtang Pass and winds through narrow gorges and steep gradients, is particularly susceptible to flash floods triggered by cloudbursts, glacial melt, and heavy rainfall. Recent years have experienced a rise in both the occurrence and intensity of floods in this area, with the devastating flash floods of 2022 and 2025 inflicting significant damage on lives, livelihoods, and infrastructure. Human encroachment along riverbanks, inadequate drainage systems, and the absence of efficient early warning mechanisms further heighten this threat. As remotely sensed datasets become increasingly available and advanced modeling software improves, the application of high-resolution physically based modeling by the scientific community has increased. Among these tools, the River Analysis System of the Hydrologic Engineering Center (HEC-RAS), developed by the U.S. Army Corps of Engineers, has remained one of the most widely applied hydraulic models. Initially developed for 1D steady flow simulation, the release of HEC-RAS 5.0 introduced full 2D capability that enabled advanced floodplain modeling over complicated topography using unstructured computational meshes. Availability in open source and the ability to interface with GIS have rendered it widely used in flood-prone regions of South Asia.

A key element in HEC-RAS modeling is the use of digital elevation models (DEMs) to define terrain. DEM resolution and vertical precision directly influence hydraulic simulation reality. For mountainous basins, medium-resolution DEMs such as the Shuttle Radar Topography Mission (SRTM, 30 m), ASTER-GDEM (30 m), and ALOS World 3D (30 m) are most widely used. Among these, ALOS DEM has been most widely used in Indian flood research due to its higher vertical accuracy and lower noise over mountains. Pande et al. (2020) proved that floodplain geometry derived from ALOS DEM, especially if enhanced by field-based cross-section correction, generated more realistic flood maps in Beas and Sutlej River basins. Bera et al. (2021) supported the observation using SRTM and ASTER versus ALOS in the Teesta Basin, where ALOS best preserved hydrological flow paths and minimized elevation artifacts.

However, DEMs by themselves are insufficient to simulate flood dynamics. Land cover and land use (LULC) significantly control surface roughness, infiltration, and runoff generation. Surface resistance in hydraulic models can be quantified using

Manning's roughness coefficients, which correlate strongly with land cover type. Therefore, by employing spatially high-resolution LULC data, differential roughness can be assigned across the floodplain, which improves model performance. Sentinel-2 MSI satellite images with 10-m resolution and wide spectral range have proven to be successful in generating high-resolution maps of LULC. Studies conducted by Belgiu and Drăguţ (2016) had also indicated the robustness of machine learning classifiers like random forest in Sentinel-2-based LULC classification. Das and Maiti (2022) employed random forest classification of Sentinel-2 to develop urban LULC maps for Bhubaneswar flood modeling in Indian scenarios. Accuracy of the classifications has been shown to be crucial in parameterizing Manning's values for urban, agriculture, forest, and bare surfaces.

Zonal statistics play an extremely crucial role in carrying out this exercise. Through the superimposition of LULC maps over a mesh or grid utilized in HEC-RAS 2D, zonal statistics allow for the derivation of dominant land cover class within a computational cell. This makes automatic assignment of Manning's roughness values possible based on the most frequent or average class in a zone. The approach not only eliminates subjectivity but also allows for repeatability and calibration. Chakraborty et al. (2023) utilized a zonal approach for ranking roughness values in the Brahmani delta based on ALOS DEM and Sentinel-based LULC, leading to effective flood hazard zonation.

The same has also been utilized in the Yamuna and Gandak basins, where Kumar et al. (2020) and Srivastava et al. (2020) demonstrated that land cover spatial variability affects flood extent and velocity outputs significantly.

In the foothills of the Himalayas, where available observational data are limited and the rainfall- runoff response is instantaneous, such integration of DEM and LULC databases with HEC-RAS 2D would serve as a practical but viable alternate. Bose et al. (2022) mapped flood hazard in the Beas River Basin from ALOS DEM and Sentinel-2 LULC. Their study emphasized incorporating fine-scale land use heterogeneity in the definition of roughness coefficients, particularly in land-converting areas for environmental changes such as urbanization or agriculture. Inundation maps based on derived hazard areas were strongly consistent with historical inundation areas, validating the use of high-resolution inputs in data-limited mountainous settings.

Model output verification is another crucial step in flood modeling, especially in areas with few detailed hydrological records. Remote sensing, and in particular Sentinel-1 SAR data, has become more important in this respect. While optical sensors are cloud-shaded, SAR images can pass through clouds and capture surface water extents during monsoon-related floods. Nasirzadehdizaji et al. (2019) utilized Sentinel-1 time series data for the delineation of flood extents in Iran and compared them with HEC-RAS results to calibrate the roughness coefficients and optimize model accuracy. Similarly, in the case of Indian studies, post-event flood maps obtained from SAR images are utilized to cross-check the simulated extents received from HEC-RAS.

Rainfall, boundary conditions, and soil type are other parameters that influence the accuracy of flood simulation. Terrain and LULC are, however, the most critical spatial inputs for hydraulic modeling. Urbanizing watersheds where developed areas alter the patterns of runoff need to be modeled for impervious surface expansion. Jain et al. (2018) used HEC-RAS combined with GIS-based change detection to evaluate flood vulnerability in the Ganga Basin. They determined the areas of high urban growth were exposed to much higher flood risk, which could be quantified only after adding layers with newer LULC data into the model.

Some research has also started to integrate HEC-RAS with emerging technologies like Google Earth Engine (GEE), artificial intelligence (AI), and cloud computing to automate the preparation of inputs. GEE enables near-real-time extraction of LULC maps and DEM derivatives, downloadable and consumable in desktop-based flood modeling systems. Gorelick et al. (2017) justified the advantages of GEE toward planetary-scale geospatial analytics, offering new avenues to model in data-poor regions.

Other researchers also emphasized the necessity of DEM preprocessing before flood modeling. Basic DEMs tend to have sink errors, artificial depressions, or unrealistic ridges, and this would not allow for correct simulation of overland flow. Hydrological conditioning techniques such as depression filling, stream burning, and slope correction are now being used extensively to improve DEM hydrological realism. Hirt et al. (2010) showed that even freely available DEMs like ASTER and SRTM can be prepared to be ready for hydraulic modeling by adequate preprocessing.

The assignment of Manning's n-values is always an art and a science. While Chow's handbook and regional guides to simulation provide standard values for each land use category, in practice, calibration is necessary. Modelers often adjust roughness coefficients using observed velocity profiles, inundation area, or flow depth. Chakraborty et al. (2023) performed sensitivity analysis of n-values for each LULC zone and found that slight changes might lead to a considerable alteration in the modeled flood extents. Thus, zonal assignment of roughness values obtained from LULC requires careful inspection and calibration.

Moreover, model resolution makes a contribution to sensitivity of flood outputs. Coarse meshes also smooth out the small-scale elements like roads, embankments, and canals, whereas finer meshes possess higher spatial resolution but at the expense of increased computational demand. A compromise has to be made between model detail and efficient processing, particularly when large basins are being simulated. Overall, the use of HEC-RAS 2D in flood modeling has developed into a reliable practice when coupled with high-fidelity DEMs, high-resolution LULC maps, and selective utilization of zonal statistics. In the Indian subcontinent and the Himalayas, where topography is complex and in situ data availability is limited, the above tools are a way to generate reliable flood hazard maps. The integration of Sentinel-2 images for LULC, ALOS DEM for height, and SAR for validation together provides a solid foundation for physically based data-intensive flood modeling. With climatic variability still not done changing precipitation patterns and intensifying flood frequency, the demand for such an integrated high-resolution approach will continue only to grow. To understand flood dynamics in such sensitive regions, it is crucial to gather high-resolution topographic information and employ advanced hydrodynamic modeling techniques, along with a thorough watershed management approach. Accurate flood modeling not only helps assess the extent of flooding and vulnerability but also aids in formulating effective mitigation strategies and land-use policies. In the context of the Beas River, two-dimensional hydrodynamic modeling using tools like HEC-RAS is essential for simulating flood behaviors and analyzing the spatial and temporal impacts of these events. This study, as an application of flood dynamics in Beas River basin using high-resolution geospatial information such as ALOS DEM and seasonal Land use/land cover (LULC) maps, is of utmost significance due to the following reasons. Firstly, Himachal Pradesh's Mandi district, which is under the catchment of the Beas River, is very vulnerable to flash floods and landslides, as was seen during the 2022 and 2025 disasters. The uncontrolled growth, the hilly terrain, and the vulnerable geology of the region increase its vulnerability to intense rainfall events. For prevention of the recurring socioeconomic loss, disruption of livelihood, and loss of human lives, the interplay between topography, river hydrodynamics, and dynamic land use has to be comprehended. Based on high-resolution ALOS Digital Elevation Model (DEM) data, the current study correctly maps flood-risk areas, identifies areas of quick generation of runoff, and maps river stream paths in detail. The topography data of this study are the backbone of solid hydrological modeling, early warning, and hazard maps.

Secondly, the seasonal LULC data (pre-monsoon, monsoon, and post-monsoon 2022) introduce a time dimension to the assessment of flood impacts. Pre-monsoon LULC is a reference baseline for land vulnerability to identify low-infiltration surfaces like barren lands and parts of agricultural lands that generate surface runoff. Monsoon LULC is representative of land condition at the time of peak rainfall, with saturated zones and real-time water body boundary. Most critically, post-monsoon LULC contains observable evidence of land cover change due to flood damage, like croplands that have been affected, scoured channels, and displaced drainage lines. Season contrast requires quantifying the flood damage to facilitate an estimation of economic loss, ecological disturbance, and infrastructure vulnerabilities. Third, the study improves the utilization of 2D hydrodynamic modeling (HEC-RAS) in a Himalayan complex basin. Modeling flood extent, depth, and velocity from actual terrain and land cover data, the study offers usable data on high-risk zones, impact zones, and flow characteristics during disaster phases. Lastly, this study offers great contributions for emergency managers, planners, and policymakers. It assists urban planners to zone development away from floodplains, engineers to design flood-resistant structures, and disaster managers to create specific mitigation measures. Through enhancing the scientific understanding of flooding in the Beas Basin, the study directly assists climate adaptation planning, disaster risk reduction, and sustainable development in a changing Himalayan context.

6.1.1 Research Gaps

Despite the considerable contributions of this research in modeling flood dynamics and evaluating the impacts of land use and land cover (LULC), several areas of inquiry remain unaddressed.

Bridging these gaps could further improve our comprehension and management of flood risks within the Beas River basin and other similar mountainous regions:

Enhanced Hydraulic Modeling for Vulnerability Assessment: Although the study effectively employs digital elevation models (DEMs) to analyze terrain and assesses impacts through LULC, there exists a possibility for more detailed hydraulic modeling (e.g., utilizing software like HEC-RAS or FLO-2D) to accurately simulate flood inundation depths, velocities, and durations for various return periods. Such models, informed by high-resolution DEM, could yield more precise and actionable insights for vulnerability assessments than relying solely on LULC changes. This would enable a more accurate appraisal of direct physical damage to infrastructure and buildings predicated on specific water levels and flow forces, moving from generalized affected regions to detailed damage evaluations.

- Integration of Socioeconomic Vulnerability Factors: Although the study highlights urban areas in flood-prone zones, it provides limited incorporation of socioeconomic vulnerability aspects beyond merely geographical location. Future studies could explore demographic details (e.g., age, income brackets, marginalized groups), accessibility to resources, adaptive capacities, and the existing social networks of impacted communities in greater depth. Grasping these socioeconomic factors is vital for formulating equitable and effective disaster preparedness, response, and recovery initiatives that cater to the varied impacts of floods across different community segments, ensuring that interventions are customized for those most vulnerable.
- Projection of Future Flood Scenarios Under Climate Change: The research employs LULC data from 2022, presenting a snapshot of circumstances related to specific flooding incidents during that year. Nevertheless, a significant void exists in forecasting future flood scenarios in the context of climate change. Incorporating climate models to predict changes in monsoon rainfall intensity, duration, and frequency, along with snowmelt patterns (considering the aspect analysis), could yield probabilistic flood hazard maps for upcoming decades. This proactive approach is crucial for long-term strategic planning and implementing adaptive strategies that account for changing flood risks rather than solely depending on historical data.

So, the study is designed to gather and preprocess comprehensive spatial (DEM, LULC, river geometry) and hydrological (rainfall, discharge) data essential for detailed hydrodynamic flood modeling of the Beas River basin, and to characterize the terrain and seasonal land cover of the basin. The objectives are: to establish and operate a two-dimensional (2D) hydraulic model utilizing HEC-RAS 6.6 to simulate flood inundation across three distinct scenarios: a normal Flood (2022 baseline), a flood with a 20% increase, and a flood with a 50% increase, incorporating LULC-based infiltration and Manning's roughness coefficients; to examine and illustrate the spatial distributions of flood characteristics (maximum inundation depth, peak flow velocity, and water surface elevation) for each simulated flood scenario; to create composite flood hazard maps by amalgamating simulated flood depths and velocities, identifying locations with low, moderate, and high hazard levels; and to conduct zonal statistics using GIS to quantify the extent of various land use/land cover (LULC) types (built-up, herbaceous, shrubland, cultivation, forest, barren) that are impacted by floodwaters during both pre-monsoon and post-monsoon conditions for each flood scenario.

6.1.2 Regional Geological Setup

Understanding the local geologic setup is essential for a comprehensive flood hazard assessment because geological and geomorphological conditions play a substantial role in hydrological processes, drainage patterns, and flood hydrodynamics within a river basin. The Beas River basin, situated in Lesser Himalyas and some parts of Shiwalik Province the first geological region near Mandi is the one where the rock types found are the phyllites and quartzites of Chail Group with Mandi Granite including the Tira Sujanpur and Dehra Gopipur areas, we enter the Siwalik zone of the Himalayas [provide general geological province, e.g., the Lesser Himalaya, or if available, specific location], has a geologically sophisticated history that influences its current topography and ground conditions. The basin is commonly characterized by a heterogeneous lithology, which may include sedimentary, metamorphic and igneous rocks like shale, sandstone, quartzite, granite and shale [cite common types of rock like sedimentary, metamorphic, or igneous rocks like shales, sandstones, limestones, granites, or gneisses]. The varying permeability and erodibility of the different rocks have direct impact on infiltration rates, recharge to groundwater, and susceptibility of the slope to erosion and landslides, which in return may add to sediment load in the river and alter channel morphology during flood. Structurally, the region is typically defined by the geological structures of this area depend on the prominent Himalayan geological structures, namely the presence of fault, fold, thrust, and lineaments, including the Main Boundary Thrust (MBT) [cite common structural elements like faults, folds, thrusts, or lineaments, such as major thrust faults or complicated folding]. The structural elements can affect the direction of river channels, tributary network evolution, and the presence of geological weaknesses that may promote flood-driven erosion or channel migration. Seismicity, wherever prevalent in the region, can also potentially induce ground deformation or landslides that change the topography and affect flood routing. The presence of unconsolidated alluvial deposits in the floodplains and river terraces, which can cover older bedrock, is central to controlling the hydrological response of the basin. The deposits are more porous and permeable than bedrock and influence groundwater–surface water interactions as well as the storage capacity of the floodplain during flooding. The thickness and structure of these surface deposits also control soil types, which in turn influence infiltration and runoff generation. In summary, the intricate interaction between lithology, structural geology, and surficial deposits of the Beas River basin's geological configuration strongly governs its hydrological properties; hence, a complete grasp of all these factors is required to model accurately its hydrodynamics and flood risk.

6.1.3 Study Area

The study site for the current study is the Beas River basin of Mandi district of Himachal Pradesh in the western Himalayas of northern India. The Beas River, one of the five great rivers of the Indus river system, originates at the Rohtang Pass at an elevation of approximately 3,978 m above sea level and flows for approximately 470 km before merging with the Sutlej River in Punjab. It runs mostly southwest, cutting through long, deep gorges and narrow valleys, and is crucial for maintaining the ecology, hydrology, and economy of the area. Mandi district, between about 31°13′N to 32°04′N latitude and 76°37′E to 77°23′E longitude, is placed in the central region of Himachal Pradesh and covers a highly dissected mountainous area. The altitude in the district ranges from 600 m to over 3000 m, yielding steep slopes and complex drainage. This altitudinal variation coupled with intense monsoonal rainfall and large-scale land-use change makes the area very susceptible to flash floods and landslides. The Beas River in Mandi is fed by various tributaries like the Suketi Khad, Uhl River, and Rana Khad, which significantly contribute to the hydrological processes of the basin. Tributaries respond to high-intensity rainfall events and, in most instances, cause sudden flow variations and flash flood conditions. The region is monsoonal in nature, where the majority of the rainfall for a year is experienced between June and September. Heavy and short-duration rainfalls during monsoon season have a propensity to lead to overbank flow, lowland flooding, and extensive damage to infrastructure and settlements.

6.2 Methodology

This flowchart is a graphical and structured depiction of the methodology followed for undertaking the flood hazard assessment study of the Beas River basin Farr et al. (2007). It is a chronological depiction of the interdependent steps, ranging from simple data collection and preprocessing to hydrodynamic modeling, scenario simulation, and final output analysis and interpretation. The diagram is a holistic overview, with the extensive range of high-resolution hydrological and geospatial data utilized, the extensive computational software (such as HEC-RAS 2D, Google Earth Engine, and ArcGIS Pro), and the wide range of outputs derived for flood inundation and velocity estimations and overall hazard. All phases are designed

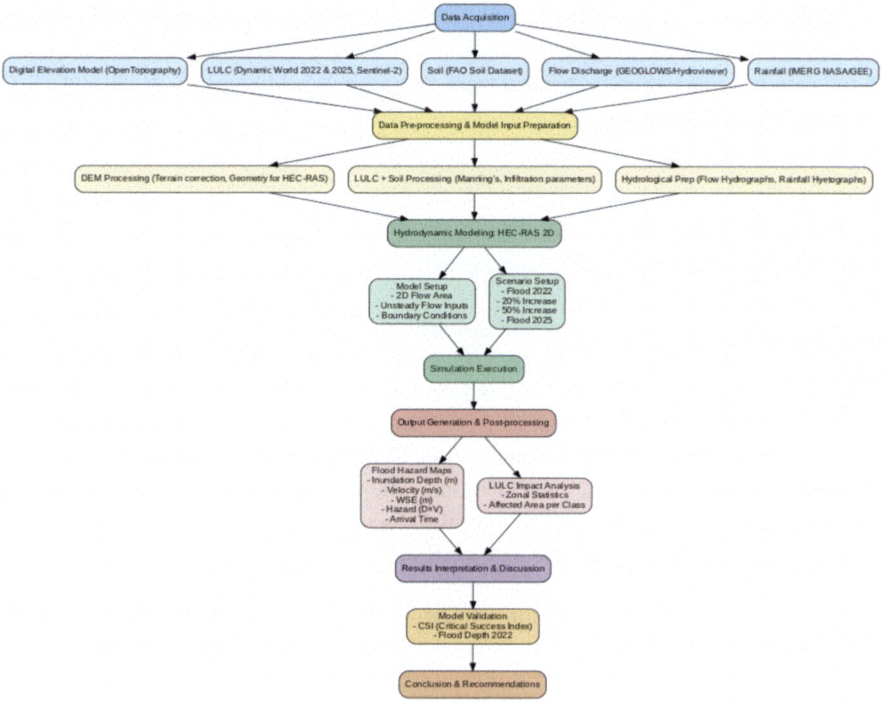

Fig. 6.1 Detailed chart of methodology

to complement each other magnificently, to give a firm and reliable basis for the understanding and mapping of flood risks in current and future anticipated scenarios (Fig. 6.1).

The key steps are listed below:

6.2.1 Data Acquisition and Preprocessing

The cornerstone of flood modeling revolved around the acquisition and preprocessing of several critical datasets:

Digital Elevation Model (DEM) A 30-m high-resolution ALOS PALSAR DEM was acquired from OpenTopography and preprocessed to remove sinks and depressions. Terrain data were utilized in an effort to demarcate the computational domain within HEC-RAS as well as to effectively represent surface elevations and river channels (Kulp & Strauss, 2019; Yamazaki et al., 2011).

Land Use/Land Cover (LULC) Data Seasonal LULC maps for pre-monsoon 2022, monsoon 2022, post-monsoon 2022, and pre-monsoon 2025 were produced from

Sentinel-2 Copernicus satellite imagery by supervised classification in Google Earth Engine (Hamunyela et al., 2018; Loveland & Belward, 1997). Flood 2025 scenario LULC data were specifically derived from Dynamic World Sentinel-2. These LULC classes were significant for:

6.3 Hydrological Data

- Flow Discharge Data: Upstream boundary conditions for the hydrodynamic model were retrieved from GEOGLOWS Hydroviewer data. This was historical flow discharge data for the 2022 flood event that was used to construct the flow hydrograph.
- Rainfall Data: Hourly precipitation data were retrieved from IMERG (NASA) through the use of Google Earth Engine. These rainfall data were applied in the model to simulate direct rainfall over the floodplain and contribute to runoff.
- Manning's n-values of roughness coefficients were assigned to different land cover categories in the hydraulic model in terms of flow resistance modification. Infiltration parameters (maximum deficit, initial deficit, and potential percolation rate) were set by LULC class impacting rainfall–runoff transformation.

6.4 Hydrodynamic Model Setup (HEC-RAS 2D)

- Model Geometry: The preprocessed DEM was imported to HEC-RAS 2D to develop a computational mesh of varying cell sizes (10–30 m) with refinement of river and urban zones (Brunner, 2016; Zheng et al., 2018).
- Surface Roughness Assignment: Manning's n-values were distributed to each mesh cell using the classified LULC map for each season, reflecting the varying surface roughness of different land cover classes (Qi et al., 2020; Aboelnour & Engel, 2018).
- Boundary Conditions: Upstream: GEOGLOWS data inflow, hydrographs. Downstream: Normal depth or rating curves. Rainfall Input: IMERG-based rainfall data were utilized using the rain-on-grid methodology across the computational domain. Infiltration Environments: In rainfall simulations, Green–Ampt infiltration properties were utilized based on the LULC classes.

6.4.1 Simulation Scenarios of Flood

Three individual flood events were simulated under different LULC conditions to assess the flood hazard in an integrated manner: observed 2022 flood event, 20% increased discharge scenario, 50% increased discharge scenario, and projected flood

2025 scenario. All scenarios were run for both pre-monsoon and post-monsoon LULC conditions to analyze temporal variations in flood response.

6.4.2 Output Analysis and Spatial Processing

The hydrodynamic model generated multiple raster outputs required for flood hazard estimation: maximum flood depth, maximum flow velocity, water surface elevation (WSE), and time of flood arrival (Robinson et al., 2017). The flood hazard index (FHI) was calculated using the following formula:

$$\text{FHI} = d * (v + 0.5) \tag{6.1}$$

Here, d is the depth, and v is the velocity. All these results were post-processed using ArcGIS Pro for computation of zonal statistics based on LULC classes, mapping of affected zones, generation of detailed flood inundation maps, and compilation of comprehensive flood hazard and risk zones for prioritization.

6.4.3 LULC Risk and Impact Mapping

Inundation maps for every one of the scenarios were overlaid with classified maps of LULC to compute the affected area (in hectares) per land cover class. This analysis identified seasonal flood exposure and vulnerability variations, dentification of flood-prone, and high-hazard zones, which are crucial for giving precedence to adaptation and mitigation strategies.

6.5 Results

The Beas River, marked here by the intense blue channels, cuts through this varied landscape, its course determined exactly by underlying topography. The river originates in the high Himalayas and then flows roughly south through hilly and mountainous country before potentially opening out into a wide reservoir in the west of the map. Though its precise range of flow in this given map is not quantified, the overall length of the Beas River is approximately 470 km (290 miles) long, with a vast majority of the 256 km (159 miles) traversing solely within Himachal Pradesh before entering Punjab. Such a long extension allows for flood occurrences to be deep upstream, with accumulated water flowing down over extensive distances to strike downstream areas (Zhang et al., 2020). It allows for the provision of data necessary to delineate precisely the river network of tributaries, as well as their slopes. Steep river sections will transport water quickly, whereas the flat sections, especially in lower valleys, will have slow-moving water with broader inundation, directly influencing the duration and extent of flooding. The compilation of the high-resolution DEM data with the flow properties of the river is therefore important for forecasting the flood pathway, estimating inundation depth and ultimately for general flood risk assessment and management within the Beas River basin. A high-resolution ALOS (Advanced Land Observing Satellite) DEM was applied to acquire the intricate topographic variations of the basin. Prior to modeling, the DEM (Fig. 6.2) was preceded by preprocessing methods, including the removal of spurious sinks (depressions) to yield hydrologically correct flow directions. The DEM was precisely projected to the Universal Transverse Mercator (UTM) Zone 43N projection within the WGS 1984 datum (EPSG:32643). Shaded relief analysis, derived from the ALOS DEM (e.g., Fig. 6.2), outlines the striking terrain features of the basin with crystal clarity. Steep and mountainous terrain dominates at elevated points, with focus in the Mandi district through steep elevational gradients. In contrast, the lower parts, or the Una and Hamirpur valleys, possess relatively little flat terrain. These elevational gradients are critical, pointing to the fact that the inherent rapid runoff generation in upstream mountainous terrain significantly and quickly contributes to the area of flooding in the downstream flat region (Bhuyian & Kalyanapu, 2020; Demir & Kisi, 2016).

The slope map (Fig. 6.3) is a simple geographical derivative, showing the measured steepness of the land surface across the area of interest. Usually quantified in terms of degrees or percent and precisely sorted in many ranges, this map provides instant and

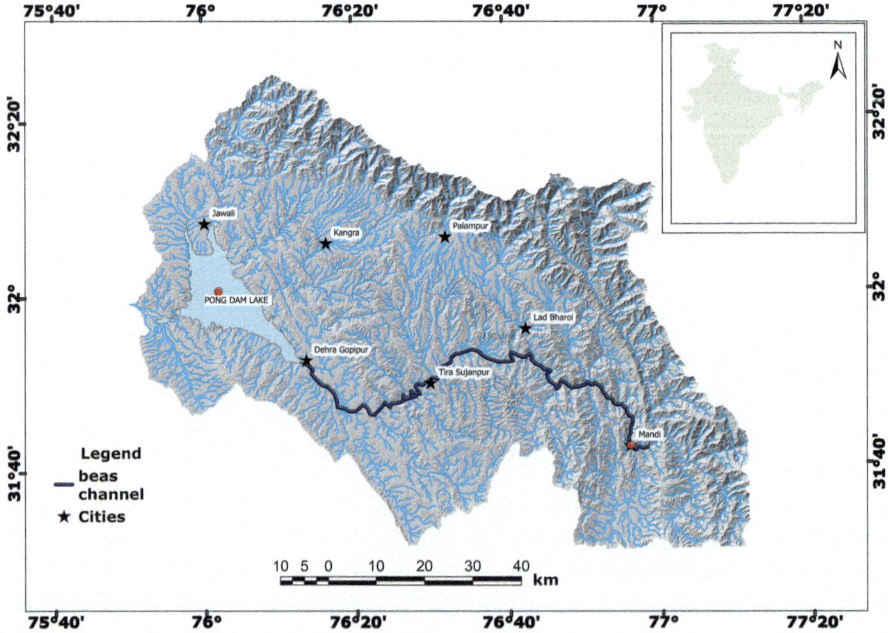

Fig. 6.2 Location map of the study area with drainage details

priceless information regarding the hydrologic response of the landscape. It yields critical information on how fast surface runoff will occur, where the floodwaters will tend to be concentrated and gain destructive kinetic energy, and where they are likely to dissipate and persist. Areas depicted with highly rising inclines, usually graphically represented in darker red or brown on the map key (e.g., representing mountain ranges like 32.657°–74.351°), are of utmost relevance for flood research. When it rains upon these steep surfaces, gravity forces drive water to travel quickly downhill. This rapid acceleration diminishes the time of infiltration into the ground, producing a sudden and massive conversion of rain to runoff. A great quantity of water is instantaneously concentrated and channeled into the drainage system, greatly accelerating the increase in river stages and immensely increasing the potential for flash floods, particularly for narrow and confined downstream valleys. These zones of high gradient are thus defined as regions of initial high-runoff production in the catchment headwaters. Conversely, regions bordered by lower slopes (e.g., the light yellow regions for 0.001°–9.622°) are defined as having much slower moving water. This slower velocity has more room for storing water, greater infiltration into the ground, and less overland flow. But a crucial consequence for flood modeling is that when eventually the floodwaters spill over riverbanks and flow onto these flat areas, they spread out far over vast distances. The far-reaching inundation leads to more widespread and prolonged flooding, reaching more areas of land, buildings, and human habitation. Accordingly, slope map is not just vital for the identification of areas of intensified runoff but also for accurate delineation of locations most vulnerable to extensive and prolonged overtopping and therefore an essential and invaluable part of robust flood modeling and hazard assessment.

The aspect map (Fig. 6.4), a significant topographic derivative, shows the compass direction in which the slope is orientated. While its role in the forming of flood dynamics is typically described as indirect, it still plays a significant role, particularly over mountain and snow-conducive areas such as the Beas basin, by influencing microclimatic regimes and vegetation regimes. Different aspects receive varying degrees of solar radiation throughout the day and year, leading to varying temperature and moisture regimes. For instance, in the Northern Hemisphere, south slopes (traditionally depicted as blue in the map key, indicating 157.5°–202.5°) are directly exposed to more sunlight and for longer hours. This increased insolation is to blame for quicker snowmelt during hotter periods of the year, such as late spring or early summer, and even during some winter thaw. The abrupt dissolution of snowpack within the higher catchment contributes significant regional, often sudden, water volumes to the Beas River system. Such influx will powerfully magnify already high river flows, increase baseflow rates, or even generate extreme flood events, particularly when augmented with rainfall, even in the absence of excessive rain. Furthermore, warmer south-facing slopes could bear alternate types of vegetation that have distinct hydrological properties. Conversely, north-facing slopes (customarily red on the map, indicating 0°–22.5° or 337.5°–360°) get less direct sunlight. Thus, they typically retain snow longer during the melt period and often support denser, more water-tolerant vegetation (e.g., forests of conifers) due to cooler, more

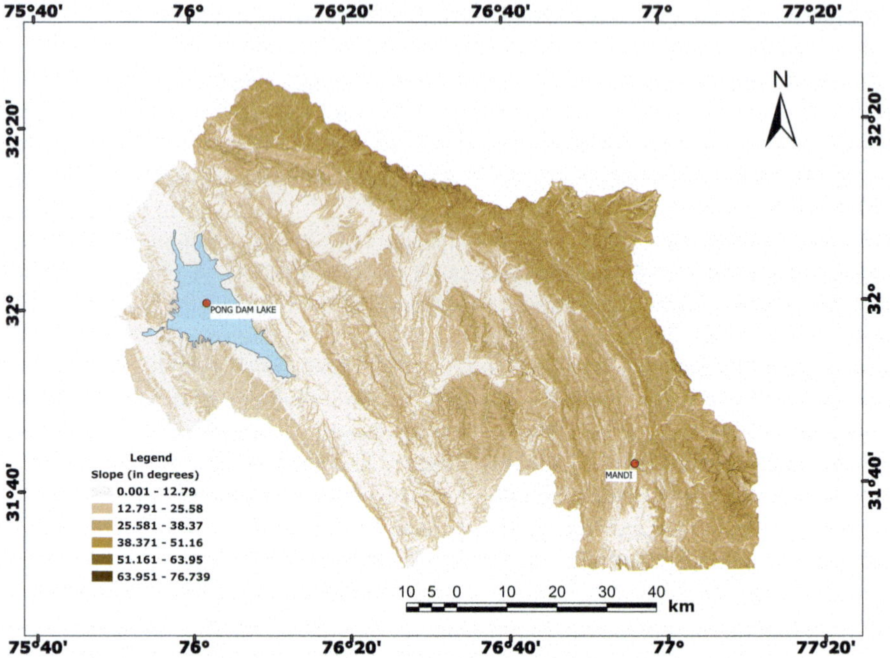

Fig. 6.3 Slope map of the study area

humid conditions and lower evapotranspiration. Higher density vegetation can significantly improve greater interception of higher water by canopies, greater infiltration of water into the soil, and reduced surface runoff with lower hydraulic resistance, thereby potentially offering some natural attenuation of floods and regulation of flow.

Therefore, analyzing the aspect map provides abundant information on the spatial heterogeneity of snowmelt velocity, soil moisture characteristics, and vegetation cover characteristics over the basin. This complements the large-scale hydrological understanding obtained from other topographic data and provides valuable information for improved flood prediction and risk assessment, especially in areas where the snowmelt is a key contributor to river discharge and the microclimatic heterogeneity affects the hydrological response.

Land use/land cover (LULC) classes are essential in modeling flood hazards since they represent the physical characteristics of the Earth's surface, which play a significant role in controlling hydrological processes such as runoff generation, infiltration, and hydraulic roughness. The LULC classes used in this study were diligently derived from high-resolution Sentinel-2 satellite imagery, representing three distinct and hydrologically meaningful seasonal phases: pre-monsoon, monsoon, and post-monsoon. The land cover and land use classes divided the landscape into basic features such as urban, agricultural lands, forest, bare soil, and water bodies with specific hydraulic roughness and infiltration properties that were translated into model parameters (for example, Manning's n-values and Green–Ampt infiltration

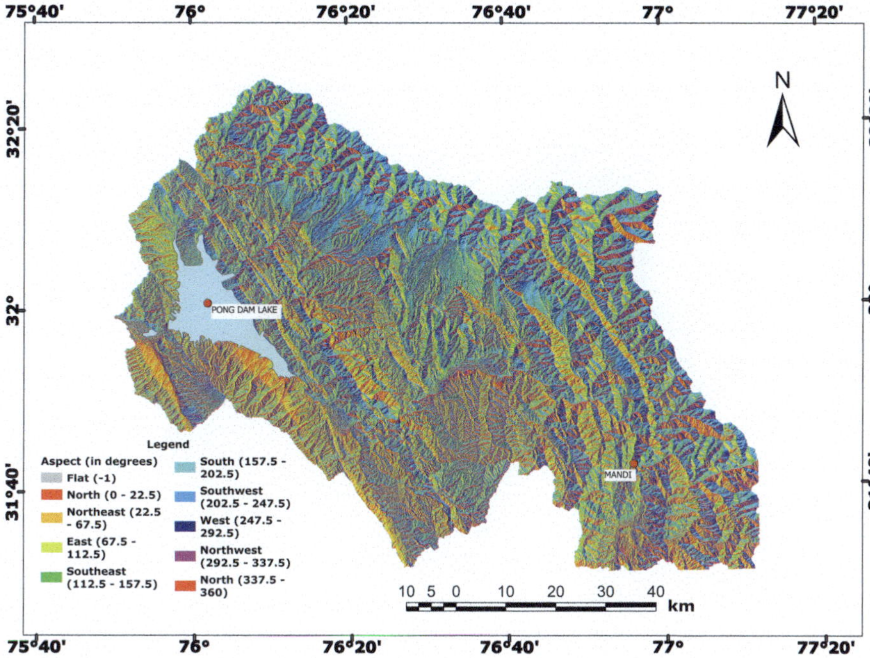

Fig. 6.4 Aspect slope direction showing the slope direction of the study area

parameters). The seasonal LULC maps provided dynamic information regarding the surface conditions of the basin, including the variation in vegetation density, soil moisture, and human activity over seasons.

Pre-monsoon The season was typically characterized by a prevalence of dry, often crusted, bare soil and sparse vegetation, easily apparent in cropland and naturally bare surfaces that have not yet experienced significant rainfall. This condition inevitably (Fig. 6.5) gives rise to high surface runoff and lower infiltration capacity due to low antecedent moisture content in the soil and perhaps crusty ground surfaces. Therefore, the terrain becomes more prone to rapid development of flow and instantaneous overland flow at the onset of monsoon rains, reinforcing initial peak flooding.

Monsoon The LULC maps during monsoon times (Fig. 6.6) indicated a sudden increase in vegetation cover density and typically saturated or close-to-saturated soil conditions for the entire basin. Although local vegetation density slows flow, increases interception, and promotes infiltration to some extent, the large quantity and intensity of rainfalls characteristic of this period are likely to swamp natural capacity for uptake. This generates extensive saturation-excess runoff, increased surface water storage, and large contributions to river discharge. Altered hydraulic roughness by dense vegetation also greatly affects regulation of the speed of flow in floodplains.

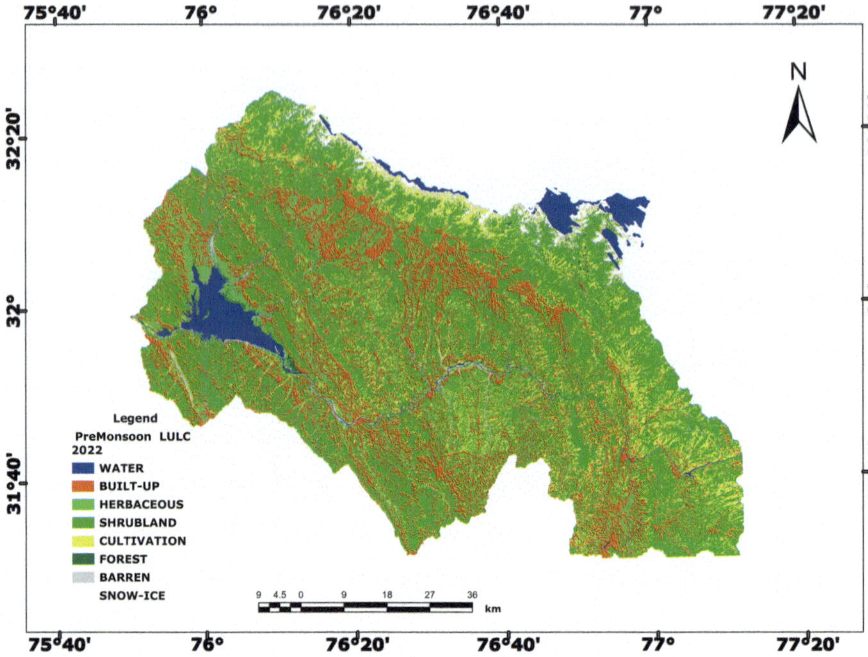

Fig. 6.5 Land use/land cover (LULC) map for pre-monsoon 2022

Post-monsoon Post-monsoon LULC maps were employed to identify areas that remained inundated or exhibited clear evidence of prolonged flood impact even after the main monsoon event and retreat of the peak flows. These signs comprised significant ground-truth information for model calibration and validation of the output from the HEC-RAS simulation, particularly in terms of the accurate spatial extent and duration of inundation. Post-event LULC comparison with pre-event situations allows for assessing remaining flood hazard and the identification of possible long-term recovery or targeted mitigation zones.

Further, for the future scenario (flood 2025 model), LULC data were retrieved from Dynamic World (Sentinel-2) so that it is possible to assess flood hazards in terms of future anticipated land cover situations, thereby enabling a plan for a forward-looking flood risk management. The dynamic character as illustrated in these seasonal and future LULC maps represents an entire scenario of the hydrological response variability of the basin.

Rainfall Data Hourly rainfall data, which was acquired from the IMERG (Integrated Multi-satellitE Retrievals for GPM) product and processed with great care using Google Earth Engine, was employed as a principal input to the simulation period June 28–July 2, 2022. The data provided the observed precipitation patterns, which served as spatially uniform precipitation input over the 2D simulation domain

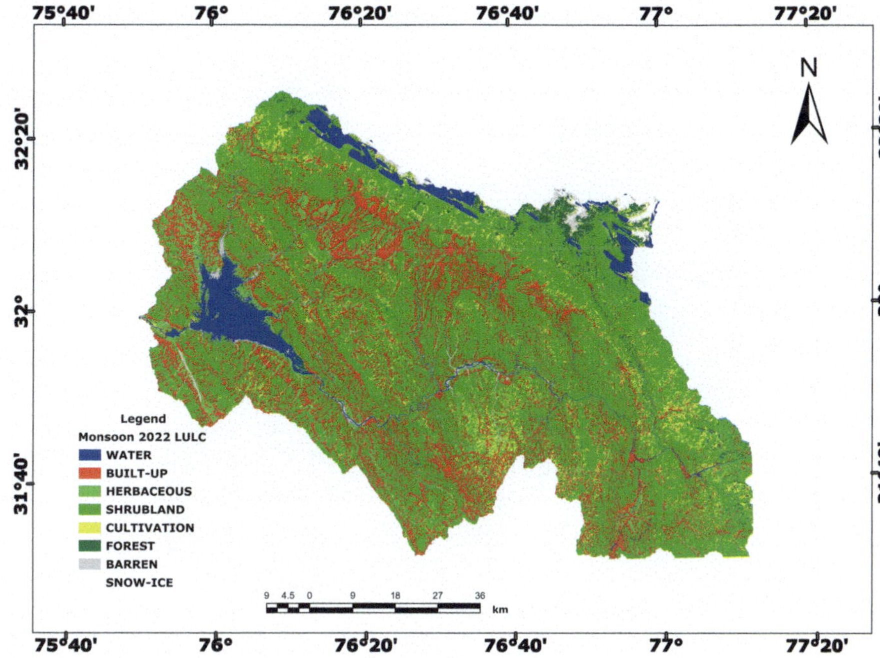

Fig. 6.6 Land use/land cover (LULC) map for monsoon 2022

to accurately record the direct rainfall contribution toward the flood event. The magnitude and temporal pattern of the rainfall, as observed by IMERG, were critical to be in order to replicate the ensuing hydrological response of the basin. The hourly rainfall data utilized for simulations of the 2022 flood event included the standard observed values and progressively higher surmised scenarios (20% and 50% rise). This table is highlighted on the peak intensities of rainfall and cumulative precipitation totals on the most significant days, providing the quantitative basis for the hydrological inputs. Application of this spatiotemporally disambiguated rainfall data in the HEC-RAS 2D model was pivotal toward understanding the basin's response to direct precipitation, particularly for areas prone to localized runoff generation and flash flooding (Fig. 6.7 and Table 6.1).

6.5.1 Observed River Discharge Data Analysis

Accurate representation of upstream river discharge and water levels is the key to controlling strong hydrodynamic simulations and interpreting flood propagation. These data were responsible for setting up the upstream inflow boundary conditions for the HEC-RAS model and served as a critical benchmark for subsequent model

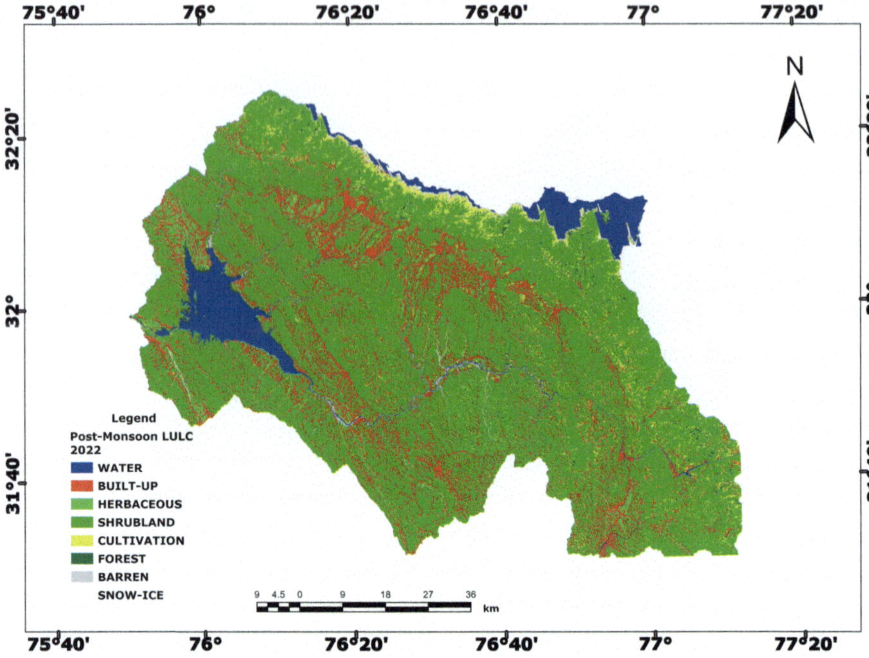

Fig. 6.7 Land use/land cover (LULC) map for post-monsoon 2022

Table 6.1 Percentage of post-monsoon affected land cover area

Value	Flood 2022	20%	50%
Built-up	20.29541626	19.76237793	20.29541626
Herbaceous	8.88949585	8.17411499	8.88949585
Shrubland	8.359899902	8.24624939	8.359899902
Cultivation	13.25495605	13.11911621	13.25495605
Forest	22.1944458	22.19464111	22.1944458
Barren	12.37031555	11.38283691	12.37031555

calibration and validation. The primary hydrological input used for the simulation of the 2022 flood event was a detailed flow hydrograph, which showed the observed discharge at the time. Besides the flood 2022 historical observed flow, two future or extreme scenarios, which are deemed to be highly significant, were developed: a 20% greater discharge and a 50% greater discharge. These step increments were applied in the modeled 2022 hydrograph to check the sensitivity of flood extent and depth to heightened flow magnitudes, which may represent potential future scenarios because of higher precipitation or snowmelt contribution, or to check how much existing flood defenses can withstand more intense events. Table 6.2 shows an example of the hourly discharge data used in the simulations with the observed "flood 2022"

Table 6.2 Percentage of post-monsoon affected land cover area

Value	Flood 2022	20%	50%
BUILT-UP	21.48955688	20.99093018	21.48955688
Herbaceous	22.09562988	22.14244385	22.09562988
Shrubland	12.69456787	12.71238403	12.69456787
Cultivation	7.986224365	7.388928223	7.986224365
Forest	25.11636963	25.1972168	25.11636963
Barren	12.99144592	12.04669189	12.99144592

values plotted against the corresponding 20% and 50% increased discharge scenarios. This table illustrates how the maximum flows and overall hydrograph shape were adjusted for the different scenarios, which had a direct influence on volume and water velocity entering the modeled floodplain. The delineation of the boundary conditions individually ensures that the hydrodynamic model accurately represents the diverse regimes of flow and their potential influences on inundation distributions and flood risks downstream.

6.5.2 HEC-RAS 2D Flood Simulation

To undertake a comprehensive and realistic flood hazard analysis, 2D hydrodynamic flood modeling (Fig. 6.8) was conducted using the process of HEC-RAS version 6.6. The advanced simulation technique has the capability to accurately forecast the intricate movement of floodwater over sophisticated topography by simulating flood inundation zones, water surface elevation, and flow velocity under varied rainfall and discharge conditions.

6.5.3 HEC-RAS 2D Model Setup and Parameters

The HEC-RAS 2D model was carefully set up with the following essential components and parameters:

- **Preparation of Terrain**: The ALOS DEM was utilized as the underlying terrain layer. To enable meaningful hydraulic representation, terrain was clipped to the study region of interest and conditioned more by including breaklines, which force the computation mesh to stay along significant linear features such as riverbanks and main roads. Terrain resolution was optimized so that computational stability was weighed against detail needed.

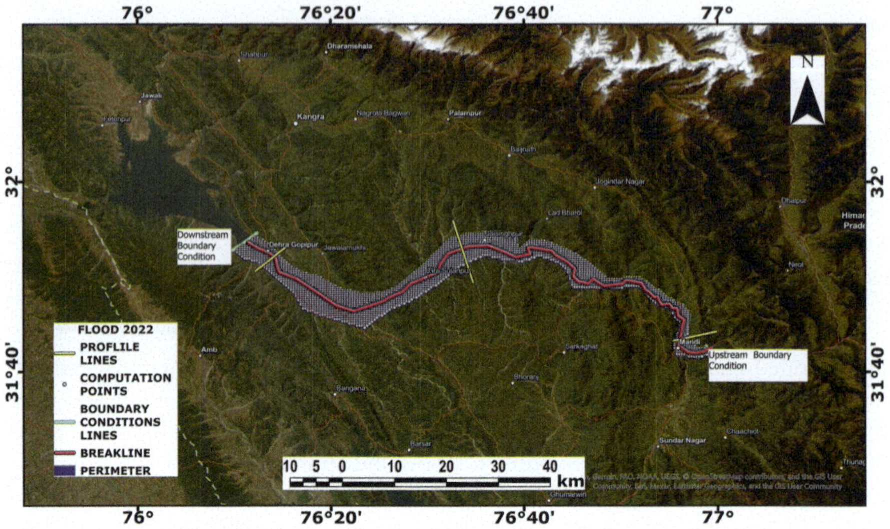

Fig. 6.8 2D area for flood 2022 created in HEC-RUS for flood simulation

- **Mesh Generation**: A 2D Flow Area was defined for the computational domain. Within this area, a deformable computation mesh was formulated with an approximate cell size of 400 cells. The mesh was highly refined along river channels, confluence points, and near hydraulic structures through placement of breaklines in key locations, thereby improving flow routing and localized hydraulic calculations accuracy.
- **Boundary Conditions**: Appropriate boundary conditions were applied to trigger the hydraulic simulation:

Inflow Boundary Upstream inflow was defined in terms of a discharge hydrograph according to past experience and rainfall–runoff prediction, which characterized the temporal variation of water into the model domain.

Outflow Boundary The typical depth boundary condition was applied at the downstream end of the model. The condition calculates the water surface elevation in terms of a mean downstream slope, allowing water to exit the model in a natural way without the application of unnatural backwater effects.

Rainfall Data Input Hourly IMERG precipitation data, spanning from June 28 to July 2, 2022 (Table), were directly incorporated into the model as a spatially uniform rainfall input over the entire 2D computational domain, enabling direct rainfall–runoff simulation within HEC-RAS.

6.6 Infiltration Parameters and Land Cover-Based Assignment

A proper simulation of rainfall transformation into runoff within a hydrological model entails a reliable description of infiltration processes. In this research, infiltration was directly simulated in HEC-RAS 2D through its land use/land cover (LULC)-based infiltration method. The method enables the spatial heterogeneity of infiltration properties to be directly attributed to various LULC classes and their varying conditions through seasons. The approach used in HEC-RAS employs critical parameters Infiltration Parameters Assigned per LULC Class (Pre / Post Monsoon)—namely maximum deficit and percolation rate—which were carefully allocated according to detailed literature review and modified for every LULC category and dominant seasonal conditions. The maximum deficit is the initial storage capacity of the unsaturated zone of the ground, reflecting how much water the earth can hold before saturation and major runoff occur. The percolation rate specifies the constant rate at which water may percolate into the soil after saturation, indicating the ability of the soil to sustain vertical water movement. Dynamic incorporation of these season-specific and LULC-dependent parameters greatly improves the model's capability to represent the spatial and temporal patterns of runoff generation within the Beas River basin. Infiltration parameters were assigned to all LULC classes with care, considering the unique hydrological attributes of pre-monsoon and post-monsoon periods in order to capture the fluctuating nature of vegetation and soil moisture.

6.7 Analysis of Parameter Assignment and Seasonal Variability

As can be seen in Table X, the infiltration parameters assigned carefully capture anticipated hydrological response of each land cover class, conforming to rates determined in the hydrological literature. For example: Urban built-up areas were given very low maximum deficit and percolation rates (5 mm and 0.1 mm/h, respectively). This is representative of their highly impervious condition, wherein most of the rainfall is immediate surface runoff with little infiltration, in line with general understanding of urban hydrology. On the other hand, natural vegetative covers like forest and shrubland have the highest maximum deficit (150/140 and 120/100 mm) and percolation rates (3.0/2.5 and 1.8/1.4 mm/h). This can be explained by their well-developed root system, increased organic matter, and undisturbed soil structure, all of which, in combination, facilitate high water absorption and vertical flow, significantly decreasing surface runoff volumes. Herbaceous and cultivation lands are in an intermediate category, and their values represent their most general soil structure and vegetative cover, which may differ with agricultural operations and growth cycles of a season. Barren lands exhibit lower infiltration rates than vegetated sites (max deficit

30/25 mm, percolation rate 0.5/0.3 mm/h), and these typically consist of hard-packed soils with little organic material, resulting in higher potential for direct runoff.

Water bodies are given zero values for both parameters because they are already full of water or infiltration is not a process in these areas. One of the key aspects of this parameterization is its seasonal adjustment. For the majority of categories, especially where there is extensive vegetation or exposed ground, maximum deficit values tend to be higher in the pre-monsoon season compared to the post-monsoon. This is for the drier antecedent conditions and perhaps more parched soil profiles that prevail ahead of the monsoon rains setting in, to enable a higher initial infiltration capacity. In contrast, percolation rates also are slightly greater in the pre-monsoon, indicating soils that are yet not entirely saturated, to support faster initial vertical flow of water. During the post-monsoon period, since soils would be wetter and vegetation possibly denser, the slightly lower maximum deficit and percolation rates indicate a lesser capability for additional infiltration and a higher chance of saturation-excess runoff.

These well-chosen and seasonally corrected infiltration parameters (Table 4) are essential to the HEC-RAS 2D model for partitioning rainfall into infiltration and surface runoff components accurately. This accurate parameterization is crucial in providing a more realistic and dependable simulation of flood generation and propagation along the Beas River basin under changing seasonal conditions.

6.8 Land Roughness (Manning's n-values)

Land roughness, defined as Manning's roughness coefficients (n-values), is an elementary hydrodynamic modeling parameter that represents the resistance experienced by flowing water over different surfaces. These n-values play a critical role in simulating realistically flow velocities, depths, and floodwater propagation overall within the 2D HEC-RAS model mesh. The spatial assignment of these coefficients was performed meticulously with respect to the land use/land cover (LULC) raster that had already been classified such that hydraulic resistance was based on the specific surface condition of each cell. In general, values ranged from 0.02 for very smooth, non-blocked surfaces (e.g., channels lined with concrete) to 0.10 for very resistive surfaces (e.g., heavy vegetation or urbanized urban surfaces).

6.9 Pre-monsoon Season (Dry Conditions)

It is of particular significance (Mannings n-value assigned to the various land cover classes for the pre-monsoon season) to employ appropriate Manning's n-values during pre-monsoon season since it possesses unique hydrological and land surface conditions. In the pre-monsoon season, the surface will be mostly dry, cover vegetation will be sparse, and agricultural crop lands will be freshly ploughed or even bare. Such ubiquitous conditions automatically lead to lower mean surface roughness over the terrain with the result of facilitating higher surface runoff flow speeds. Less friction enables water to move at faster speeds, thus leading to quicker runoff production with the onset of rain.

Low Manning's n of 0.030 was assigned to "water" bodies since it characterizes the typically clean and open flow conditions of river courses and ponds during the season. Conversely, "built-up" areas were assigned a significantly higher value of 0.100 to account for the strong resistance to flow presented by buildings, roads, and other non-permeable structures. Proper application of such specific n-values of pre-monsoon LULC is essential for the hydrodynamic model to effectively simulate the initial rapid runoff and subsequent flood spreading in a realistic way, yielding reliable simulation results in accordance with this arid season's uniqueness.

6.9.1 Post-monsoon Season (Wet/Veg. Conditions)

After the heavy monsoon season, the post-monsoon season is characterized by another unique set of land surface conditions that have a strong impact on hydraulic roughness. Under this stage, surfaces are usually saturated with water from residual moisture left after the previous heavy rainfall, and vegetation is generally at its most densely grown and mature. In addition, this stage can also observe an expansion of natural barriers like debris accumulated plant biomass, dropped waste, or leftover standing water, mostly in low-lying terrain.

All these conditions together—high vegetation cover, saturated soils, and physical obstructions—together contribute to higher resistance in water flow over the land surface. The increased friction results in slower and more diffused surface flow, where water moves along a longer path and at lower velocities in contrast to dry or sparse conditions. As such, these conditions require assignment of greater Manning's n-values in the hydrodynamic model to properly capture this increased resistance. These higher n-values (Manning n-value assigned to the various land cover classes for the post-monsoon season) are important for realistically simulating the reduced recession of floodwaters, possible extended inundation in more flat areas, and the hydrologic response to any future rainfall events that may fall on an already saturated and vegetated terrain. The proper depiction of post-monsoon roughness guarantees the model's accuracy in predicting not just the immediate flood event, but also the subsequent hydrological response of the basin during its recovery process.

6.9.2 Simulation Settings: Temporal Resolution and Duration

The reliability and accuracy of hydrodynamic flood simulations in HEC-RAS 2D are significantly conditioned by the simulation settings configured, especially the duration and temporal resolution. For the overall flood hazard appraisal of the Beas

River basin, the HEC-RAS 2D model was carefully configured to achieve the best balance between computational efficiency and the accurate depiction of dynamic flood processes.

Computation Interval and Time Step: Temporal accuracy of the simulation was established through a one-minute computation interval within the HEC-RAS Unsteady Flow Analysis interface. This high temporal resolution is paramount in capturing sharp variations in flow conditions, which are typical of flash flood events or abrupt peak discharges commonly encountered in upland river basins like the Beas. Such high temporal resolution enables the numerical engine of the model to capture detailed hydraulic processes, such as sudden rise and fall of water levels, surges, and sudden changes in direction or velocity of the flow, thus improving the accuracy of the results of the simulation.

Simulation Period and Intervals of Output: The model was run for a long duration from June 29, 2022, to October 5, 2022. This period was precisely selected to cover the whole critical monsoon phase, such that all the peak flood occurrences and their subsequent declining phases germane for a complete flood evaluation in the Beas River basin were well covered. Though the internal computation was done at one-minute time intervals, output data, including key parameters like hydrographs, maximum flood extent maps, and detailed hydraulic parameters like water surface elevation, flow depth, and velocity, were produced at one-hour time intervals. This output interval strategic selection establishes a necessary equilibrium: It offers enough temporal detail to adequately analyze the course and effects of flooding during the modeled time frame, while at the same time controlling the computational burden, file sizes, and processing time needed for post-processing and visualization. All of these settings together allow for accurate simulation of flood behavior, providing sufficient resolution for valid analysis and sound data storage.

6.10 Analysis of Affected Land Use/Land Cover (LULC) Zones

It was necessary, in the course of the very detailed and fine-grained assessment of flood impacts in the Beas River basin, to quantify overlap between simulated floods and current land uses. This was achieved through performing detailed zonal statistics, overlaying the simulated flood inundation boundaries of the hydrodynamic model onto the high-resolution seasonally classified land use/land cover (LULC) data. This operation is essential for diverging from inundation mapping into simple mappings toward a tangible quantification of potential damage, loss estimation, and vulnerability mapping. By identifying the most affected specific land cover types, the study offers valuable information for targeted mitigation efforts, resource allocation to counter disasters, and effective policymaking for sustainable development in affected flood areas. This approach allows for direct mapping of hydrological products and socioeconomic or environmental impacts, rendering it a sound basis for management of flood risk.

6.11 Zonal Statistics for Affected LULC

The calculation of conducting zonal statistics on affected LULC was carried out with great care through a series of interconnected geospatial procedures: Generation of Flood Inundation Boundary: The first critical step was the production of reliable digital models of the flooded zones for each event simulation. The maximum flood depth rasters, representing the extent of maximum inundation and depth as per the HEC-RAS 2D hydrodynamic simulations for the normal flood (2022), 20% increased flood, and 50% increased flood scenario, were utilized for that. These raster outputs were then converted into vector polygon boundaries. This vector conversion is crucial because polygon features do not have continuous, measurable areas whose boundaries delineate clearly and unambiguously the maximum inundation area for every flood event, and hence, they define "zones" on which the subsequent statistical analysis occurs.

LULC Data Integration The second operation was the fusion of the spatial flood boundaries with the high-resolution land use/land cover (LULC) data. Pre-monsoon and post-monsoon LULC rasters, which were labeled cautiously into detailed land cover classes (e.g., built-up, herbaceous vegetation, shrubland, cultivation, forest, and barren lands) from Sentinel-2 imagery, were used as the base or "zonal input" layers. The use of seasonal LULC data is important because the vulnerability and land-use effect may vary significantly over time (e.g., agricultural areas in various growth phases, or varying vegetation density affecting hydraulic resistance and flood extension).

Such integration enabled an overlay analysis to represent the underlying land attributes for each flood zone.

Calculation of Zonal Statistics The functions of zonal statistics were done by overlaying each flood inundation polygon boundary (the prescribed zones) over the corresponding LULC raster. For each individual flood inundation boundary (e.g., normal flood 2022, 20% increase, 50% increase), the area (in hectares or square kilometers) of each underlying LULC category was determined computationally. This allows for precisely how much of each individual land cover type—whether hectares of agricultural fields, square kilometers of urban settlements, or acres of woodland—was directly affected or flooded by the simulated floodwaters.

The outcome of this calculation provides valuable quantitative data for establishing the direct impacts of flooding on various land uses, enabling thorough damage calculation, establishing principal areas of vulnerability, and eventually informing flood mitigation and adaptation planning focused on these areas.

6.12 Flood Hazard Assessment

This chapter presents a comprehensive analysis of the HEC-RAS 2D flood simulation results for the Beas River basin, focusing on the impacts of varying flood magnitudes under three distinct hydrological scenarios: (1) a **normal flood condition (2022 baseline)**, (2) a **20% increased flood scenario**, and (3) a **50% increased flood scenario**. These simulations provide quantitative data on flood behavior in terms of inundation depth, flow velocity, water surface elevation (WSE), and composite flood hazard across spatial extents. Additionally, the analysis quantifies the areas of different land use/land cover (LULC) types affected under these flood conditions, enabling critical insights for regional flood risk-management (Zhang et al., 2012).

The HEC-RAS 2D hydrodynamic modeling performed for the Beas River basin was a good foundation for approximating flood hazards. The simulations generated a series of extremely high-resolution raster outputs, such as maximum flood depth, maximum flow velocity, maximum water surface elevation, and a composite flood hazard index. These figures were computed systematically for every simulated scenario (normal flood 2022, 20% increased flood, and 50% increased flood) in order to create a detailed and spatially explicit interpretation of potential flood impacts across the study area. All parameters of output yield unique information critical to end-to-end risk analysis, vulnerability mapping, and application of targeted flood management strategies.

6.12.1 Flood Depth Analysis

Definition: An elevation between the simulated floodwater surface and the base bare ground elevation is referred to as flood depth. Flood depth is a most critical hydrological parameter that represents the intensity and potential impact of flooding at specific geographical points. In direct proportion with the extent of damage to infrastructure, agricultural lands, and human settlements, flood depth is a key indicator for physical vulnerability assessment and emergency response planning.

6.12.2 Normal Flood Condition (2022 Baseline)

When the simulations were run for the 2022 baseline flood event, which was taken as a reference condition and representing observed conditions, the simulations revealed intense inundation patterns.

Maximum flood depths exceeding 3 m were outstandingly observed in low-lying, flat floodplain regions, particularly in (Fig. 6.9) the Dehra Gopipur towns and in Hamirpur valleys. They are geomorphologically most vulnerable to water accumulation due to their broad alluvial plain topography, gentle slopes, and naturally restricted

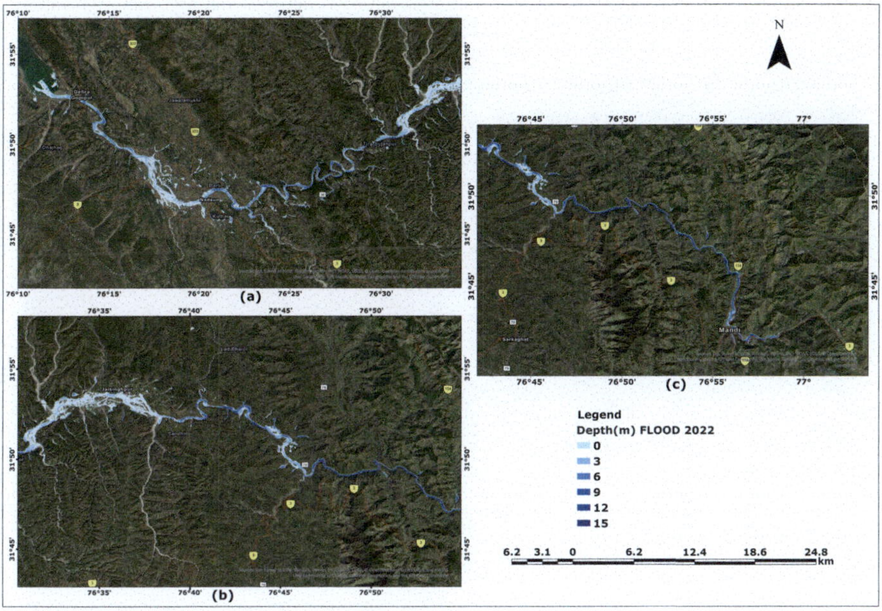

Fig 6.9 Flood depth. **a** Downstream (west): shallow flooding on flat plains. **b** Mid-stream (middle): moderate depths in mixed terrain. **c** Upstream (east): deep flows in narrow valleys

drainage capability. The worst-affected areas also extend to Una city fringes, the fertile agricultural belts of Nadaun, and parts of rural Kangra district. In these areas, more than 3 m of depth indicates widespread structural damage to single-story buildings, complete submersion of lifeline infrastructure, and rendering vast agricultural fields useless. These depths pose a serious threat to human lives, requiring massive evacuations and long-term reconstruction.

- **20% Flood Scenario Increase:** The simulation with a 20% increase from the baseline discharge produced a dramatic increase in both the areal extent and peak flooding depths, proving the basin to be highly sensitive to even marginal rises in flow. In the significant areas (Fig. 6.10) already identified as being at risk, such as parts of Dehra Gopipur and Hamirpur, inundation depth rose dramatically to between 6 and 8 m. This rise in deeper water inundated more widespread zones of settlement, intruding on communities previously considered marginally safe. Particular attention is drawn to areas around flood-risk roads and low-lying school playing fields, which experienced much greater flooding, seriously undermining emergency access, evacuation, and the maintainability of public services.

Depths within this category may result in megadamage to multi-story structures, render most vehicles worthless, and rescue attempts extremely difficult and hazardous due to pervasive, deep water.

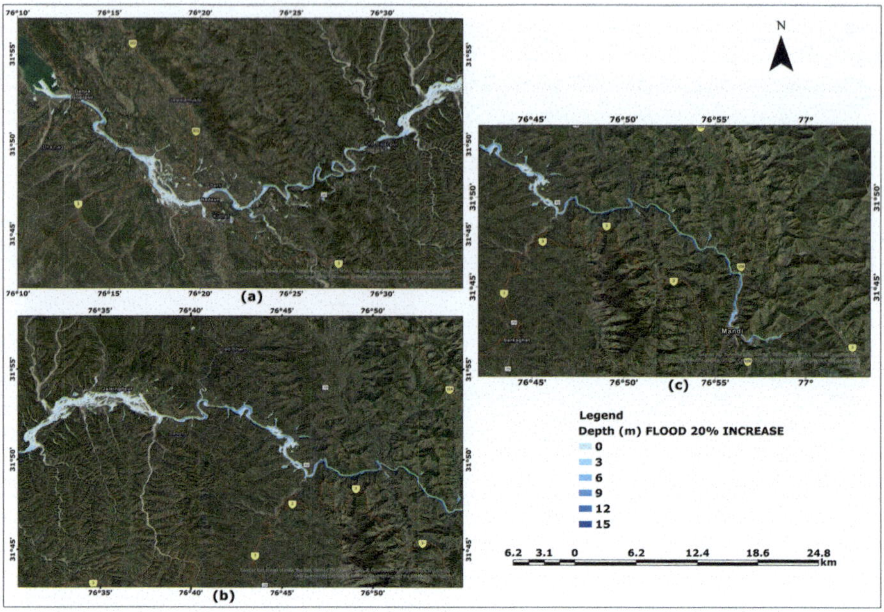

Fig. 6.10 Flood depth. **a** Downstream: wider and deeper spread. **b** Mid-stream: noticeable increase in depth. **c** Upstream: deeper channelized flows

- **50% Increased Flood Scenario**: The 50% augmented discharge scenario represents an exceptionally hydrometeorological event, to assess the basin response to a very severe flood. Severely damaging depths, over 15 m, were observed in the most constricted zones surrounding Dehra Gopipur and Hamirpur valley areas, indicating virtual complete immersion of local land and structures. In this worst-case situation, (Fig. 6.11) entire belts of crops were completely flooded, resulting in complete crop loss and severe long-term impacts on livelihoods. Widespread and record-breaking depths in large areas of the floodplain constitute a regional-scale disaster, demanding widespread emergency relief, potential mass displacement, and long-term recovery measures. Such excessive depths highlight the pressing need for effective flood mitigation structures and land-use management in order to enable community resilience to future extreme events (Figs. 6.12 and 6.13).

6.12.3 Velocity Analysis

6.12.3.1 Normal Flood Condition (2022 Baseline)

The maximum velocities of 6–10 m/s were predominantly experienced in the confined, steeper channels of the Beas River, particularly in the Mandi and Bhuntar

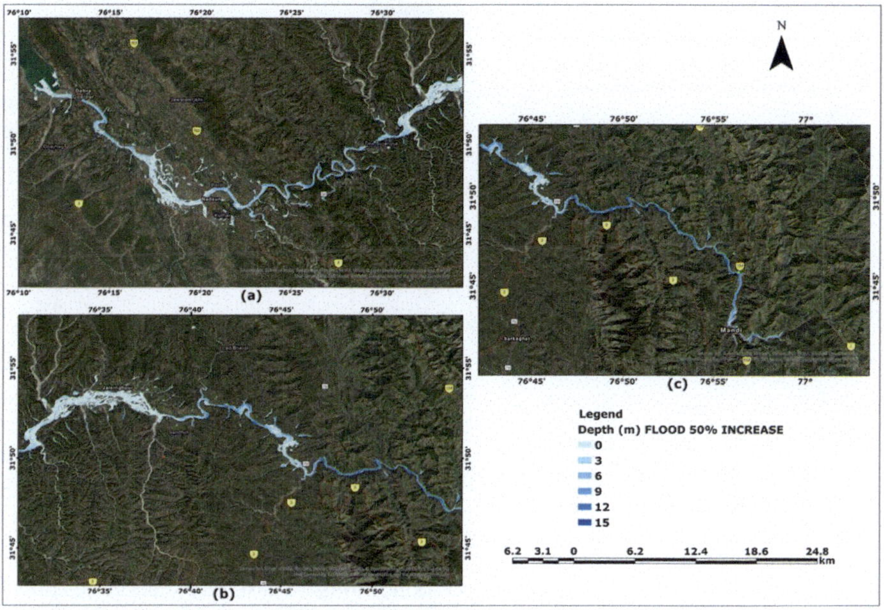

Fig. 6.11 Flood depth. **a** Downstream: widespread, deep flooding. **b** Mid-stream: high flood levels basin-wide. **c** Upstream: catastrophic depth in channels

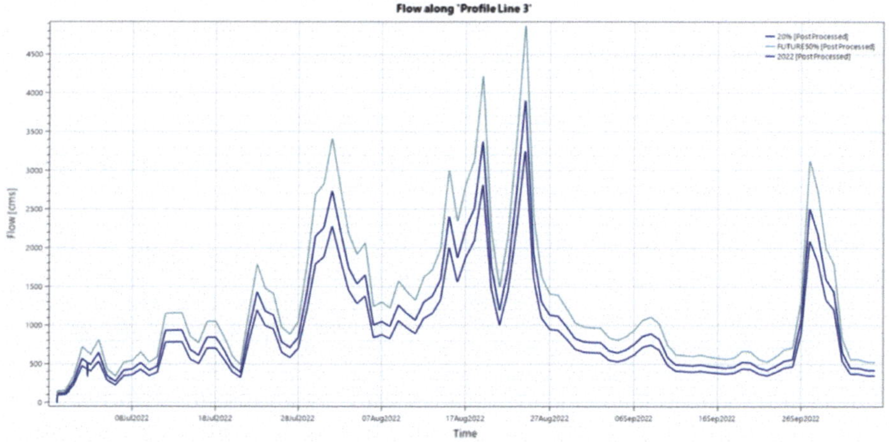

Fig. 6.12 Flow discharge graph showing upstream discharge comparatively for flood 2020, 20% increase, and 50% increase scenarios

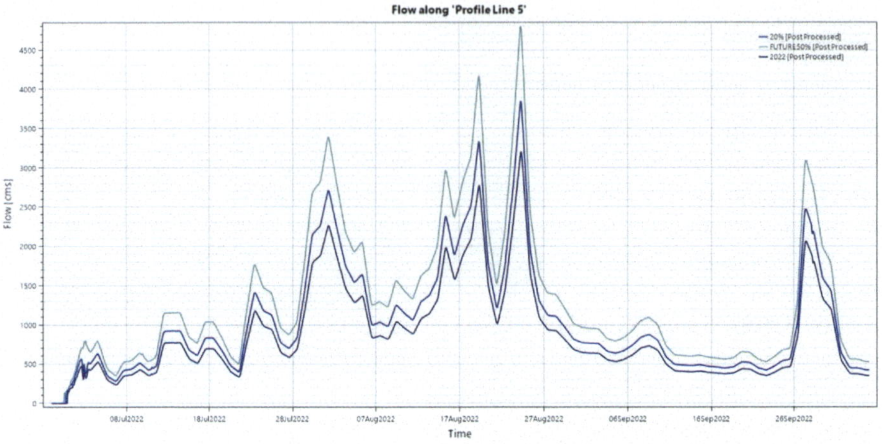

Fig. 6.13 Flow discharge graph showing downstream discharge comparatively for flood 2020, 20% increase, and 50% increase scenarios

valleys, under the 2022 baseline flood conditions. These are energetic flows (Fig. 6.14) that are a natural character of rivers flowing along incised, geomorphologically confined reaches. Such high velocities, especially on steeper river bends and constrictions, are a significant source of bank erosion, localized scouring around foundations and bridge piers, and possible undermining of riverbank structures. Even though these flows were largely limited to the expressed channel, their intensity signaled the erosional hazards at even traditional flood events.

- **20% Increased Flood Scenario**: At 20% increase (Fig. 6.15) above base discharge, the velocity map showed conspicuous expansion of high velocities. High-energy flows began to develop from the main channel to subsidiary channels and early floodplain areas with larger flood areas and a broader spread of hydraulic energy across the landscape.Most notably, medium to high speeds (4–6 m/s) now threatened existing road and bridge construction that had previously been less exposed to these forces. These speeds are powerful enough to impose significant lateral pressure on bridge abutments, destabilize road foundations, and haul large debris, leading to bridge collapses or making critical transport routes impassable and unsafe for emergency response or evacuation.
- **50% Increased Flood Scenario**: For the 50% increase (Fig. 6.16) in flood scenario, an extreme flooding scenario, velocities were higher than 12 m/s even for regions of the floodplain well beyond the center of the river channel. This highest ever recorded velocity in the overbank regions is an extremely hazardous condition. These high speeds are important from the standpoint of engineering design since they are larger than the design standards for most traditional flood defense structures and buildings. The consequences are widespread structural failure, extensive land and infrastructure erosion, and very high human life hazard. This

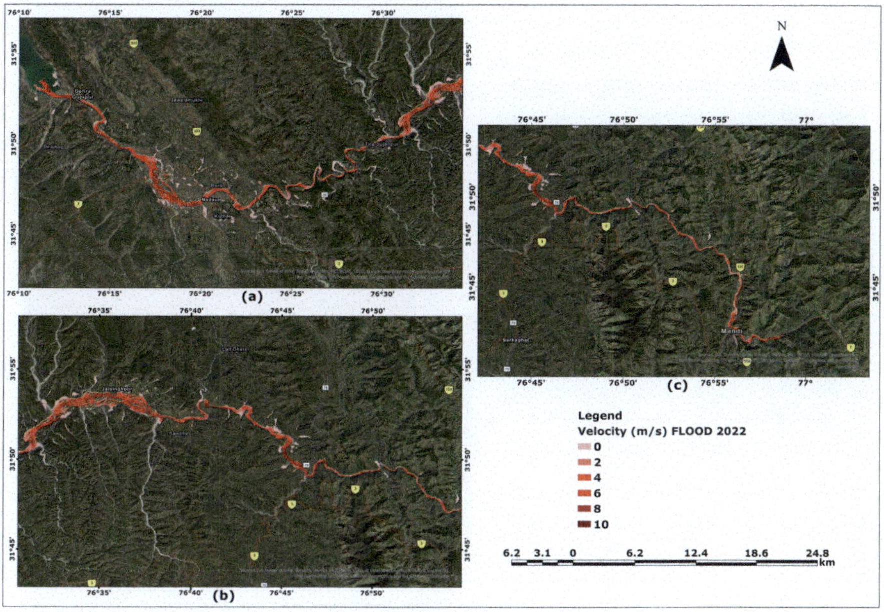

Fig. 6.14 Flood Velocity. **a** Downstream: low speeds in broad areas. **b** Mid-stream: mixed velocities across floodplain. **c** Upstream: high flows in confined reaches

situation requires highly resilient, specialized engineering solutions or rethinking land use in such high-velocity zones.

6.12.4 Water Surface Elevation (WSE) Analysis

Normal Flood Condition (2022 Baseline) When under normal flood conditions (Fig. 6.17), the numerically computed WSE slopes followed the natural topographic gradients of the Beas River basin quite closely and gradually decreased from higher to successively lower elevations from upstream regions close to Mandi toward downstream direction toward Una. The artificial WSE values were well congruent with historical flood heights and observed high-water marks throughout the region, providing a convenient gauge by which the hydrodynamic model can first be validated. This reference provides a robust reference point for determining representative flood characteristics and for testing the model's ability to replicate known hydrology.

- **20% Increased Flood Scenario**: The 20% flood scenario (Fig. 6.18) increase led to a really large mean WSE increase of around 0.8 m in low-lying settlements and floodplains compared to the baseline. This really big water level rise has a direct equivalent in a really big lateral increase in the extent of floodwaters. Built-up areas not so far impacted or only mildly impacted by flood, particularly the ones near the

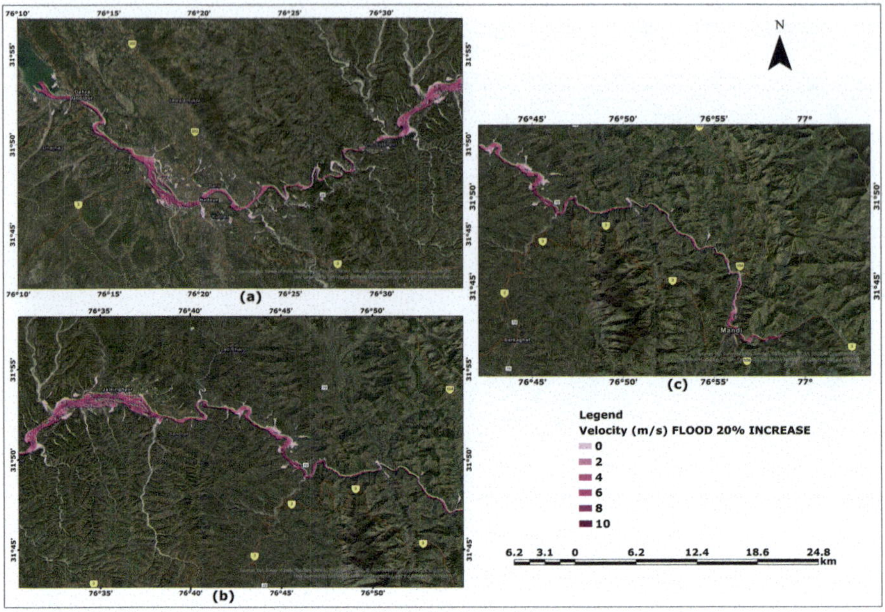

Fig. 6.15 Flow velocity (20% increase). **a** Downstream (west): low velocities over broad floodplains. **b.** Mid-stream (middle): variable flows across mixed topography. **c** Upstream (east): High velocities in steep, narrow channels

floodplain edge, were prominently submerged. This rise translates to ground floors of buildings in such large areas being inundated, critical infrastructure including roads and utilities being located, and general flood impact being felt by a much larger population base and assets.

- 50% Increased Flood Scenario**:** For the 50% increase (Fig. 6.19) in floods scenario, there was a substantial and severe rise in WSE of 1.5–2.0 m happening uniformly along vast stretches of the basin. The large increase raises the effective floodplain boundary far beyond past levels. Such an increase indicates a high likelihood of overtopping existing levees, embankments, and even impacting the operational integrity of reservoirs in case of being surpassed by their design freeboard. Such a scenario portrays the ruinous potential of rare events to bring about record depths and areal coverage of inundations and necessitates re-examination of local flood protection planning and emergency response protocols.

6.13 Arrival Time Analysis

Flood arrival time maps are a representation of the time between the onset of a flood process (e.g., onset of excessive rainfall or upstream discharge rise) and the occurrence of floods at specific points along the floodplain. This parameter is critical

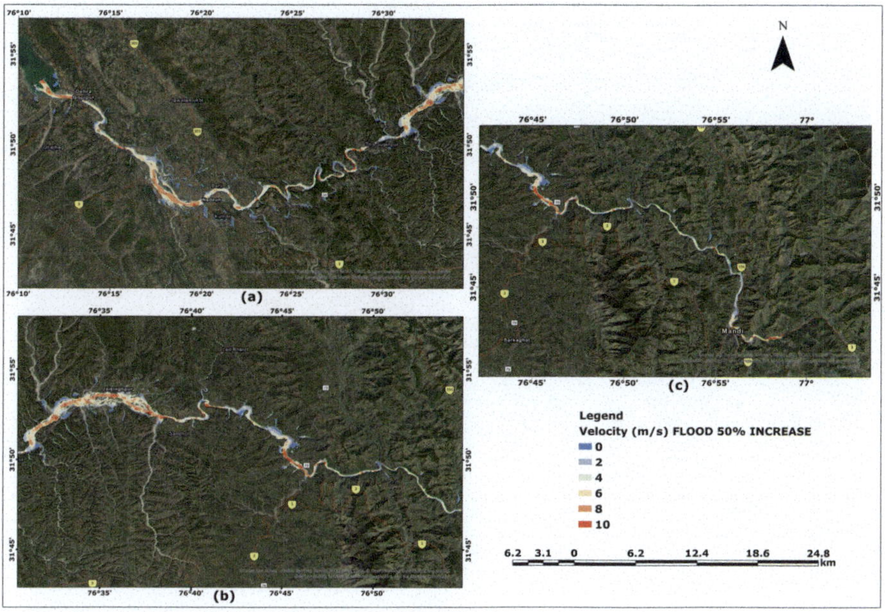

Fig. 6.16 Velocity. **a** Downstream: fast flows over wide areas. **b** Mid-stream: damaging velocities in core areas. **c** Upstream: peak speeds causing erosion

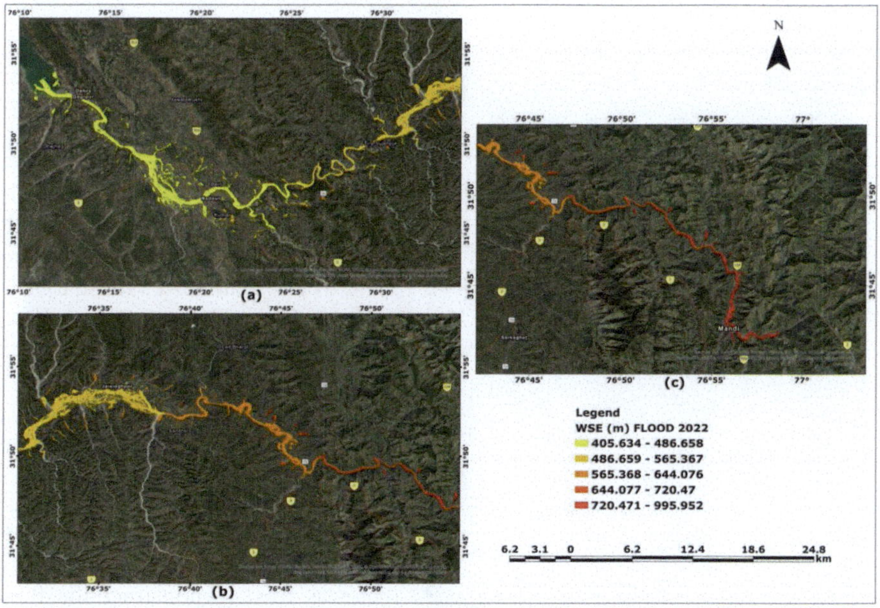

Fig. 6.17 Water surface elevation (WSE). **a** Downstream: low, terrain-following levels. **b** Mid-stream: balanced elevations across slopes. **c** Upstream: elevated WSE in steep channels

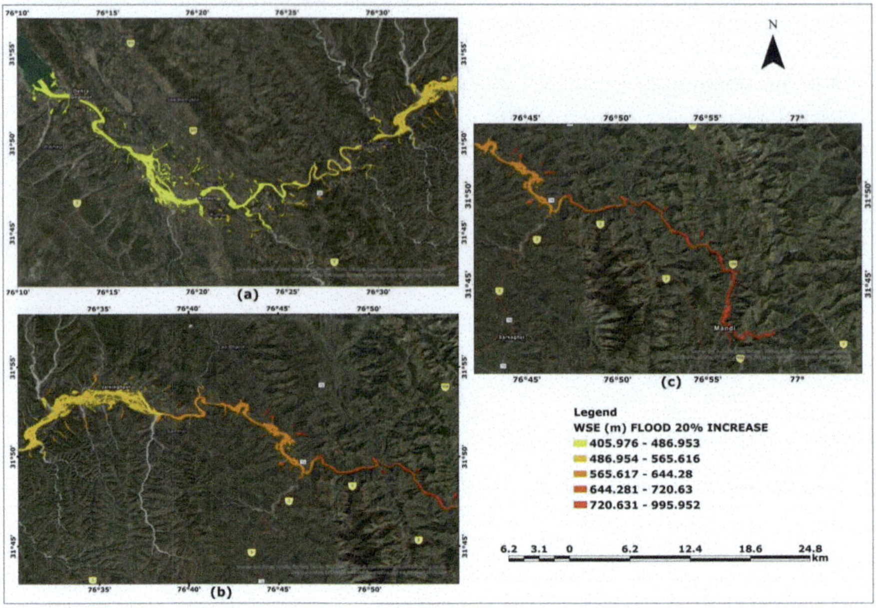

Fig. 6.18 Water Surface Elevation (20% Increase). **a** Downstream (West): elevated WSE over flat floodplains. **b** Mid-stream (Middle): uneven WSE from terrain variation. **c** Upstream (east): rising WSE along steep valleys

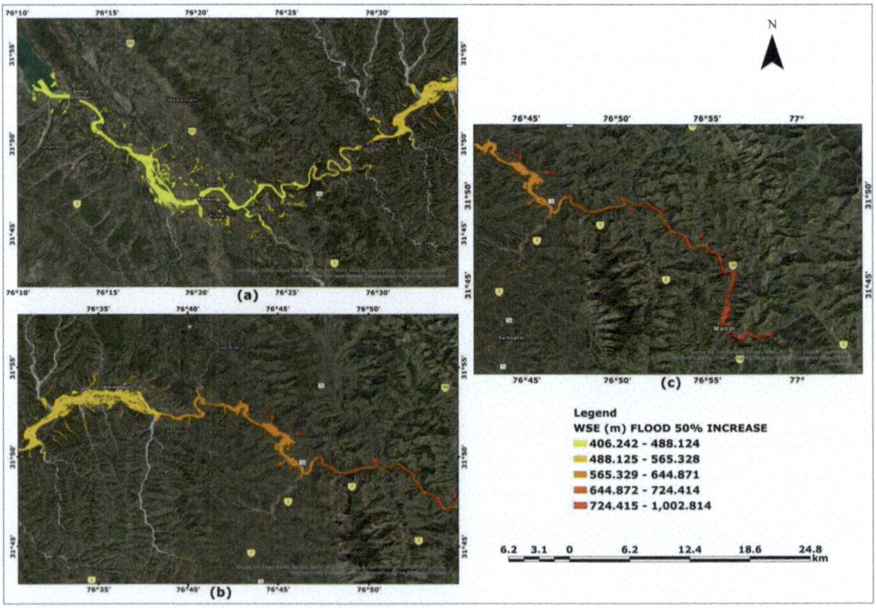

Fig. 6.19 WSE. **a** Downstream: water overtopping defenses. **b** Mid-stream: severe rise in elevations. **c** Upstream: critical WSE in upper basin

to effective early warning systems, enabling emergency managers and the public to provide an estimate of the time available for evacuation, deployment of resources, and protective actions.

6.13.1 Flood Arrival Under 20% Increased Flow

Regions closest (Fig. 6.20) to the main river channel—namely in the area of upstream localities such as Pandoh, Gutkar, and Jhatingri—experienced floodwater arrival within a short time frame of 0 to 6 hours.This rapidity indicates that these areas are highly sensitive and must be protected at once during the onset of a flood event. As the flood wave propagated outward and downstream, areas with lower topography or at a greater distance from the river, which have slower water dispersal and aggradation times, had considerably later times of arrival, ranging from 12 to 24 hours. This pattern reflects a relatively sluggish growth of the flood under conditions of moderate increased flow, providing a brief but beneficial window of response time for downstream communities.

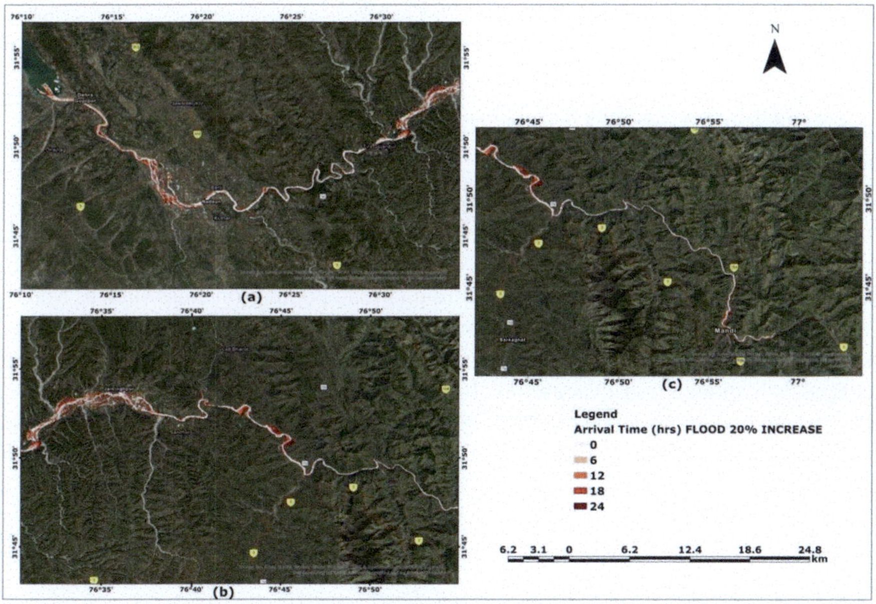

Fig. 6.20 Arrival time—20% increased flood. **a** Downstream: faster flow, reduced warning time. **b** Mid-stream: quicker wave propagation. **c** upstream: rapid inundation within hours

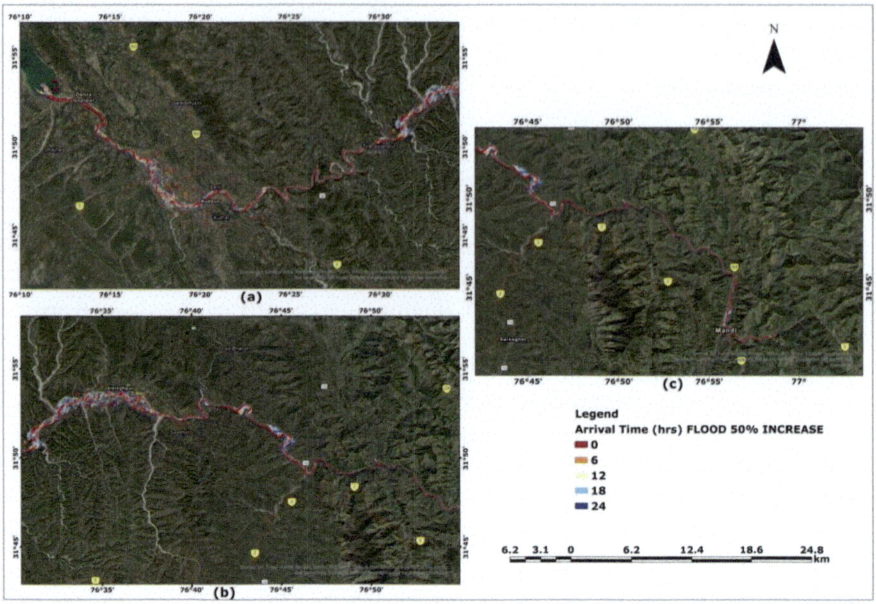

Fig. 6.21 Arrival time—50% Increased flood. **a** Downstream: minimal warning across basin. **b** Mid-stream: swift and widespread inundation. **c** Upstream: instant flooding in narrow channels

6.13.2 Flood Arrival Under 50% Increased Flow

Conversely, in sharp contrast to the above, the scenario of a 50% increase (Fig. 6.21) showed a sharp acceleration in flood arrival times across the basin. Under this extreme flow scenario, many more geographic regions were inundated within the first 6 h, also critically involving important downstream urban population areas such as Mandi town and Bhuntar. The strengthened flood wave traveled faster and covered a wider distance within a much reduced duration, drastically reducing the time for successful emergency response and evacuation. In such a catastrophic situation, barely any location remained untouched over 12–18 h back, exemplifying the strengthened and widespread nature of catastrophic floods, posing behemothic challenges for disaster management.

6.13.3 Comparative Analysis and Risk Implications

This comparative analysis categorically indicates that increased flood size significantly shortens the flood arrival time throughout the basin. This severe reduction of the response interval makes preemptive evacuation strategies and early warning systems much more indispensable and difficult to deploy effectively. Areas repeatedly in the 0–6-h inundation zone under both the 20 and 50% increase scenarios are those at greatest risk and must receive priority and ongoing attention. These areas must be prioritized for structural mitigation measures (e.g., levees, floodwalls), strategic relocation of hazard-exposed persons, and establishment of highly responsive, real-time flood monitoring and early warning systems to optimize community safety and resilience against growing flood threat.

6.13.4 Flood Hazard Analysis (Depth × Velocity Product)

The flood hazard map is a cumulative index resulting from the multiplication of the simulated flood depth (m) and the corresponding flow velocity (m/s), producing a product typically expressed in m^2/s. This one metric captures both the severity of flooding and floodwater energy, providing a more holistic and robust measure of the risk of damage to structure and to life through flooding than depth or velocity alone. Higher $D \times V$ values correspond to greater potential for structural damage, erosion, and instability for those subject to the floodwaters.

6.13.5 Normal Flood Condition (2022 Baseline)

Under normal flood conditions (Fig. 6.22), the flood hazard map indicated distinct areas of moderate hazard located primarily in the main channel of the Beas River, where moderate depth and enhanced velocities came together to generate the $D \times V$ product. Uniquely, some high-hazard areas were noted to be encircling large road intersections and high-density market areas in Dehra Gopipur. These areas, while maybe not experiencing the absolute maximum depths, showed a convergence of significant flood dynamics (moderate speed) and a relatively high concentration of exposed assets (infrastructure, business, population), implying a relatively greater risk even in typical floods.

- **20% Increased Flood Scenario**: The 20% increase (Fig. 6.23) in flood scenario resulted in the expansion of such hazard locations significantly, extending beyond the principal channel along river branches and into what were once comparatively less exposed parts of the floodplain. This expansion implied that a greater geographical area was now subject to not only inundation, but also to potentially erosive force of flows. Mid-density population areas, which under baseline conditions might have been barely safe, are now facing major flood hazard for the first time under this scenario, and an urgent review of regional flood preparedness and mitigation strategies was required. The $D \times V$ map accurately identified these new at-risk areas.
- **50% Increased Flood Scenario:** In the 50% (Fig. 6.24) augmented flood scenario, the scale of catastrophic levels of hazard extends across nearly the

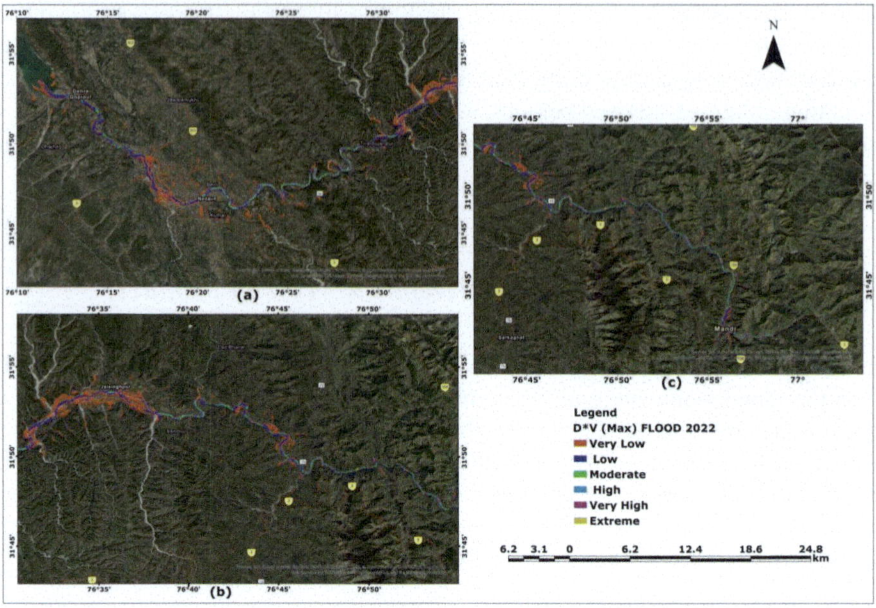

Fig. 6.22 Flood hazard (depth × velocity). **a** Downstream: localized moderate hazard. **b** Mid-stream: Moderate zones near riverbanks. **c.** Upstream: High hazard along narrow channels

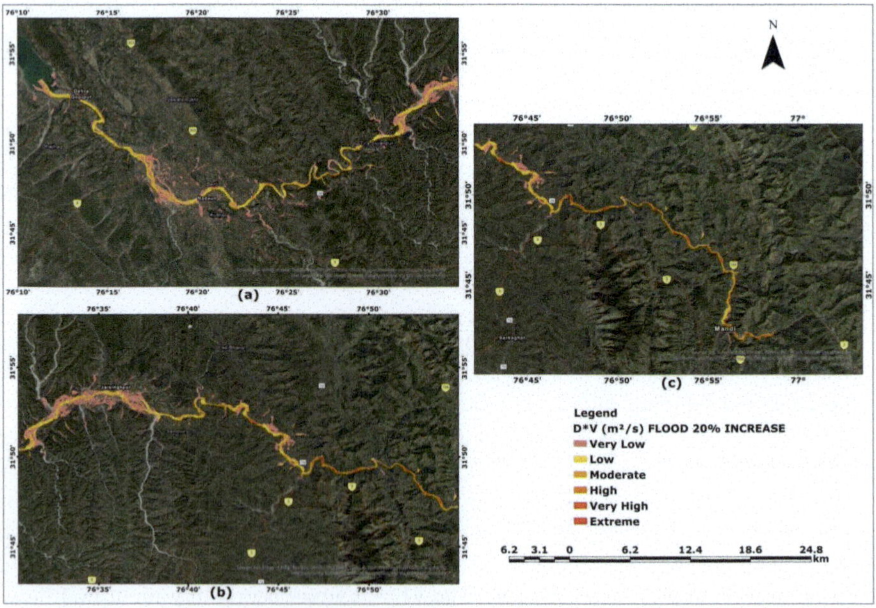

Fig. 6.23 Flood hazard (20% Increase). **a** Downstream (west): Moderate hazard in wide lowlands. **b** Mid-stream (Middle): mixed hazard in sloped terrain. **c** Upstream (East): high hazard in steep, fast-flowing zones

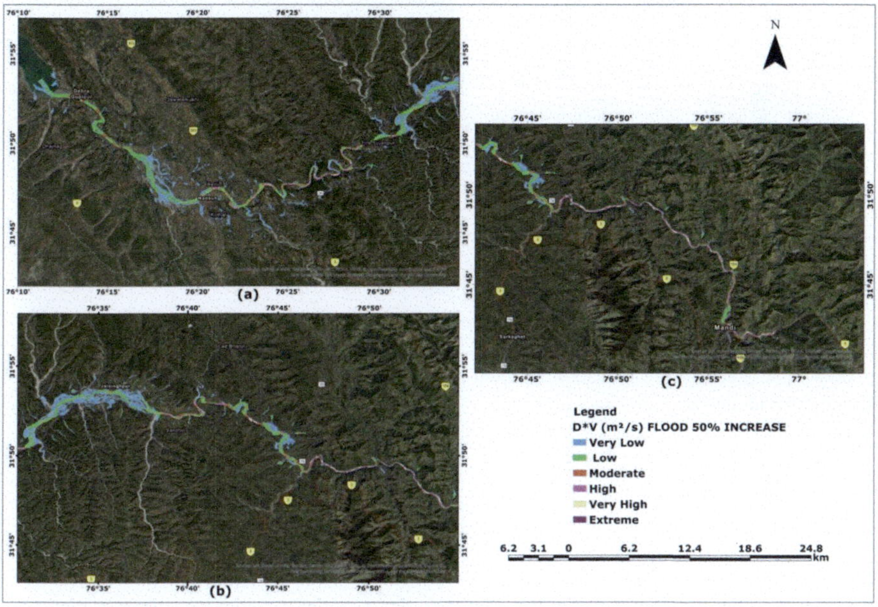

Fig. 6.24 Hazard. **a** Downstream: widespread extreme hazard. **b** Mid-stream: extensive high-risk zones. **c** Upstream: critical hazard levels throughout

entire floodplain with very large sections exhibiting severely high values of $D \times V$. Specifically, populous locations such as Hamirpur, Gagret, and Nadaun witnessed a serious threat to human life as well as property, indicating a significant likelihood of severe structural damage, widespread erosion, and unsafe conditions for human survival. The widespread occurrence of such high-hazard values under this case reflects the need for firm, extensive flood resilience planning, including potential land-use regulation, relocation programs, and the realization of highly resilient infrastructure that can withstand severe hydrodynamic forces (Fig. 6.25).

6.14 Affected Land Use/Land Cover (LULC) Regions Analysis

The zonal statistics performed on the LULC data combined with flood inundation boundaries quantified the impact of different flood scenarios on various land cover types

- **Affected LULC—Normal Flood (2022)**: Fig. 6.26 visually represents the LULC categories affected by the 2022 normal flood event under pre-monsoon conditions. Quantitatively, it shows that built-up areas experienced an affected area of 20.295 units (likely km^2 or hectares), while forest areas had the largest affected extent

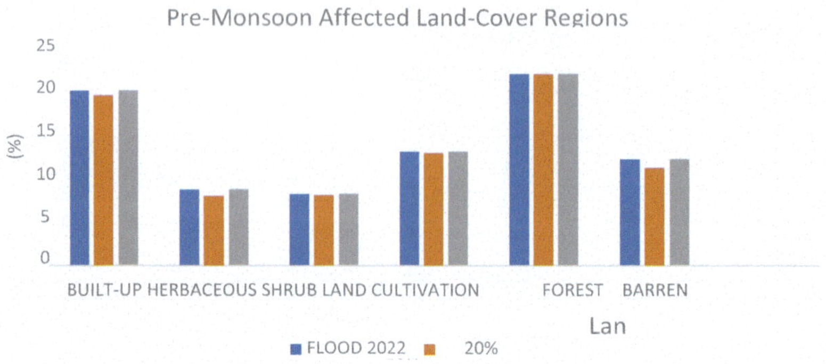

Fig. 6.25 Bar graph of pre-monsoon affected regions of flood 2022, including 20% increase and 50% increase scenarios

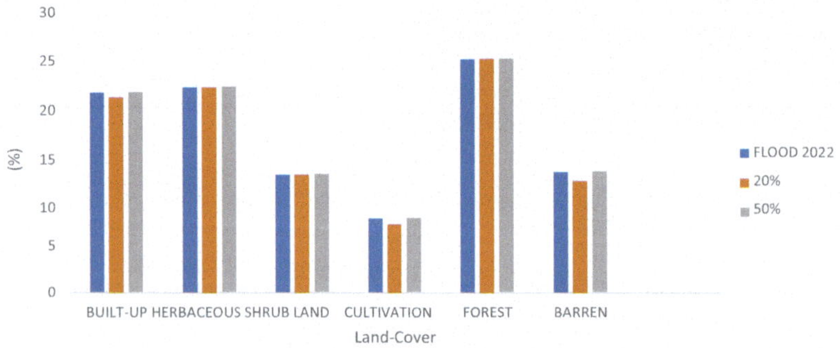

Fig. 6.26 Post-monsoon affected regions of flood 2022, including 20% increase and 50% increase scenarios

at 22.194 units. This highlights the vulnerability of both urban infrastructure and natural ecosystems even during typical flood events (Fig. 6.27).

- **Affected LULC—20% Increased Flood**: For the 20% increased flood scenario, Fig. 6.28 visually depicts the expanded affected LULC regions under pre-monsoon conditions. Table 8 indicates an affected built-up area of 19.762 units and forest by 22.194 units. The overall trend, however, shows a broader spread of inundation into various LULC types, suggesting increased pressure on agricultural lands (13.119 units) and natural vegetation.
- **Affected LULC—50% Increased Flood:** The 50% increased flood scenario, both in pre-monsoon and post-monsoon conditions, demonstrates widespread impacts across all LULC categories. Built-up areas are affected by 20.295 units and forest by 22.194 units (Percentage of Post-Monsoon Affected Area Land-Cover:). The visual representation in for post-monsoon conditions highlights that even after the peak, substantial areas of cultivation, herbaceous, and shrubland remain within the inundation boundary, indicating prolonged waterlogging and potential for significant agricultural losses (Ghorbani et al., 2022).

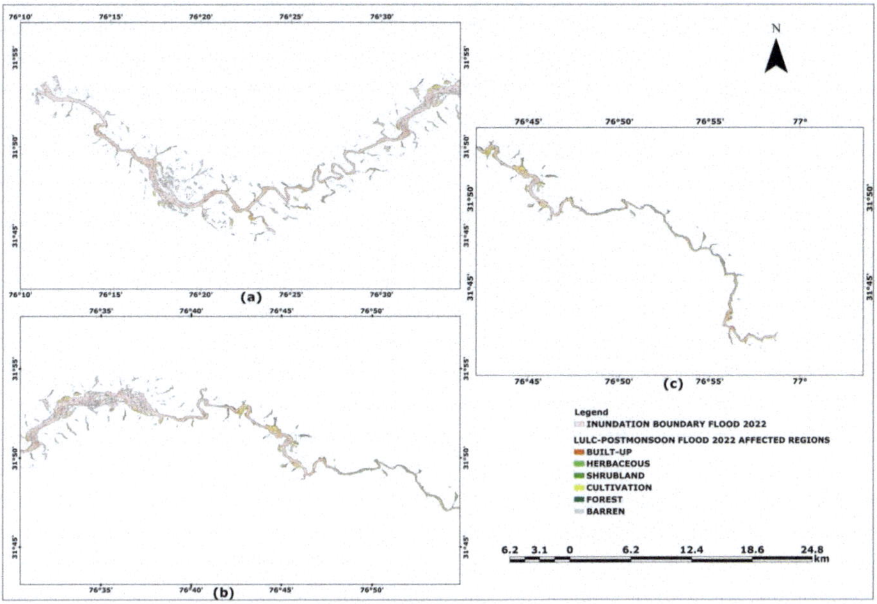

Fig. 6.27 Post-monsoon LULC affected by flood 2022

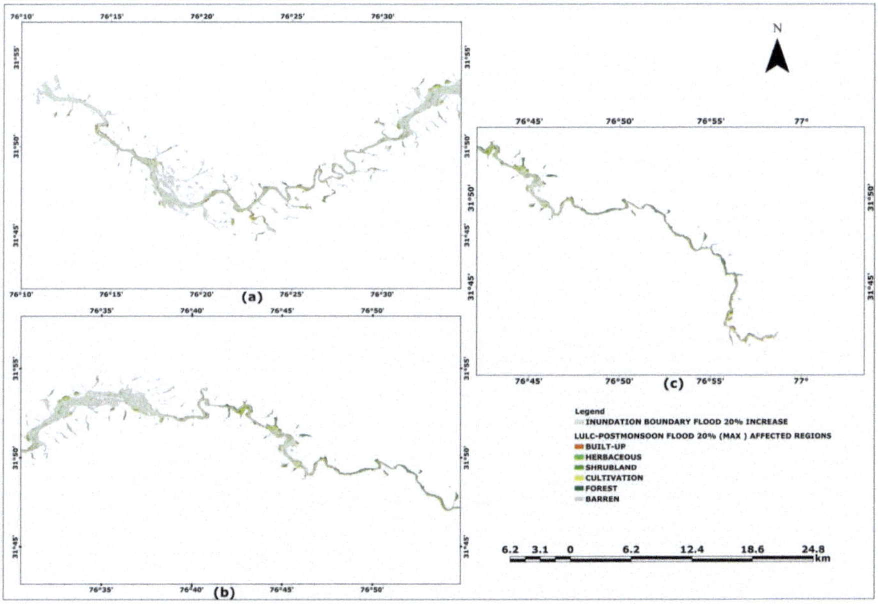

Fig. 6.28 Post-monsoon LULC affected by 20% increased flood (2022)

The **LULC affected area analysis** further refines our understanding of flood impacts. While the affected areas for built-up and forest categories remained relatively consistent across scenarios in the provided pre-monsoon data (Percentage of Pre-Monsoon Affected Area Land-Cover:), it is crucial to interpret these numbers in context. Minor fluctuations might reflect the precision of spatial intersection rather than a decrease in overall risk. The visual evidence (e.g., Fig. 6.29: for post-monsoon 50% flood) consistently shows significant inundation across various land cover types, particularly agricultural land and natural vegetation, which can lead to extensive economic losses and ecological disruption. The seasonal LULC comparison highlights the dynamic nature of the flood-affected landscape; post-monsoon conditions, with saturated soils and mature vegetation, can influence both the flow resistance and the duration of inundation, impacting recovery efforts (Ghorbani et al., 2022).

The integration of high-resolution DEM, seasonal LULC, and detailed hydrological inputs within the HEC-RAS 2D framework has proven effective in capturing the complex hydraulic behavior of the Beas River. The outputs provide a robust foundation for developing detailed flood hazard maps and serve as a crucial decision support tool. Validation against 2022 flood extents enhances the reliability of these simulations. However, the accuracy of the predictions is inherently dependent on the quality and resolution of input data, particularly future climate projections for more precise long-term flood forecasting. Further research could integrate socioeconomic vulnerability data with these hazard maps to perform a comprehensive flood risk assessment.

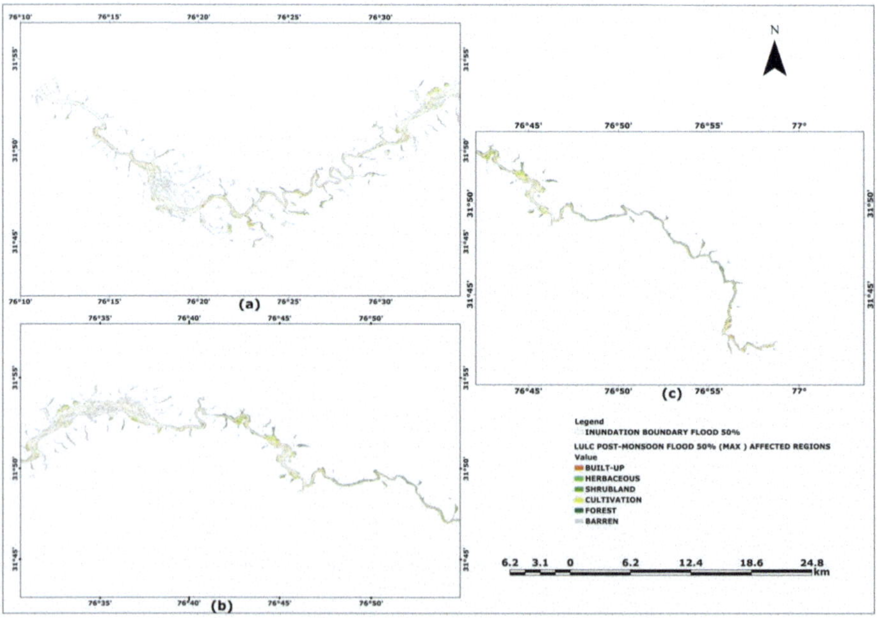

Fig. 6.29 Post-monsoon LULC affected by 50% increased flood (2022)

6.14.1 Discussion

6.14.1.1 The following discussion Incorporates the 2025 flood scenario analysis

This map displays the spatial extent of the flooded area for the flood 2025 scenario, represented by a purple boundary. The inundation boundary defines the maximum lateral spread of floodwaters. Analyzing this boundary helps in identifying areas at risk of flooding and provides a visual representation of the flood footprint. Comparing this boundary with existing infrastructure and populated areas can highlight potential impacts. The velocity map shows the flood water velocity (in meters per second) for the flood 2025 scenario. The legend categorizes velocities into different ranges, from 0 m/s (stagnant or very slow) up to 10 m/s. Higher velocities (indicated by darker shades of purple/magenta in the legend, reaching 8–10 m/s) are typically observed in the main river channel or constricted areas where water flow is concentrated. Lower velocities (lighter shades) are found in overbank areas or floodplains, where the water spreads out and slows down. Flood velocity is a critical parameter for assessing flood hazard, as higher velocities can lead to increased erosive power, structural damage to buildings, and pose greater risks to human life (Fig. 6.30).

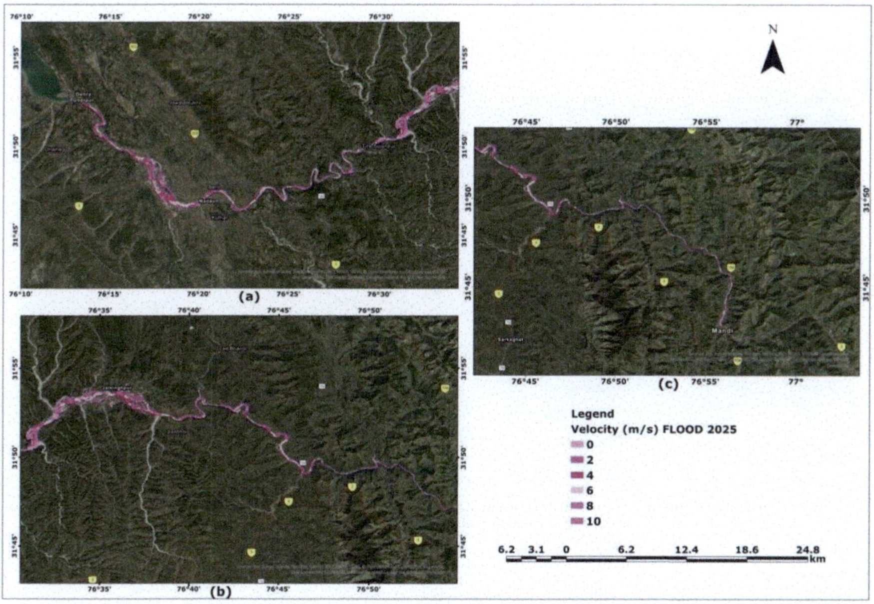

Fig. 6.30 Flood velocity: flood 2025 peak scenario. **a** Moderate velocities in the downstream with channelized flow. **b** Mixed velocities across mid-stream terrain. **c** High-velocity zones in steep upstream channels

6.14.2 WSE (m) Flood 2025

The water surface elevation (WSE) map for flood 2025 indicates the elevation of the water surface above a certain datum. The legend provides ranges of WSE in meters, from 414.936 m to 734.262 m. The WSE values generally increase upstream (toward the top left of map "a" and "b") and along the river course, reflecting the natural topography and the flow of water (Chen et al., 2015). The variations in WSE across the floodplain can reveal how high the water level will reach in different locations, which is essential for flood preparedness and designing flood protection measures (Figs. 6.31 and 6.32).

The $D * V$ (Depth multiplied by Velocity) map (Fig. 6.32a-c) provides a combined measure of flood intensity, often used as an indicator of flood hazard. The legend classifies $D * V$ values into categories ranging from "very low" to "extreme." Higher $D * V$ values (represented by warmer colors like red/orange) suggest areas with both significant depth and velocity, indicating a very high hazard due to strong currents and deep water. Lower $D * V$ values (cooler colors like green/yellow) indicate less hazardous conditions. This parameter is particularly useful for assessing the potential for structural damage, erosion, and danger to life.

The flood depth map for the flood 2025 scenario is measured in meters. The legend shows depth categories from 0 up to 15 m. Areas with greater depths (darker

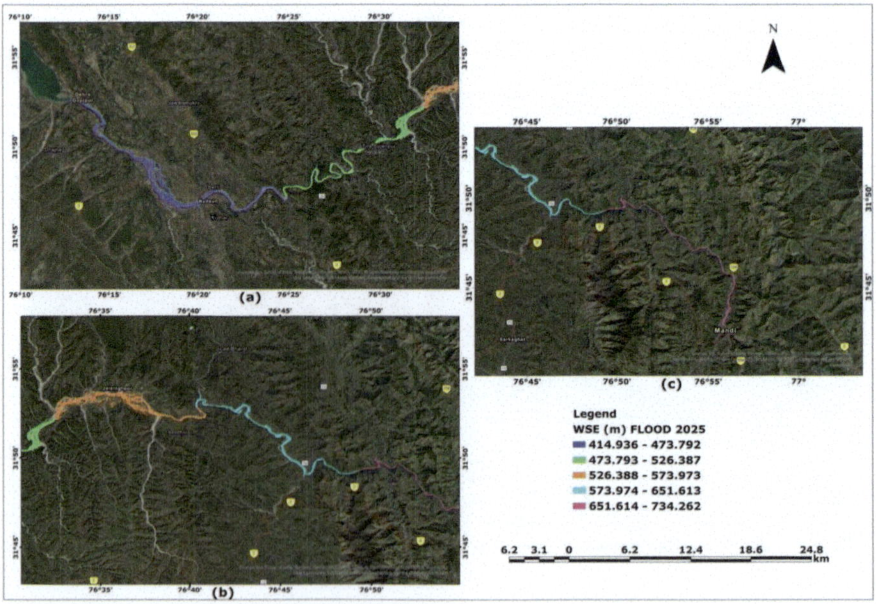

Fig. 6.31 Water surface elevation flood 2025 scenario. **a** Elevated WSE across wide floodplains in the downstream. **b** Significant rise in mid-stream WSE. **c** High WSE in confined upstream valleys

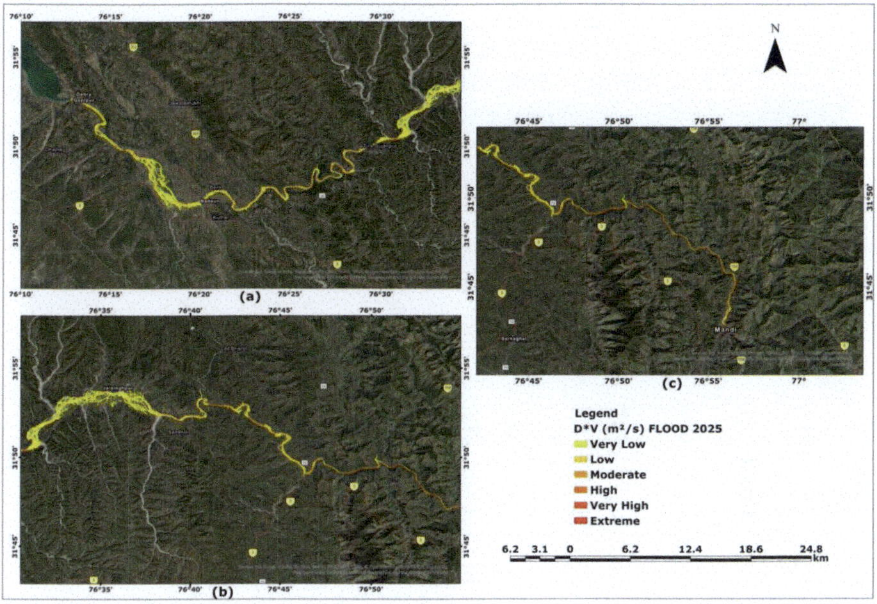

Fig. 6.32 Flood hazard ($D \times V$)—Flood 2025 scenario. **a** Varying hazard levels across the lower basin. **b** Moderate to high-hazard zones in the middle basin. **c** Intense localized hazards in the upper reaches

blue shades) are typically found within the main channel and adjacent low-lying floodplains. Shallower depths (lighter blue shades) are observed at the fringes of the inundated area. Flood depth is a primary indicator of flood impact on infrastructure, agriculture, and human safety. Understanding depth distribution is crucial for evacuation planning, damage assessment, and designing flood-resilient structures (Figs. 6.33 and 6.34).

The percentage distribution of various land cover types within the pre-monsoon flood-affected areas for the year 2025. Among all categories, forest areas account for the highest proportion at 25.54%, indicating that dense vegetative regions were significantly impacted during the flood event. Barren lands, which typically have minimal vegetation cover, make up 24.57% of the affected area. These regions are more susceptible to surface runoff and erosion, contributing to their high vulnerability during flooding. Cultivated lands represent 21.98% of the flood-impacted zone, reflecting the potential impact on agricultural productivity and local livelihoods. Built-up areas contribute 18.55%, suggesting that urban settlements and infrastructure were also notably affected. Shrublands, forming 10.96%, are moderately impacted and often act as transitional zones between more densely vegetated and cleared areas. Herbaceous cover, the least affected at 5.01%, may indicate sparsely vegetated lands or grasslands that were either at higher elevations or less exposed.

Overall, the data show that both natural landscapes (forests, barren, shrubland) and human-modified land uses (agriculture, built-up) were significantly affected.

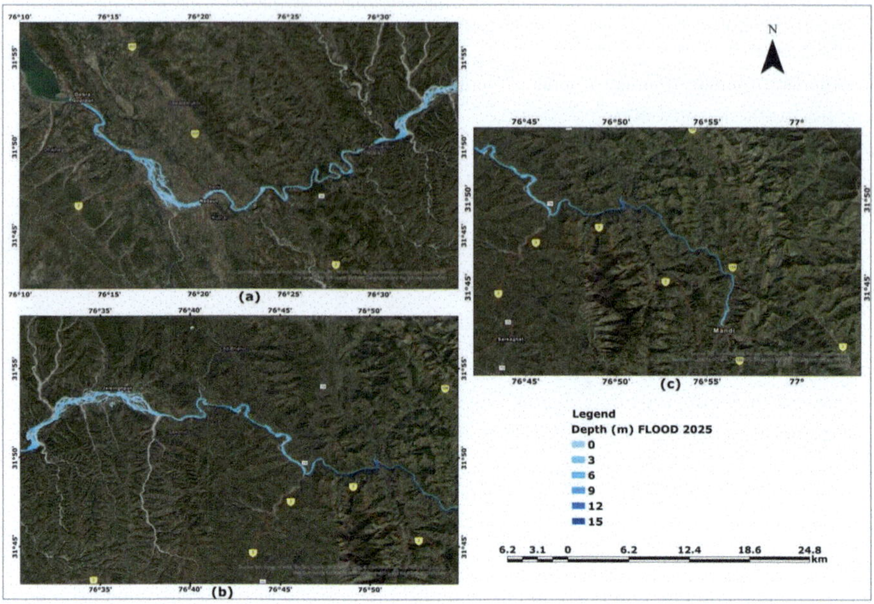

Fig. 6.33 Flood depth: flood 2025 peak scenario. **a** Widespread shallow flooding in the downstream. **b** varied depths in the mid-stream region. **c** Deep, channelized flows in the upstream segment

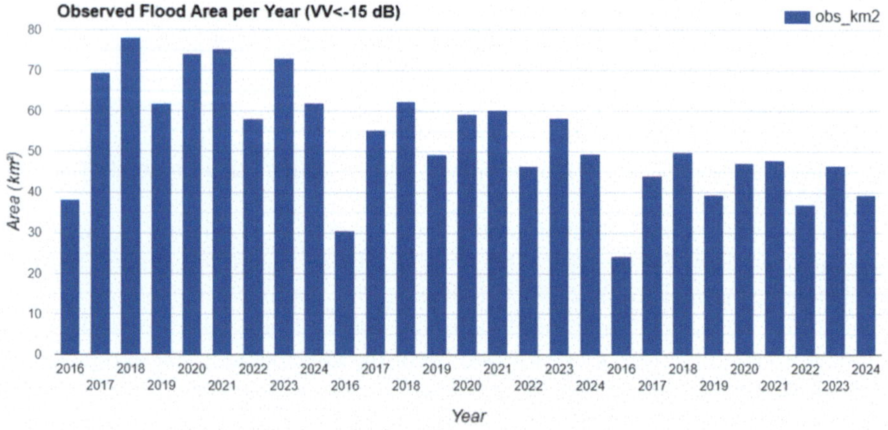

Fig. 6.34 Observed flood area analysis

The high percentage of forests and barren areas suggests that topographical and land management factors play a crucial role in flood vulnerability, while the substantial share of cultivated and built-up lands underscores the socioeconomic risks associated with such flood events.

6.15 Flood Model Performance and Validation Analysis

The graphs provided offer a detailed look at the performance and validation of a flood model over time. The comparison is directed toward understanding modeled versus observed flood areas, assessing the accuracy and stability of the model across the different years (Bhuyian & Kalyanapu, 2020; Demir & Kisi, 2016).

1. Observed Flood Area Analysis

The bar graph "Observed Flood Area per Year (VV < − 15 dB)" shows the observed flood area in square kilometers (km^2) as delineated from remote sensing imagery, presumably from SAR data at backscatter level − 15 dB. This graph shows significant year-to-year variability in modeled flood extent, with peak flood areas in 2018 and 2020. The variability corresponds to the dynamic nature of flooding based on rainfall, snowmelt, and antecedent soil moisture status.

2. Modeled Flood Area Analysis

The line chart "Modeled Flood Area per Year" shows the total area of flooding as modeled.

Each line is one distinct model setup, possibly for varying input parameters like Manning's roughness factor or some other model threshold. Worth pointing out here is the way that the area of flooding as modeled is constant from year to year for any particular configuration. This means that the model is applying a static input, such as a single flood event or a given set of boundary conditions, to generate its output, rather than simulating the year-to-year variability in the observed data. This is a significant difference from the observed data and suggests that the model is a hazard-based model rather than a predictive model for real-time events.

3. Overlap Analysis and Model Accuracy

The "Overlap Area per Year (Observed ∩ Modeled)" plot quantifies the spatial accuracy with which the modeled flood area overlaps the observed flood area. The plot shows an evolving pattern of increasing overlap from 2016 through a peak in 2021 and then a decrease.

This indicates that the model has higher accuracy in detecting the observed flood extent in some years than in others. Even though the modeled flood extent is static, its intersection with the observed data is variable because the flood extent observed itself is very variable. This is the critical measure of the model's spatial accuracy over time that reflects its inability to replicate the whole range of observed flood extents (Fig. 6.35).

4. Critical Success Index (CSI) and Model Performance

The "Flood CSI (%)" line chart is a direct measure of model performance as a whole. The CSI is a mean validation measure that combines the concepts of precision and recall to establish how closely the model flood area approximates the observed flood area. A high CSI (almost 100%) outcome is a good model–observation agreement.

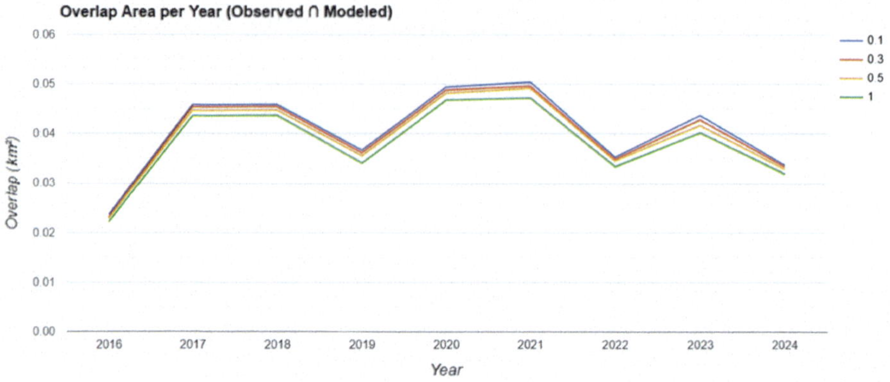

Fig. 6.35 Overlap analysis and model accuracy

The plot of a uniformly high CSI from 2022 to 2025 shows that the model is highly accurate for the validation years selected. The tiny drop with increasing time could be because of land cover change or some other environmental factor not fully accounted for in the model. The bar plots titled "2022 Data" and "2025 Data" likely provide a graphical breakdown of the CSI computation. The green bar shows the correctly modeled region (overlap), and orange bar shows false positives, false negatives, or both. In the CSI, the plot would most likely be showing the percentage of the modeled region that is correctly overlapping with the observed region (Fig. 6.36).

The high green proportion in both the years validates the high CSI scores and shows the good performance of the model during those specific periods.

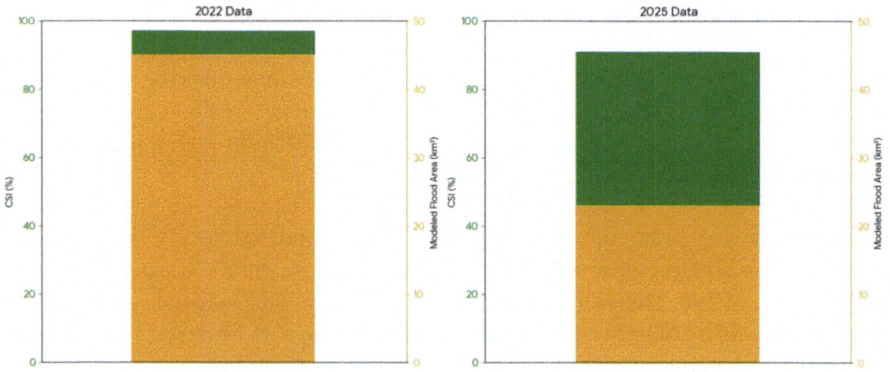

Fig. 6.36 Critical success index (CSI) for 2022 and 2025 flood

6.16 Conclusion

This study successfully leveraged HEC-RAS 2D hydrodynamic modeling (Zhang et al. 2020), to perform a comprehensive flood hazard assessment for the Beas River basin under current and projected flood scenarios. By meticulously integrating high-resolution spatial data, seasonal land use/land cover classifications, and detailed hydrological inputs (including flow discharge and precipitation data), the study has generated robust simulations of flood depth, velocity, and water surface elevation.

The findings clearly indicate a significant and escalating flood hazard across the basin. Even under the 2022 normal flood conditions, substantial areas, including built-up zones and critical infrastructure, are vulnerable. The simulated 20 and 50% increased flood scenarios, alongside the projected flood 2025 event, reveal an alarming expansion of inundation extents, increased flood depths reaching up to 15 m in extreme cases, and velocities reaching up to 10 m/s in various parts of the floodplain. Water surface elevations (WSEs) also show significant changes across these scenarios, reaching over 734 m in the flood 2025 scenario. Furthermore, the $D * V$ (depth–velocity product) maps for these increased and projected scenarios highlight regions of very high to extreme flood hazard. These extreme conditions pose catastrophic threats to both human life and property, necessitating urgent and proactive flood risk management strategies. The analysis of affected land use/land cover zones further quantifies these impacts, highlighting areas where agricultural productivity and urban development are severely compromised across all simulated flood conditions.

Abbreviations

ALOS PALSAR	Advanced Land Observing Satellite Phased Array L-band Synthetic Aperture Radar
DEM	Digital Elevation Model
$D * V$	Depth-Velocity product
FHI	Flood Hazard Index
GEE	Google Earth Engine
GEOGLOWS	Group on Earth Observations Global Water Sustainability
HEC-RAS	Hydrologic Engineering Center's River Analysis System
IMERG	Integrated Multi-satellitE Retrievals for GPM (Global Precipitation Measurement)
LULC	Land Use/Land Cover
NASA	National Aeronautics and Space Administration
WSE	Water Surface Elevation

References

Aboelnour M, Engel B (2018) Integrating land use change and hydrological modeling for flood risk assessment. Environ Model Softw 100:1–15. https://doi.org/10.1016/j.envsoft.2017.11.017

Belgiu M, Drăguţ L (2016) Random Forest in remote sensing: A review of applications and recent advances. ISPRS J Photogram Remote Sens 114:24–31. https://doi.org/10.1016/j.isprsjprs.2016.01.011

Bera, A., Singh, R., & Ghosh, S. (2021). A hydrological study on the Teesta River Basin using ALOS PALSAR DEM and remote sensing data. Master's thesis. Indian Institute of Technology

Bhuyian MNM, Kalyanapu AJ (2020) Application of HEC-RAS 2D for floodplain modeling and mapping in data-scarce regions. Water 12(5):1341. https://doi.org/10.3390/w12051341

Bose, P., Sharma, R., & Raj, K. (2022). Flood hazard assessment in the Beas River Basin using ALOS PALSAR DEM and Sentinel-2 derived LULC. Master's thesis. National Institute of Technology

Brunner GW (2016) HEC-RAS, river analysis system: hydraulic reference manual (version 5.0). U.S. Army Corps of Engineers, Hydrologic Engineering Center

Chakraborty S, Mishra A, Pal S (2023) Estimation of Manning's roughness in the Brahmani delta using remote sensing-derived LULC and ALOS DEM. J Hydrol Eng *28*(2):05022055

Chen J, Liao A, Cao X, Chen X, Mills J (2015) Global land cover mapping at 30 m resolution: a POK-based operational approach. ISPRS J Photogram Remote Sens 103:7–27. https://doi.org/10.1016/j.isprsjprs.2014.09.002

Das SK, Maiti R (2022) Probabilistic simulation of soil moisture using hydrometeorological data and infiltration models. Hydrol Res 53(6):745–758

Demir G, Kisi O (2016) Flood hazard mapping by using geographic information system and hydraulic model. KSCE J Civil Eng 20(1):115–124

Farr TG, Rosen PA, Caro E, Crippen R, Duren R, Hensley S, Kobrick M, Paller M, Rodriguez E, Roth L, Seal D, Alsdorf D (2007) The shuttle radar topography mission. Rev Geophys 45(2). https://doi.org/10.1029/2005RG000183

Ghorbani M, Darvishi M, Niknejad M (2022) Land use land cover classification using Sentinel-2 imagery and machine learning algorithms: A comparative study. Environ Monitor Assess 194(8):1–17. https://doi.org/10.1007/s10661-022-10260-5

Gorelick N, Hancher M, Dixon M, Ilyushchenko S, Thau D, Moore R (2017) Google Earth engine: Planetary-scale geospatial analysis for everyone. Remote Sens Environ 202:18–27. https://doi.org/10.1016/j.rse.2017.06.031

Hamunyela E, Reiche J, Verbesselt J, Herold M (2018) Detecting tropical deforestation using Sentinel-1 radar data. Remote Sens Environ 210:444–454

Hirt C, Filmer MS, Featherstone WE (2010) Comparison and validation of recent freely available global DEMs and SRTM-based terrain corrections. Earth Surface Processes Landforms *35*(7):803–820

Jain SK, Singh VP, Gupta A (2018) Urbanization impacts on surface runoff in small catchments using SCS-CN and Green-Ampt methods. J Hydrol Hydromech 66(4):362–371

Kulp SA, Strauss BH (2019) New elevation data triple estimates of global vulnerability to sea-level rise and coastal flooding. Nat Commun *10*:4844. https://doi.org/10.1038/s41467-019-12808-z

Kumar R, Bhattacharjya RK, Pandey A (2020) Two-dimensional flood simulation using HEC-RAS in parts of the Yamuna River Basin. Water Practice Technol *15*(2):465–475

Kundzewicz ZW, Kanae S, Seneviratne SI, Handmer J, Nicholls N, Peduzzi P, Mechler R, Bouwer M, Amell N, Mach K, Muir-Wood R, Sherstyukov B (2014) Flood risk and climate change: Global and regional perspectives. Hydrol Sci J *59*(1):1–28. https://doi.org/10.1080/02626667.2013.857411

Loveland TR, Belward AS (1997) The international geosphere–biosphere programme data and information system global land cover data set (DISCover). Acta Astronautica 41(4–10):681–689

Mahdianpari M, Salehi B, Rezaee M, Mohammadimanesh F, Zhang Y (2019) Very deep convolutional neural networks for complex land cover mapping using multispectral remote sensing imagery. Remote Sens 11(24):3041. https://doi.org/10.3390/rs11243041

Nasirzadehdizaji R, Homayouni S, Safari A, Mahdavi S (2019) Flood detection and susceptibility mapping using Sentinel-1 and a machine learning approach: Hybrid bagging ensemble of k-NN classifiers. Remote Sens 12(2):266. https://doi.org/10.3390/rs12020266

Pandey AC, Nathawat MS, Gaur MK (2013) Land use land cover mapping through satellite data and GIS: A case study of the Dehradun region, Uttarakhand, India. Int J Remote Sens 34(22):8293–8307

Pande CB, Moharir KN, Patel K (2020) Analysis of floodplain morphology using ALOS DEM in Beas and Sutlej River basins. J Environ Manage 267:110642

Qi W, Li B, Zhu L, Wang J (2020) Assessing the impacts of LULC changes on flood risk using remote sensing and hydrological modeling: a case study in the Yangtze River Basin. J Hydrol 584:124702. https://doi.org/10.1016/j.jhydrol.2020.124702

Robinson NP, Allred BW, Jones MO, Moreno A, Kimball JS, Naugle DE, Erickson TA (2017) A dynamic Landsat-derived normalized difference vegetation index (NDVI) product for the conterminous United States. Remote Sens 9(8):863

Sanyal J, Lu XX (2004) Application of remote sensing in flood management with special reference to monsoon Asia: a review. Nat Hazards 33(2):283–301

Srivastava PK, Singh SK, Han D, Szilard S (2020) Integrated assessment of extreme events and hydrological responses in the Indo-Nepal Gandak River Basin. Environ Dev Sustain 23(6):8643–8668. https://doi.org/10.1007/s10668-020-00956-5

Yamazaki D, Sato T, Kanae S, Hirabayashi Y, Oki T (2011) FLO-2D flood modeling with improved global topography using Hydrosheds and SRTM. Hydrological processes 25(7):1129–1135

Zhang H, Su Z, Lu L, Yang K (2012) Assessment of land surface models in estimating surface turbulent fluxes over heterogeneous landscape. J Geophys Res: Atmos *117*(D6). https://doi.org/10.1029/2011JD017130

Zhang Y, Wang D, Li W, Li M, Zhang L (2020) Comparative study of flood simulation using HEC-RAS 1D and 2D models in urban areas. Water 12(3):787. https://doi.org/10.3390/w12030787

Zheng X, Tarboton DG, Maidment DR, Liu YY (2018) Terrain-based parameterization of physically based distributed hydrologic models using HEC-HMS. J Hydrol Eng 23(2):04017060. https://doi.org/10.1061/(ASCE)HE.1943-5584.0001596

Chapter 7
Application of HEC-RAS Flood Modeling of a Himalayan Foothills River: A Case Study of a Part of the River Torsa

Adrija Raha and Kaustav Baidya

Abstract Floods in the Himalayan river are a concerning natural event, especially in the foothill zone. The Torsa River is a prominent river of eastern Himalaya, known for being highly prone to seasonal flooding due to intense monsoonal precipitation and complex topographic conditions. This work is purposed to quantify the flood inundation and risk perception of the Torsa River in the foothill area near the Cooch Behar town. HEC-RAS 1D flood inundation method is applied on high-resolution terrain data (Digital Elevation Model) and historical unsteady flow data, to simulate the flood dynamics and delineate the inundation zone along the Torsa River. This application is very important for understanding flood-prone areas, land-use planning, and risk zonation.

Keywords HEC-RAS · Flood inundation · Torsa River

7.1 Introduction

A flood is considered to be a hydro-geomorphological and meteorological process of a drainage basin. It has an impact on altering the existing landscape both in the plain land and in the deltaic part of the basin (Bera and Sahu 2023). The concept of flood encompasses multiple dimensions, each reflecting different aspects of this natural hazard. Broadly, it may be understood in three contexts: (1) hydrological floods characterized by varying magnitudes and irregular flow frequencies, often linked with climate change; (2) floods as hazards that threaten human settlements, infrastructure, and livelihoods, thereby creating vulnerability; and (3) the environmental and geomorphological consequences of floods, including their role in reshaping landscapes and altering ecological components. Flooding typically occurs when a sudden surge of streamflow, generated by intense or prolonged rainfall, exceeds the carrying

A. Raha (✉) · K. Baidya
Department of Geography, Presidency University, Kolkata, India
e-mail: adrijaraha1997@gmail.com

M. Biswas et al. (eds.), *Himalayan Flood Hazards*, Advances in Geographical and Environmental Sciences, https://doi.org/10.1007/978-981-92-0702-2_7

capacity of a river channel at a particular location. In India, this situation is a recurring phenomenon during the monsoon season when the increased discharge volume of rivers often surpasses their natural thresholds, leading to widespread inundation.

The Himalayan and Sub-Himalayan regions, particularly North Bengal, are highly prone to such recurrent floods due to their climatic and topographic characteristics. This zone includes the districts of Darjeeling, Kalimpong, Jalpaiguri, Alipurduar, Cooch Behar, and the northern part of Islampur subdivision in Uttar Dinajpur. These regions experience extremely high rainfall intensity, ranging from 2000 mm to more than 4000 mm annually, of which nearly 80% is concentrated in the monsoon season. The present annual normal rainfall for the North Bengal sector stands at about 2479 mm, with districts like Darjeeling, Cooch Behar, and Jalpaiguri receiving rainfall on approximately 114, 112, and 110 days per year, respectively. The trajectory of the monsoon winds, which break against the Eastern Himalayas, results in heavy downpours, often intensified by low-pressure troughs and break monsoon conditions. Precipitation of 200–300 mm within just two hours is not uncommon in this region, creating conditions highly conducive to flash floods and severe waterlogging.

Given these recurrent hazards, flood management authorities face the crucial task of mitigating flood risks through both preventive and corrective measures. This involves a range of interventions, including infrastructural development (such as embankments, levees, and dams), river training works, and watershed management strategies (Daxikar et al. 2008). Experiences from global contexts, such as the devastating floods in Europe that caused billions in damages, underscore the urgent need for systematic flood hazard estimation and management. Innovative methodologies like "fluctuating complexity" analysis, which relies on temporal and data-based assessments, have been developed for better flood prediction (Moel et al. 2009). Furthermore, recent advancements in technology have introduced comprehensive risk reduction approaches that combine structural measures (e.g., flood barriers, drainage systems) with non-structural strategies (e.g., community preparedness, land-use planning, early warning systems). These approaches aim not merely to control floods but to minimize overall flood hazards and their cascading effects, thereby offering solutions that are scalable from local to global contexts (Kreibich et al. 2015).

In the case of the Torsa River in North Bengal, the impacts of flooding have been well documented. The river experiences its peak discharge during the monsoon, particularly in July, when heavy rainfall in the catchment area accelerates inflows. Communities residing along both banks of the Torsa frequently suffer from recurrent flooding, which has historically culminated in catastrophic events such as the floods of 1978 and 2000. Every year, authorities like the Central Water Commission (CWC) and the Irrigation & Waterways Department (I&WD) issue flood warnings for the Torsa, highlighting the vulnerability of agricultural fields and settlements adjacent to its banks. Beyond inundation, one of the most severe consequences of these floods is bank erosion. Since the alluvial plains are composed of unconsolidated sandy deposits, water recession during the post-monsoon period leads to large-scale erosion, threatening nearby habitations and anthropogenic structures.

The geomorphology of the region further exacerbates flood risk. The piedmont zone, characterized by a steep surface gradient, reduces the residence time of river water, which in turn increases the force of flood discharge. Embankment breaches are common, allowing floodwaters to spill into agricultural lands, causing crop destruction and livelihood loss. On the alluvial plains, the impact intensifies due to the dense network of drainage channels—comprising active river courses, tributaries, and abandoned channels—which allows floodwaters to spread rapidly (Saha and Bhattacharya 2016). Downstream areas, especially around Cooch Behar and near the Torsa's confluence with the Kaljani River, have been recognized as highly flood-prone zones (Chakraborty and Dutta 2013).

Additionally, sedimentation plays a crucial role in shaping the hydrological dynamics of the Torsa. During floods, the river's sediment load rises dramatically, often clogging parts of the active channel. The flashy nature of floods, driven by sudden inflows from the mountain catchments, not only exacerbates sediment deposition but also promotes channel instability. Annual low-intensity floods, although less destructive than major ones, often trigger bifurcation of the river's active course, particularly on the piedmont surface (Saha and Bhattacharya 2021). This constant reworking of channels disrupts local ecosystems and complicates flood management further.

In summary, floods in North Bengal, particularly in the Torsa River basin, represent a complex interaction of climatic, hydrological, and geomorphological factors. They bring about not only immediate hazards such as inundation and erosion but also long-term transformations of the landscape. Therefore, effective flood management in such regions demands an integrated approach that combines engineering solutions, community-based preparedness, and sustainable land and water resource planning.

7.2 Study Area

The Torsa River, also known as the Machu in its upper reaches, originates from the Chumbi Valley of southern Tibet at an elevation of approximately 7,065 m above mean sea level. From its glacial-fed source, the river descends through steep mountainous terrain before entering Bhutan, where it assumes a prominent role in shaping the fluvial landscape of the Himalayan foothills. After traversing Bhutan, the river enters the Indian state of West Bengal, where its course and hydrological dynamics become particularly significant in terms of flooding and sediment transport.

Within the Indian territory, the river exhibits notable morphological changes. Below the Hasimara Bridge on National Highway 31, the Torsa bifurcates into two distributary channels—Sil Torsa and Char Torsa. These distributaries flow separately for several kilometers before reuniting in the Patla Khowa forest region. This bifurcation–reunion characteristic highlights the unstable and dynamic nature of the river's fluvial system, which is influenced by the region's geomorphology and seasonal discharge variations.

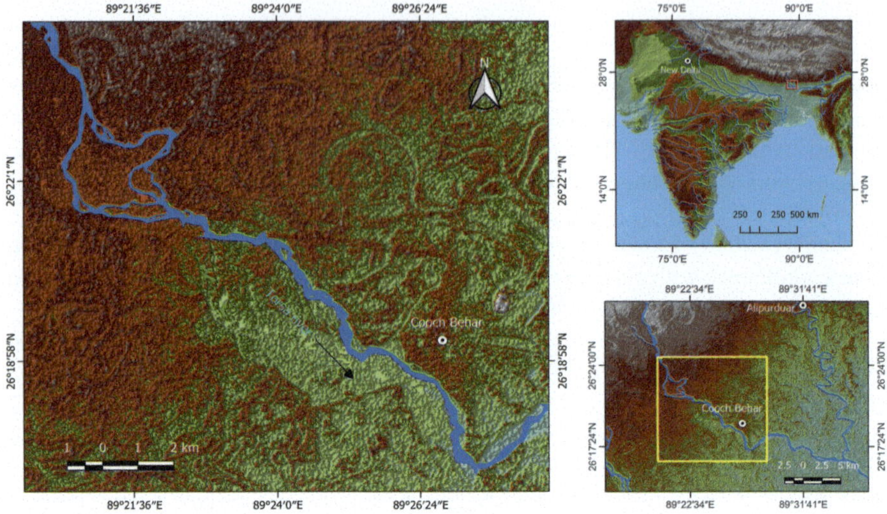

Fig. 7.1 Selected reach of the river Torsa (Starting: 26°23′54.90″N, 89°20′25.35″E Ending: 26°18′4.29″N, 89°28′19.93″E)

Continuing downstream, the river skirts past Cooch Behar town, a district headquarters with dense population and agricultural dominance. In this stretch, it is joined by two important tributaries—the Kaljani and Raidak-I rivers—thereby increasing its discharge capacity significantly. Eventually, the combined flow of the Torsa and its tributaries crosses into Bangladesh, where it joins the mighty Brahmaputra River near Nageswari in Rangpur district. This transboundary nature of the Torsa underscores its importance not only as a regional water resource but also as part of a larger international river system shared by China, Bhutan, India, and Bangladesh (Fig. 7.1).

The Torsa River extends over a total length of about 222 km, of which approximately 74 km lie within West Bengal. The total catchment area of the river basin is estimated at 7,192 square kilometers, encompassing diverse physiographic zones ranging from high-altitude glacial terrains in Tibet to low-lying floodplains in Bangladesh. This wide altitudinal range contributes to significant hydrological variability, where high-intensity rainfall in the Himalayan foothills during the monsoon season results in sudden surges of discharge and recurrent flooding in the plains.

Given its geographical setting and hydrological characteristics, the Torsa River plays a critical role in shaping the socio-economic and environmental conditions of North Bengal. However, its flashy flow regime, frequent inundations, and associated hazards such as bank erosion and sediment deposition make it one of the most flood-prone rivers in the region. This has important implications for water resource management, disaster preparedness, and transboundary river governance in the basin.

HEC-RAS as a tool:

The Hydrologic Engineering Center's River Analysis System (HEC-RAS) is a widely used hydraulic modeling tool designed to simulate water flow dynamics within channels, rivers, and floodplains (USACE 2010). Developed by the U.S. Army Corps of Engineers, HEC-RAS initially focused on one-dimensional (1D) steady flow hydraulic modeling, but over the years, it has evolved into a comprehensive platform capable of supporting a range of hydraulic and hydrodynamic analyses. Its reliability and adaptability have made it one of the most extensively adopted tools among government agencies, research institutions, and water resource managers.

The strength of HEC-RAS lies in its ability to simulate water surface profiles under steady flow conditions and evaluate the influence of hydraulic structures such as levees, floodplains, weirs, bridges, and culverts. This capability allows researchers and engineers to assess flood inundation scenarios and quantify potential impacts on surrounding landscapes. Previous studies have demonstrated its usefulness in determining inundation extents and assessing the hydrological behavior of reported flood events (Manandhar 2010; Al-Qudah 2011).

In addition to 1D steady flow modeling, the current versions of HEC-RAS support more advanced simulations, including one- and two-dimensional (1D and 2D) unsteady flow analysis, sediment transport modeling, water quality and temperature modeling, and flood risk mapping. The core computational framework is based on the hydrodynamic approach, with the shallow water equations forming the foundation for 2D unsteady flow simulations. This enables the system to represent complex hydraulic processes with greater accuracy and to evaluate flood propagation under diverse climatic and geomorphological conditions.

To develop a robust HEC-RAS model, several types of input data are required. These include the topography of the study area (generally represented in the form of digital elevation models), Manning's roughness coefficients to account for surface resistance, hydrological inputs such as precipitation or discharge data, and a comprehensive understanding of the stream network and hydraulic structures present within the basin. In this study, Shuttle Radar Topography Mission (SRTM) data (NASA JPL 2013) were used to construct the terrain model, providing the necessary elevation framework for hydraulic computations. The spatial distribution of the study area was preliminarily visualized using Google Maps imagery (Fig. 7.1), which offered a general overview of the basin's geographical context.

Thus, by integrating hydrological inputs with topographic and structural data, HEC-RAS provides a powerful framework for simulating flood inundation and assessing the hydraulic response of river systems under varying conditions. The methodology outlined here forms the foundation for analyzing flood dynamics in the selected study area and for evaluating the effectiveness of flood management strategies.

7.3 Methodology

The **main steps of HEC-RAS modeling:**

1. Creating a new project in HEC-RAS

2. Setting terrain with DEM and projection in RAS Mapper (Fig. 7.2)
3. Creating flow paths, bank lines, and Cross Sections (Fig. 7.3)
4. Inputting unsteady flow data (Fig. 7.4)
5. Running unsteady flow analysis
6. Checking results

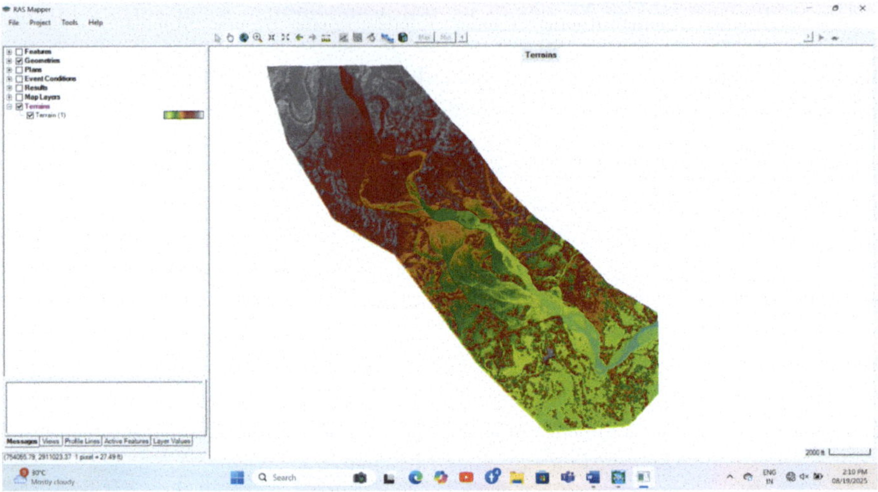

Fig. 7.2 Creating terrain file on HEC-RAS using SRTM DEM file

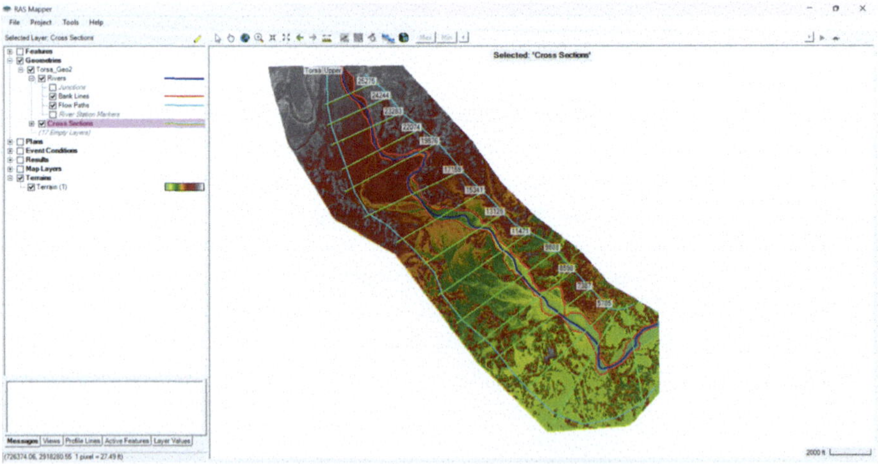

Fig. 7.3 Creating the flow paths, bank lines and cross sections

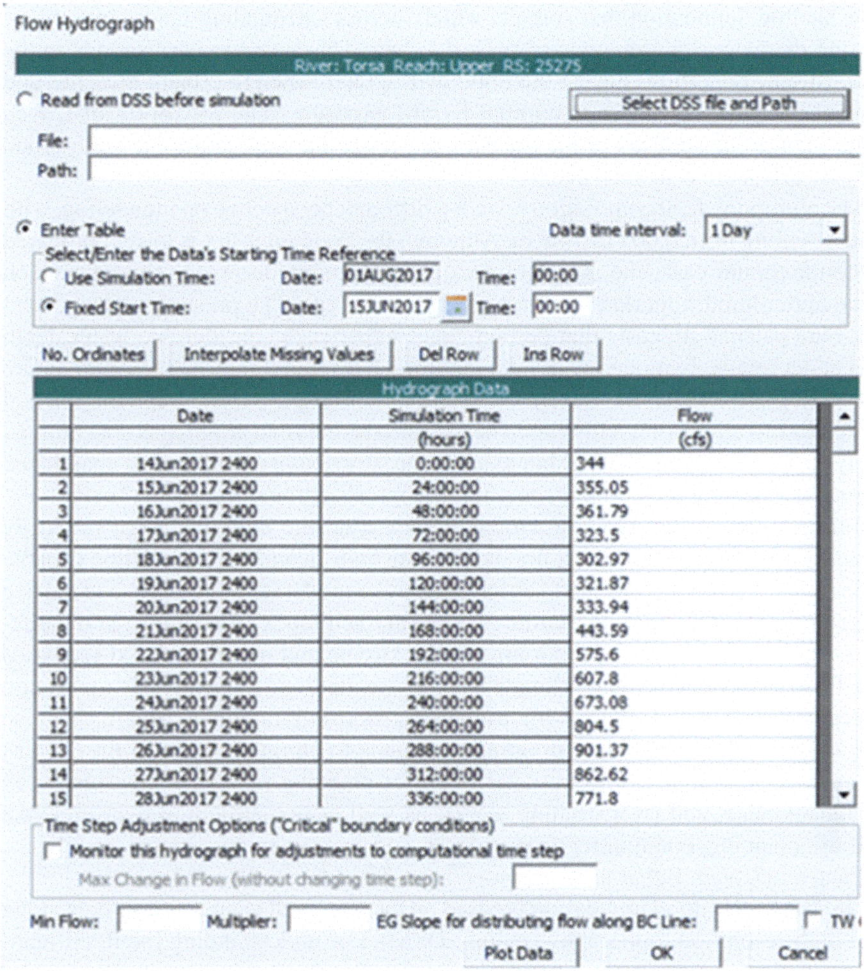

Fig. 7.4 Inserting unsteady flow data to simulate the flood model on river Torsa

7.4 Results and Discussion

The flood depth simulations of the Torsa River basin near Cooch Behar, generated through HEC-RAS 2D hydraulic modeling, provide a comprehensive picture of the spatial and hydrological dynamics of flooding in the region. The maximum depth map demonstrates water depths ranging between 0.001 and 7.72 m, with the highest inundation concentrated along the active river channel and adjacent low-lying floodplains. The dark blue patches indicate zones where.

Flood waters reach critical depths, posing significant risks to nearby settlements, agricultural fields, and rural infrastructure. In contrast, the lighter cyan shades show shallow inundation that spreads widely across surrounding landscapes, highlighting the extensive influence of peak flooding. This scenario corresponds to the most extreme conditions during the flood hydrograph, when riverbank breaches and channel overflows result in maximum hazard exposure. The proximity of Cooch Behar town to these flood-prone tracts underscores the vulnerability of both urban and peri-urban areas to such events.

The minimum depth simulation reveals a different perspective on inundation, with depths ranging from 0.001 to 5.68 m. Although the flood intensity is lower compared to the maximum case, the inundation extent remains widespread, covering much of the agricultural hinterland and low-lying rural areas. The presence of persistent dark blue patches suggests that certain zones experience considerable water depth even under relatively moderate flood conditions. This highlights the recurring nature of waterlogging, which has severe implications for crop loss, soil degradation, and prolonged disruption of livelihoods. The map also reveals that flooding is not confined solely to the main river channel but extends into distributary courses and abandoned channels, reflecting the complex geomorphological setting of the Torsa basin.

Taken together, the two simulations demonstrate the dual challenge posed by floods in this region: while peak floods generate high-magnitude inundation capable of destroying infrastructure and settlements, even moderate floods cause extensive waterlogging that hampers agriculture and local livelihoods. The comparison underscores the importance of incorporating both extreme and moderate flood scenarios into flood risk management frameworks. These findings suggest that mitigation strategies must not only address high-intensity, short-duration flood hazards but also long-duration, lower-intensity flooding that leads to chronic vulnerabilities. Thus, the simulation outputs form a crucial basis for planning structural measures such as embankments and river training works, as well as non-structural measures like land-use planning, community preparedness, and early warning systems, to enhance resilience in Cooch Behar and the wider Torsa River basin (Fig. 7.5).

HEC-RAS 1D flood inundation is based on the historical unsteady flow data along with high-resolution topography (mainly DSM). For understanding the flood inundation of this area, we've reviewed several flood reports, government documents, and news articles. From these sources of literature, the whole thing is conceptualized, and the peak discharge of the monsoonal seasons is chosen during the flood period to simulate the model with validated historical data. HEC-RAS 1D flood inundation is based on the historical unsteady flow data along with high-resolution topography (mainly DSM). For understanding the flood inundation of this area, we've reviewed several flood reports, government documents, and news articles. From these sources of literature, the whole thing is conceptualized, and the peak discharge of the monsoonal season is chosen during the flood period to simulate the model with validated historical data. The annual flood report 2017, published by the Irrigation and Waterways Directorate, Government of West Bengal, mentions the occurrence of floods and an increase in stage height near the Hasimara and Cooch Behar section. It is reported that the peak flood occurred on 12th August, and the water level was 116.40 and 42.90 m at Hasimara and Cooch Behar station, respectively. It is also reported that a period of seven days of prolonged rainfall from 8 to 14th August

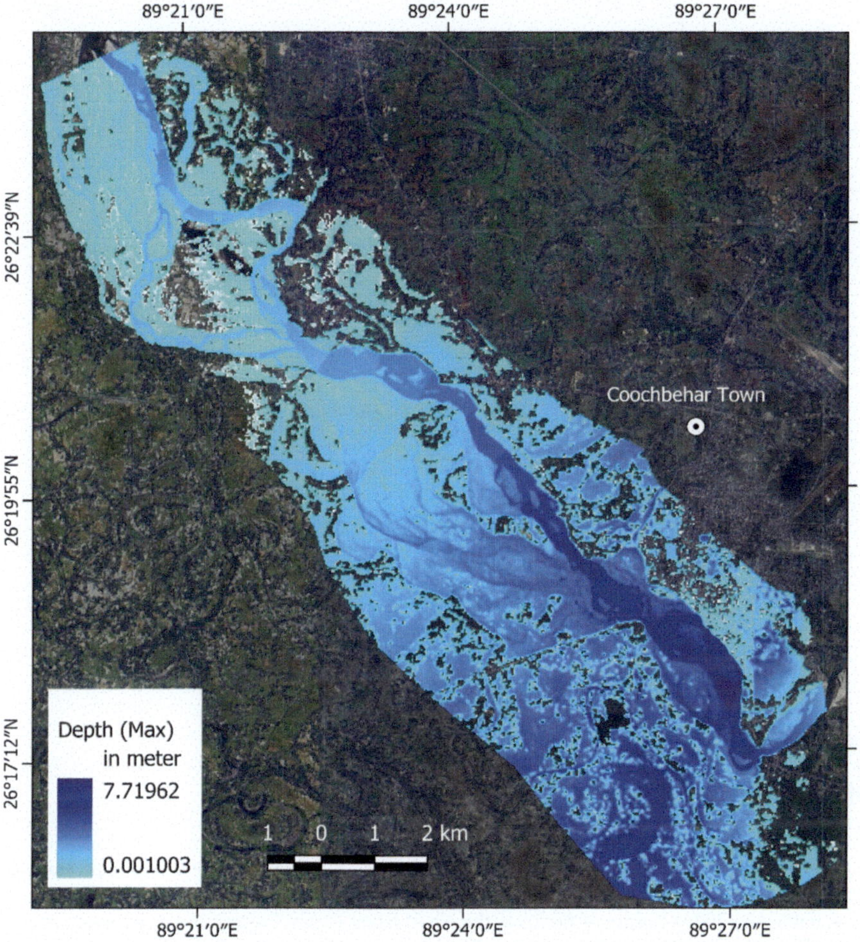

Fig. 7.5 Flood simulation model of part of the Torsa River

caused severe inundation in some major flood-prone areas. The Hasimara station received almost 1013.60 mm of total rainfall in this period (Fig. 7.6).

7.5 Concluding Remarks

HEC-RAS, being a very efficient and reliable flood modeling tool, successfully delineated the inundation extent and depths under varying discharge conditions over a period in the Torsa River. The river reach chosen for the study is highly urbanized along with highly susceptible to flooding. The key finding clearly reflects the flood hazard and long-term waterlogging issues along this stretch. It will create a new

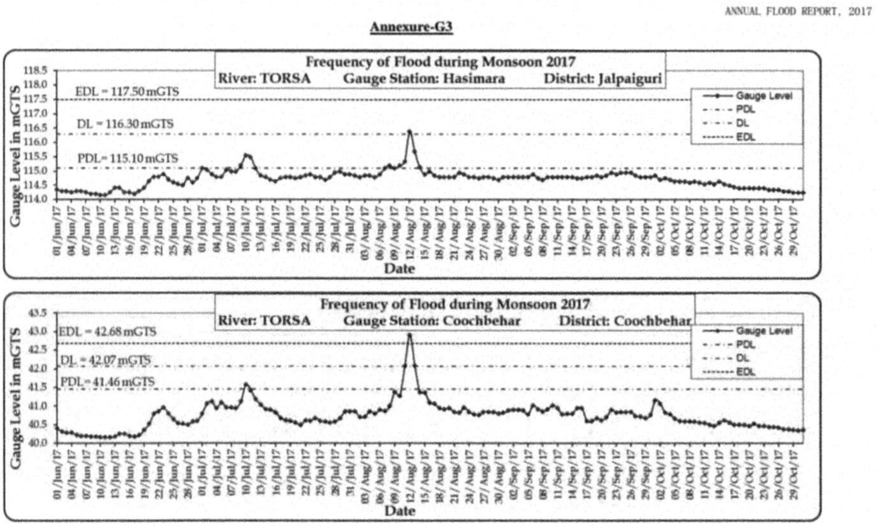

Fig. 7.6 Flood frequency during monsoon in two respected stations of Torsa River

scope of discussion for researchers and planners to identify the flood measures and mitigation necessary in this area.

References

Al-Qudah KA (2011) Floods as water resource and as a hazard in arid regions: a case study in Southern Jordan. Jordan J Civil Eng 159(2979):1–14

Bera S, Sahu AS (2023) Nature of floods in the Khari River Basin, Eastern India. In: Islam A et al (eds) Floods in the Ganga–Brahmaputra–Meghna Delta. Springer, pp 285–289

Chakraborty S, Dutta K (2013) Causes and consequences of fluvial hazard—a hydro-geomorphic analysis in Duars region India. Indian Stream Res J 2(12):1–10

Daxikar A, Hillier T, Makarem F, Stone S (2008) Application of GIS technology in coastal flood hazard mitigation. In: Solutions to coastal disasters 2008. American Society of Civil Engineers, pp 396–409

De Moel H, Van Alphen J, Aerts JCJH (2009) Flood maps in Europe—methods, availability and use. Nat Hazard 9(2):289–301

Kreibich H, Bubeck P, Van Vliet M, De Moel H (2015) A review of damage-reducing measures to manage fluvial flood risks in a changing climate. Mitig Adapt Strat Glob Change 20(6):967–989

Manandhar B (2010) Flood plain analysis and risk assessment of Lothar Khola (Master's thesis, Tribhuvan University, Institute of Forestry, Pokhara, Nepal)

NASA JPL (2013) NASA Shuttle Radar Topography Mission (SRTM) Global 1 arc second. NASA EOSDIS Land Processes DAAC

Saha UD, Bhattacharya S (2016) A study of river induced major hydro-geomorphic issues and associated problems in Torsa basin West Bengal. Geograph Rev India 78(2):132–145

Saha UD, Bhattacharya S (2021) Channel avulsion in the Torsa River course and its response to topographic and hydrological controls on the Himalayan Foreland Basin. J Earth Syst Sci 130(4):1–29. https://doi.org/10.1007/s12040-021-01641-2

United States Army Corps of Engineers. (2010). HEC-RAS river analysis system: Hydraulic reference manual (Version 4.1). Hydrologic Engineering Center, U.S. Army Corps of Engineers

Chapter 8
Discussion of Flood Disaster in Central Himalaya with Gravity Anomalies and Seismic Activities

Kutubuddin Ansari

Abstract The geological features and fault structures in the Himalayas play a major role during earthquakes or any natural disaster, such as a flood. The current discussion about the rugged topography, weak geological formations, and active seismic conditions is the major reason for flood disasters in the central Himalaya. The impact of topographic and gravity measurements on the rainfall and the change of glaciers is discussed in the study. Gravity anomalies, horizontal and vertical filter methods, and ratio tests are used to delineate the edge of geological structures in the central Himalayas. Finally, the earthquake catalog is modeled, and a quantile–quantile plot is plotted to show its relation with geological structure. These earthquake events increase the landslides, dam failures, and river blockage-related floods in the Himalaya.

Keywords Flood disaster · Gravity anomalies · Seismicity · Nepal Himalayas

8.1 Introduction

The Himalayan structure is categorized by major thrust fault, named main Himalayan thrust (MHT) (Mukul et al. 2018; Ansari 2018), coinciding with the main frontal thrust (MFT) fault by reaching the surface along the foothills (Bettinelli et al. 2006). The region surrounding the zone of rupture is crossed by additional active major faults such as the main boundary thrust (MBT) and the main central thrust (MCT) (Liu et al. 2017). Including these major faults, several known and unknown geological features and fault structures play a major role during earthquakes or any natural disaster, such as a flood. This is very important to understand this kind of structure for the assessment of natural hazards.

K. Ansari (✉)
Department of Geomatics Engineering, Karadeniz Technical University, Trabzon 61080, Türkiye
e-mail: kdansarix@gmail.com

M. Biswas et al. (eds.), *Himalayan Flood Hazards*, Advances in Geographical and Environmental Sciences, https://doi.org/10.1007/978-981-92-0702-2_8

The central Himalayas region is generally categorized by three types of rivers based on their origin. This is a standard hydrological and geomorphological classification used in the Himalaya. The first kind of river that originates from glaciers is permanent snowfields in the high Himalaya, known as snow-fed (glacial) rivers. These rivers have perennial flow throughout the year because of glacial melting and monsoon rains. They load very high sediments during the melt, especially in the summer season. For example, the Koshi river, the Gandaki river, and the Karnali river are included in this category. The second category of rivers is rain-fed rivers, which originate from the mid-hill, such as the Bagmati and Kamala rivers. The flash flood is common in these rivers because of intense rainfall. They have significant flow during the monsoon and reduced or dry flow during the winter seasons. The third category of rivers, which are mixed with snow and rain, originates from both high snow-fed mountains and rainfall-fed tributaries in the mid-hills. These rivers are Narayani River, Bheri River, and Seti River.

The seismicity of the Himalayas is well known, and there are numerous earthquakes recorded alongside the active faults. Deeper earthquakes within the range of 60–80 km are identified in the north of MCT and defined moderate and small earthquake sources in the depth of 20 km across the Himalayas (De la Torre et al. 2007). The earthquakes that occurred in the north of the MCT are strike-slip dominated, while the southern part of the MCT is measured by thrust fault (Baranowski et al. 1984). The frequency of earthquake occurrence in the region MHT is comparatively high because of its tectonic happenings. The information about historical seismicity was incorporated and implied by studies with a period of occurrence of about 1 in 700 years (Ansari et al. 2022; Rajendran et al. 2015). The studies based on instrumental, geodetic, and paleoseismological point out that there is always the possibility of a large number of events (Mw 8.5 or 8.0) or a single (Mw $\geq$ 9.0) or a single (Mw $\geq$ 9.0) incidence almost at any place of the arc of the Himalayas (Stevens and Avouac 2015). The evaluation of historical earthquake events to current time seismicity has been done by Rajendran and Rajendran (2005), where they found a spatial and temporal scale clustering in the eastern Nepal Himalayas. Geological evidence of the eastern Himalayas has been used by Bollinger et al. (2014), and they concluded the return period of great earthquakes is around 750 $\pm$ 140 and 870 $\pm$ 350 years. There are two different existing seismogenic zones of epicentral depth with the ranges of 0 to 25 km and 50 to 100 km that have been estimated in the Nepal Himalayas (Liang et al. 2008; Monsalve et al. 2006). This might be possible because of lower crust seismic events with low strength (Chen and Molnar 1983). In reality, it is very difficult to identify the future earthquake size and location in the Himalayas because most of them are controlled by various unknown constraints, such as the strike and dip of MHT structural variation. The new ideas need to be studied carefully, and the complexity of earthquake recurrences with time needs to be evaluated for the evaluation of risk.

Although space geodetic techniques, along with seismic data, are important for understanding crustal deformation caused by geological faults (Ansari and Park 2019; Ansari K and Bae, 2020), the gravity approach for detecting limits of geological features and faults is commonly used in many places worldwide (Nyalugwe et al.

2020; Yanis et al. 2020; Dewanto et al. 2022). Newton's law of gravitation, which states that the force acting between two masses is directly proportional to the product of their masses and inversely proportional to the square of their distance from one another, is the foundation of the gravity method's physical theory (Surya et al. 2019; Benyas et al. 2021). A geophysical method called the gravity method is used to evaluate changes in Earth's density and gravity with each measurement (Yanis and Marwan 2019; Saibi et al. 2019). The gravity acceleration values affected by topography are measured by the free-air gravity anomaly, which ignores the bulk of the rock beneath. The spatial variations of their horizontal and vertical derivatives are usually used to interpret and analyze gravitational data. The technique known as edge detection helps uncover important geological features such as lineaments, faults, basins, and rock formations in the graphical analysis of the gravity approach (Yanis et al. 2023).

8.2 Data and Methodology

The study discussed the number of flood events that occurred in the central Himalaya, especially in the Nepal Himalaya. The flood event data during 1971–2016 have been collected from the National Society for Earthquake Technology (NSET 2016). The bathymetry height, including the gravity measurements in the central Himalaya, is used, and the gravity anomalies that present subsurface geological fault structures are identified. The mathematical summary for bathymetry and gravity analysis is as follows:

The gravity observation data in terms of free-air anomaly (Δg_{FA}) gained from the satellite observations are given by (Yanis et al. 2023):

$$\Upsilon = \Upsilon_0 + \frac{GM}{r\gamma} \sum_{n=2}^{n_{\max}} \left(\frac{a_{ref}}{r}\right)^n \sum_{m=0}^{n} (\overline{c}^*_{nm} \cos m\lambda + \overline{s}_{nm} \sin m\lambda) \overline{P}_{nm}(\cos\theta) \tag{8.1}$$

Here, the terms Δg_0 and ζ_0 are with zero order for free-air anomaly and height data, respectively. The global gravity model scale parameter is given by a_{ref} calculated with the geometrical distance of r from the center of the Earth. M is the mass of the Earth, and G is the gravitational constant. Normalized Legendre function is $\overline{P}_{nm}(\theta)$ with m and n spherical harmonic order and degree, respectively for satellites, with the maximum degree of $n_{\max}$. $\overline{S}_{nm}$ and $\overline{C}_{nm}$ are the normalized spherical harmonic coefficients.

The residue of regional and Bouguer anomalies is turned into a residual anomaly, which can be given by the following mathematical Eq. (8.2) (Abdulrahim et al. 2022).

$$G(i, j) = c.e^{-\frac{(i-u)^2+(j-v)^2}{2\sigma^2}} \tag{8.2}$$

Here, the Gaussian kernel matrix is given by $G\ (i,\ j)$ with a constant matrix of c, and the components of the matrix are given by i, u, j, and v components.

The amplitude of the first vertical derivative (FVD) gravity anomaly (g_h) corresponding to the elevation height (h) is estimated as shown by Eq. 8.3 (Evjen 1936):

$$\mathrm{FVD} = \frac{\partial g_h}{\partial h} \tag{8.3}$$

The amplitude of total horizontal derivative (THD) uses the x-axis and y-axis directional derivatives and is calculated as shown by Eq. 8.4 (Miller and Singh 1994):

$$\mathrm{THD} = \sqrt{\left(\frac{\partial g_h}{\partial x}\right)^2 + \left(\frac{\partial g_h}{\partial y}\right)^2} \tag{8.4}$$

The arctan ratio of FVD to THD amplitudes is called tilt derivative ratio (TDR), which diverges between—π/2 and π/2 and is calculated as follows (Miller and Singh 1994):

$$\mathrm{TDR} = \tan^{-1}\left(\frac{\mathrm{FVD}}{\mathrm{THD}}\right) \tag{8.5}$$

These types of enhancement techniques are utilized for first-order derivatives and numerous joint derivatives, capable of offering rectangular, circular, and linear tectonic boundaries enhancements with depth variation up to the Moho (Kumar et al. 2022; Narayan et al. 2017).

8.3 Result and Discussion

The number of disaster events that occurred due to floods, earthquakes, and other disasters during 1971–2016, totaling 46 years are analyzed, as shown in Fig. 8.1. The number of counted events was approximately 26,000. If we took the average of these events, they would become around two events per day. The record shows that, on average, about 1,000 people have been killed or missing every year due to disaster events. The greatest number of deaths in a single year, 9442, occurred in 2015, followed by 1969 in 2001, 1885 in 1993, 1422 in 1999, 1398 in 1997, 1279 in 1988, and so on (Jimee et al. 2019). If we took the average, then three deaths or missing persons per day. The figure shows that in the early years Sastri et al. (1971), the events were low, but they kept on increasing over the years. The highest events occurred during 2011, which were almost 1800. This is a very high number, which shows that more than five events per day. The flood and earthquake from the rivers are the main cause of these events.

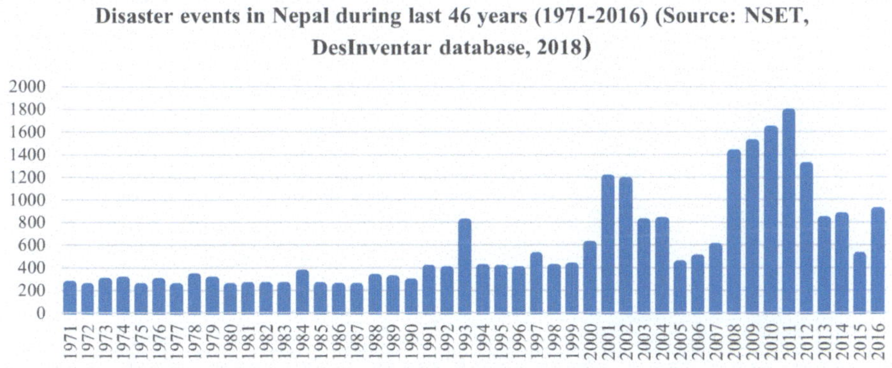

Fig. 8.1 Disaster events in Nepal during the last 46 years (1971–2016) (NSET 2016)

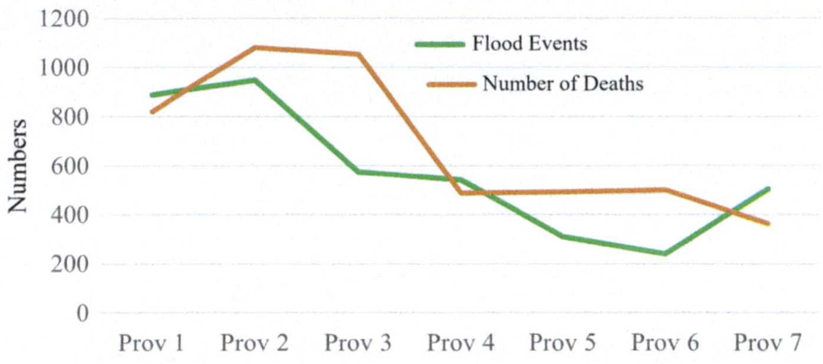

Fig. 8.2 Damages and losses by flood on the basis of provinces (Jimee et al. 2019; Shrestha et al. 2020)

The NSET (2016) shows that during 1971–2016, a total of 4012 flood events occurred (Nepal country) in these 45 years, and because of these events, around 4796 people died or are missing, as shown in Fig. 8.2. The figure shows province-wise data all over Nepal, which is the main region of the central Himalaya. The results show that the highest number of total events occurred in province 2, which was around 23.68%, and deaths were around 22.52%. The lowest number of occurring events is recorded in province 6, which was around 5.99%, and the number of total deaths or missing persons has been seen in province 7, which is around 7.57%. In terms of house damage, around 98,818 during 1971–2016 in the Nepal region. It shows that a lot of the buildings were destroyed in the flood disaster and people were affected. The estimated economic losses due to flood events are very high (Shrestha et al. 2020). The rivers that have water flow are sustained by snow melting during the summer and spring seasons and by rain during the monsoon. The flood in the central Himalaya is a common water-induced disaster and a recurring phenomenon. It causes many deaths and injuries, besides severe impacts on the vital infrastructures (Shrestha et al. 2020).

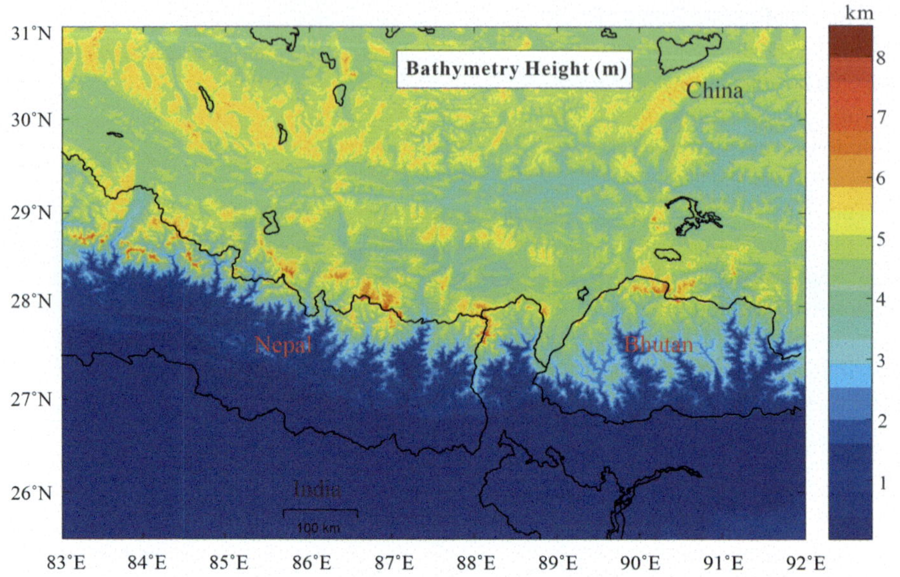

Fig. 8.3 Regional map of the selected area along the Nepal Himalayas has been plotted by using GEBCO bathymetry data. Color bar: mean sea level (MSL), height in m

The GEBCO bathymetry data were used, and a topographic contour plot using MATLAB 2017 was prepared of the given region extending from 83°E to 92°E and 25.50°N to 31° N, which covers most of the region of the central Himalaya, as shown in Fig. 8.3 (Internet ref-1). The contour plot shows that the bathymetry height gradually increases along the north–south trend, varying from a height of around 0.5 km to 8.5 km. The regional maps are quite useful to better display the relief and topography of the Himalayas. The topography and relief of the Himalayas have a great influence on the generation of rainfall floods and, as a result of erosion. Bookhagen and Burbank (2006) investigated the distribution of rainfall and its relation with relief and topography using eight years (1998–2005) of measurements over the Himalayas. They noticed two different maxima of rainfall along the Himalayan strike. The first peak of rainfall was observed in the lesser Himalayan southern margin (mean elevation of 0.9 ± 0.4 km and mean relief of 1.2 ± 0.2 km), and the second peak belonged to the greater Himalayas (both mean elevation and mean relief of 2.1 ± 0.3 km). This demonstrates that the maxima of rainfall at high locations, most likely received by relief, strengthened the phase of monsoon with the same kind of relief distribution. The impact of topographic measurement on the change of glaciers is vital because it also affects climate change. Garg et al (2017) selected 18 glaciers (1994–2015) distributed over the central Himalayas and explored the topographic influences on the glacier parameter changes using remote sensing satellite observations. They analyzed numerous topographic factors such as glacier slope, altitude, length, area, compactness ratio, hypsometry, etc., using multivariate statistical analysis. The study concludes that out of 18 glaciers, five glaciers had modest,

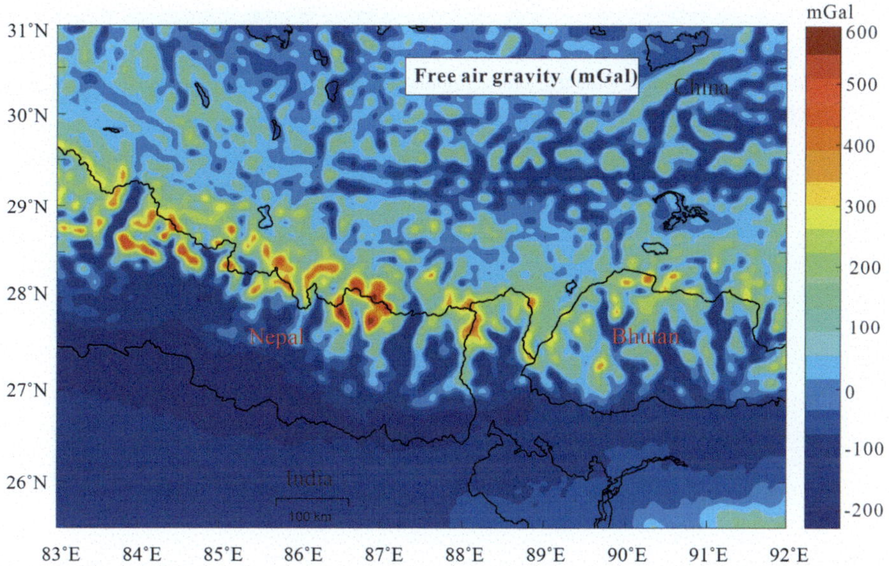

Fig. 8.4 Free-air gravity anomaly contour plot of the selected area along the Nepal Himalayas

eight glaciers had moderate, and five glaciers had strong influences of topography. Recently, the mountain of the Himalayas faced the largest continental earthquake, which included the Mw 7.8 Gorkha earthquake in 2015. It is concluded that the cause of the Gorkha earthquake is the subside of high mountains in Himalayas, and the Himalayan topography growth may mainly happen during the ongoing phase of post-seismic (Elliott et al. 2016). In this way, by analyzing several examples, we noticed that the Himalayan topography plays such an important role in natural disasters such as floods, heavy rainfall, glaciers, and earthquakes.

The free-air gravity (FAG) data is used, and a gravity map over the Nepal Himalayas from the EGM2008 data measurements was prepared as shown in Fig. 8.4 (Internet ref-2). The plot shows gradient of gravity anomalies from frontal Himalayas to the North Himalayas. The local area of higher Himalayas shows positive gravity anomalies that decline slowly negatively in the frontal part. These gravity values, which are high, range up to 600 mGal in the region, and show low values around −200 mGal. The topography of the Himalayas and gravity anomalies of the Himalayas are well correlated with each other. Tiwari et al. (2009) analyzed the lithospheric effective elastic thickness and isostatic gravity anomalies in one of the Himalayan regions (Karakoram–Pamir regions) between topography and Bouguer anomaly by using coherence analysis. They used an integrated approach of lithospheric model and finally suggested that the continental subduction from the Indian and Eurasian side leads to a complex compressional situation for large seismic events. Tiwari et al. (2006) combined magnetic and gravity measurements data from Northeast Indian (Sikkim) Himalayas with available data from India, Tibet, and Bangladesh and tried to define the structure of the crust in the region part of the Himalayas. These magnetic

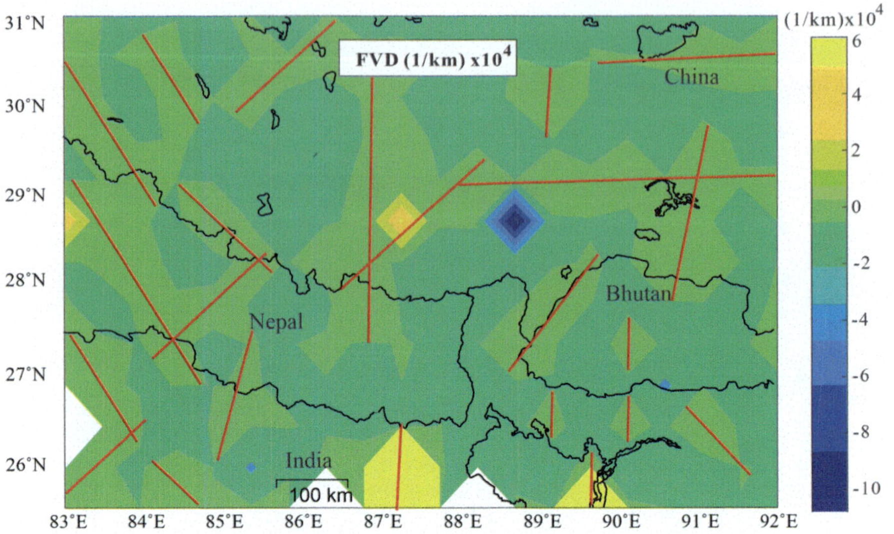

Fig. 8.5 The delineated geological faults structures estimated by the FVD with edge extension (red lines). Color bar: FVD amplitude of gravity anomaly in 10^4 km^{-1}

and gravity observation Northeast Indian Himalayas allowed for mapping the crustal structure in detail, for example, to show the under-thrusting of Indian plate, flexure support for the topography of Himalayas, and to produce the large variations in depth of Moho, which contribute maximum observed signal of gravity on the surface. The gravity anomalies variation shown in Fig. 8.2 has several reasons. For example, MHT which is mapped in the INDEPTH profile extending in south, joins the MBT and MCT at mid-crust depth. This kind of low dipping interface has dip of around 6° with small changes in density (30 kg/m^3), will comparable with the dip estimated by seismicity analysis. However, if it extends up to the upper crust with same contrast of density, it generates a significant exit in the gravity anomaly (Tiwari et al. 2006).

I used the horizontal and vertical filter methods and ratio tests of both methods to delineate the edge of geological structures in the Himalayas. The FVD values are estimated by using Eq. 8.3 and plotting their outcomes with their spatial trend in Fig. 8.5. These delineated geological fault structure plots reveal the trend of the major structural features. The red line diagrams plotted in MATLAB 2017 environment by using the FVD technique show that the structural fabrics mostly trend ~ NW–SE, and a few are horizontal in the eastern parts. The data using FVD estimate anomalies varied between −6 and 10 mGal, suggesting the presence of faults at numerous places. The low derivative values (around −10 mGal) support this stipulation and are surrounded by high values (1 mGal). In general, these derivatives show various local fault distributions that control the tectonic activity and seismicity of the Himalayan region. The tilt derivative data TDR are estimated by using Eq. 8.5, and results are shown in Fig. 8.4. The diagrams plotted in the MATLAB 2017 environment by using the TDR technique display a major NE-SW trend and a few horizontal lines

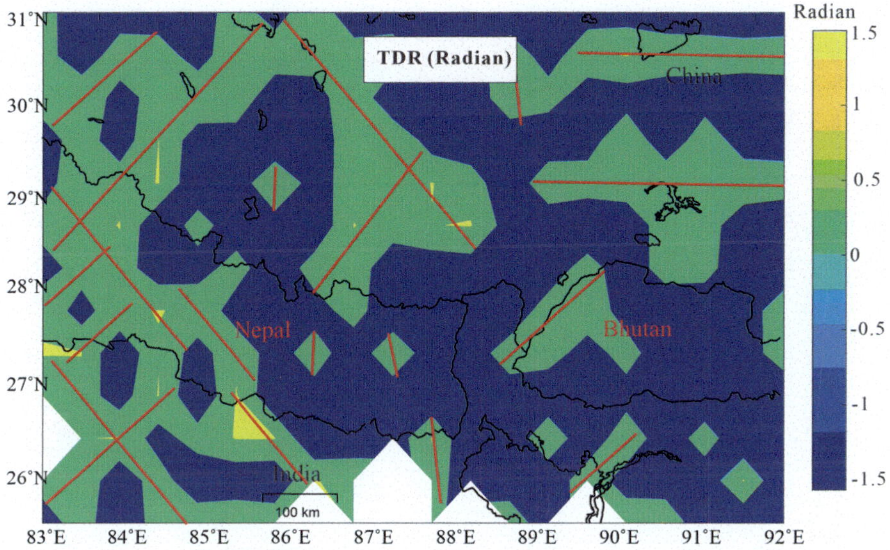

Fig. 8.6 Red lines: geological fault structures deduced from TDR. Color bar: TDR amplitude in radians

in the eastern parts. Overall, the TDR variation lies between −1.5 radians and 1.5 radians. Here, the positive value of TDR suggests that the position will be above the surface of anomalies, negative TDR values are outside the object, and if they are zero, then it means their position is on the geological structure edge. This technique is helpful because of its simple methodology in the detection of geological structure. It is assumed here that the border of the fault is between the geological structure and trace of other fault can be presented by tilted radian values (Cooper and Cowan 2006). This process can be carried out very quickly and freely relatively other conventional methods of gravity, which normally take a huge time and economic expenses (Yanis et al. 2023) (Fig. 8.6).

I used the earthquake catalog during the years 1965–2005 (accessed by De la Torre et al. 2007) and 2005–2020, accessed through the EMSC (Internet Ref-3), as shown in Fig. 8.7. We selected a lower threshold value (Mw 3) is so that the smaller earthquakes (Mw < 3) are discarded. There is a total of 804 seismic events extending from 83°E to 92°E and 25.50°N to 31°N, which includes the eastern Nepal Himalayas and the southern Tibetan Plateau. The spatial distribution of earthquakes with Magnitude size consisting to the fault locations (MCT, MBT, and MFT), plotted with different colors, indicates risky active seismic zone in the region. It can be noticed from the figure that the small-sized events are grouped near each other, and the surrounding location of their respective main shock (green circles). The quantile–quantile (QQ) plots of the seismic events' magnitudes (Mw) of diverse ranges are plotted with distinct colors as shown in Fig. 8.8. The grouping of the earthquakes is used with Mw 3, Mw 4, Mw 5, and Mw 6–7. As we can see from the

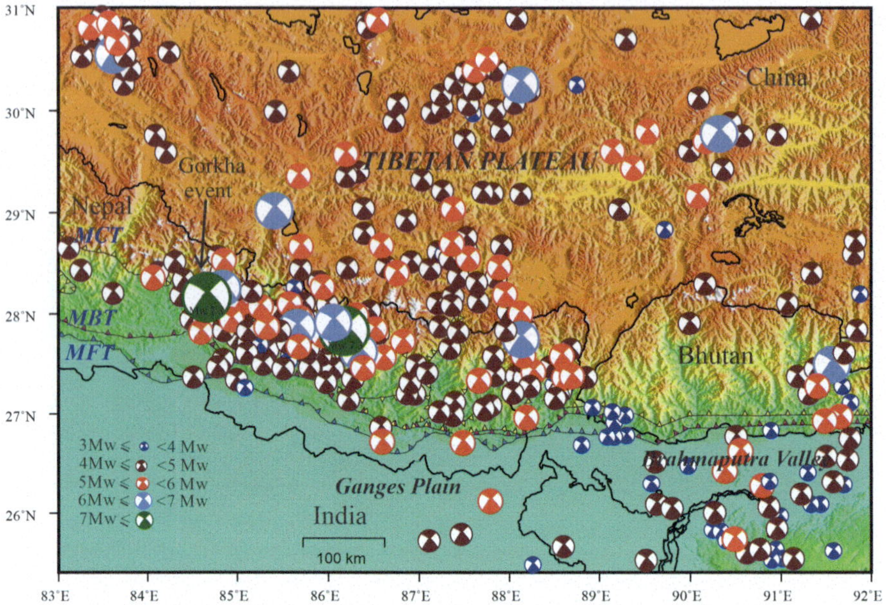

Fig. 8.7 Seismicity plot in eastern Nepal Himalayas and the southern Tibetan Plateau from the years 1965–2005 (accessed by De la Torre et al. 2007) and 2005–2020 accessed through the EMSC (Internet Ref-3) (Ansari et al, 2022)

figure, the earthquakes with Mw greater than 7 have only two events, hence they are combined with Mw 6. It is clear from the QQ plots that they are diverging from the straight line, showing that earthquake Magnitude does not estimate a linear trend, while they estimate a nonlinear trend. Probably this is possible because of unusual seismic-generated strain, which may not be accurately determined and is unstable from major active faults are larger seismic events (Ansari et al. 2017). This indicates that the generating strain because of small earthquakes includes a minor level of active geological fault structure, which we discussed before in Figs. 8.3 and 8.4. This gravity method is connected with seismicity, which must be understood for the appropriate maintenance of natural disasters.

8.4 Conclusions

The study investigates number of disaster events because of floods and seismic activities. There are number of flood events, and the loss of lives shows that the rivers passing from the mountains are the cause of flood disasters. The topographic measurements using bathymetry height in the central Himalayas showed that the height gradually increases along the north–south trend variation from a height of around 0.5 to 8.5 km. This topography of the Himalayas has a good connection with

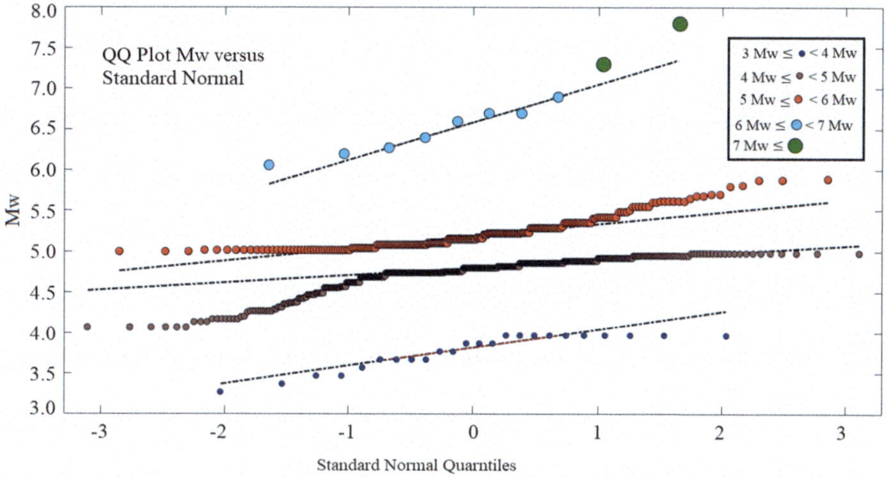

Fig. 8.8 The quantile–quantile (QQ) plots of the seismic events' magnitudes (Mw) of diverse ranges are plotted with distinct colors

rainfall and glacier changes. The gravity analysis has been carried out using Free-air gravity maps over the Nepal Himalayas. The results show that the high gravity values range up to 600 mGal in the region, and they show low values around −200 mGal. The study used the horizontal and vertical filter methods and ratio tests of both methods to delineate the edge of geological structures in the Himalayas. The data using FVD estimate anomalies varied between −6 and 10 mGal, suggesting the fault's presence at numerous places. Overall, the TDR variation lies between −1.5 radians and 1.5 radians. Here, a positive value of TDR suggests that the position will be above the surface of anomalies, negative TDR values are outside of the object, and if they are zero, then it means their position is on the geological structure edge. The quantile–quantile (QQ) plots of the seismic events' magnitudes (Mw) divert from the straight line, showing that earthquake Magnitude does not estimate a linear trend, while they estimate a nonlinear trend. Probably this is possible because of unusual seismic-generated strain, which may not be accurately determined and is unstable from major active faults are larger seismic events.

References

Abdulrahim H, Al-Naji AJ, Yousif SR, Al-Hussaini BS (2022) Determining the tectonic implication derived from geophysical data of Bahr An-Najaf Basin. SW Iraq Iraqi Geologica J 55:40–46. https://doi.org/10.46717/igj.55.1F.3Ms-2022-06-18

Ansari K, Corumluoglu O, Sharma SK (2017) Numerical Simulation of Crustal Strain in Turkey from Continuous GNSS Measurements in the Interval 2009–2017 Abstract Journal of Geodetic Science 7(1):113–129. https://doi.org/10.1515/jogs-2017-0013

Ansari K (2018) Crustal deformation and strain analysis in Nepal from GPS time-series measurement and modeling by ARMA method. Int J Earth Sci 107(8):2895–2905. https://doi.org/10.1007/s00531-018-1633-7

Ansari K, Park KD (2019) Contemporary deformation and seismicity analysis in Southwest Japan during 2010–2018 based on GNSS measurements. Int J Earth Sci 108(7): 2373–2390. https://doi.org/10.1007/s00531-019-01768-w

Ansari K, Bae TS (2020) Contemporary deformation and strain analysis in South Korea based on long-term (2000–2018) GNSS measurements. Int J Earth Sci 109(1): 391–405. https://doi.org/10.1007/s00531-019-01809-4

Ansari K, Bae TS, Singh KD, Aryal J (2022) Multivariate singular spectrum analysis of seismicity in the space–time-depth-magnitude domain: insight from eastern Nepal and the southern Tibetan Himalaya. J Seismol 26(1):147–166. https://doi.org/10.1007/s10950-021-10057-6

Baranowski J, Armbruster J, Seeber L, Molnar P (1984) Focal depths and fault plane solutions of earthquakes and active tectonics of the Himalaya. J Geophys Res: Solid Earth 89(B8):6918–6928. https://doi.org/10.1029/JB089iB08p06918

Bollinger L, Sapkota SN, Tapponnier P, Klinger Y, Rizza M, Van Der Woerd J, Tiwari DR, Pandey R, Bitri A, Bes de Berc S (2014) Estimating the return times of great Himalayan earthquakes in eastern Nepal: evidence from the Patu and Bardibas strands of the main frontal thrust. J Geophys Res Solid Earth 119(9):7123–7163. https://doi.org/10.1002/2014JB010970

Bookhagen B, Burbank DW (2006) Topography, relief, and TRMM-derived rainfall variations along the Himalaya. Geophys Res Lett 33(8)

Benyas K, Aarab A, Qarbous A, Lakhloufi A, Manar A, Amar M, Idrissi A, Elmimouni M (2021) Exploiting aeromagnetic and gravity data interpretation to delineate massif deposits of Rehamna Area (Western Meseta-Morocco). The Iraqi Geolog J 13–28. https://doi.org/10.46717/igj.54.2C.2Ms-2021-09-21

Bettinelli P, Avouac JP, Flouzat M, Jouanne F, Bollinger L, Willis P, Chitrakar GR (2006) Plate motion of India and interseismic strain in the Nepal Himalaya from GPS and DORIS measurements. J Geodesy 80(8–11):567–589. https://doi.org/10.1007/s00190-006-0030-3

Cooper GRJ, Cowan DR (2006) Enhancing potential field data using filters based on the local phase. Comput Geosci 32(10):1585–1591. https://doi.org/10.1016/j.cageo.2006.02.016

Chen WP, Molnar P (1983) Focal depths of intracontinental and intraplate earthquakes and their implications for the thermal and mechanical properties of the lithosphere. J Geophysic Res Solid Earth 88(B5):4183–4214. https://doi.org/10.1029/JB088iB05p04183

De la Torre TL, Monsalve G, Sheehan AF, Sapkota S, Wu F (2007) Earthquake processes of the Himalayan collision zone in eastern Nepal and the southern Tibetan Plateau. Geophys J Int 171(2):718–738. https://doi.org/10.1111/j.1365-246X.2007.03537.x

Dewanto BG, Priadi R, Heliani LS, Natul AS, Yanis M, Suhendro I, Julius AM (2022) The 2022 Mw 6.1 Pasaman Barat, Indonesia Earthquake, confirmed the existence of the talamau segment fault based on teleseismic and satellite gravity data. Quaternary 5(4):45. https://doi.org/10.3390/quat5040045

Evjen HM (1936) The place of the vertical gradient in gravitational interpretations. Geophysics 1:127–136

Elliott JR, Jolivet R, González PJ, Avouac JP, Hollingsworth J, Searle MP, Stevens VL (2016) Himalayan megathrust geometry and relation to topography revealed by the Gorkha earthquake. Nat Geosci 9(2):174–180

Garg PK, Shukla A, Jasrotia AS (2017) Influence of topography on glacier changes in the central Himalaya, India. Global Planet Change 155:196–212

Internet ref-1: URL: http:/www.gebco.net/data_and_products. Accessed on 12-July-2023

Internet ref-2: URL: https://bgi.obs-mip.fr/data-products/outils/egm2008-anomaly-maps-visualiza tion/. Accessed on 13-July-2023

Internet Ref-3: European–Mediterranean Seismological Centre (EMSC), URL: https://www.emsc-csem.org/Earthquake

Jimee GK, Meguro K, Dixit AM (2019) Nepal, a multi-hazard risk country: spatio-temporal analysis. J Nepal Geolog Soc 58:145–152
Kumar U, Narayan S, Pal SK (2022) Structural and tectonic interpretation of EGM2008 gravity data around the Laccadive ridge in the Western Indian Ocean: an implication to continental crust. Geocarto Int 37(11):3179–3198
Liang X, Zhou S, Chen YJ, Jin G, Xiao L, Liu P, Fu Y, Tang Y, Lou X, Ning J (2008) Earthquake distribution in southern Tibet and its tectonic implications. J Geophys Res Solid Earth 113(B12). https://doi.org/10.1029/2007JB005101
Liu C, Dong P, Shi Y (2017) Stress change from the 2015 Mw 7.8 Gorkha earthquake and increased hazard in the southern Tibetan Plateau. Phys Earth Planet Int 267:1–8. https://doi.org/10.1016/j.pepi.2017.04.002
Miller HG, Singh V (1994) Potential field tilt—a new concept for location of potential field sources. J Appl Geophys 32:213–217
Monsalve G, Sheehan A, Schulte-Pelkum V, Rajaure S, Pandey MR, Wu F (2006) Seismicity and one-dimensional velocity structure of the Himalayan collision zone: earthquakes in the crust and upper mantle. J Geophys Res Solid Earth 111(B10). https://doi.org/10.1029/2005JB004062
Mukul M, Jade S, Ansari K, Matin A, Joshi V (2018) Structural insights from geodetic global positioning system measurements in the Darjiling-Sikkim Himalaya. J Struct Geol 114:346–356. https://doi.org/10.1016/j.jsg.2018.03.007
Narayan S, Sahoo SD, Pal SK, Kumar U, Pathak VK, Majumdar TJ, Chouhan A (2017) Delineation of structural features over a part of the Bay of Bengal using total and balanced horizontal derivative techniques. Geocarto Int 32(4):351–366
NSET (2016) DesInventar Database 1971–2016. National Society for Earthquake Technology-Nepal (NSET)
Nyalugwe VN, Abdelsalam MG, Katumwehe A, Mickus KL, Atekwana EA (2020) Structure and tectonic setting of the Chingale Igneous Ring Complex, Malawi from aeromagnetic and satellite gravity data: implication for Precambrian terranes collision and Neogene-Quaternary rifting. J Afric Earth Sci 163:103760. https://doi.org/10.1016/j.jafrearsci.2020.103760
Rajendran CP, John B, Rajendran K (2015) Medieval pulse of great earthquakes in the central Himalaya: viewing past activities on the frontal thrust. J Geophys Res Solid Earth 1623–1641. https://doi.org/10.1002/2014JB011015
Rajendran CP, Rajendran K (2005) The status of central seismic gap: a perspective based on the spatial and temporal aspects of the large Himalayan earthquakes. Tectonophysics 365:19–39. https://doi.org/10.1016/j.tecto.2004.09.009
Saibi H, Mogren S, Mukhopadhyay M, Ibrahim E (2019) Subsurface imaging of the Harrat Lunayyir 2007–2009 earthquake swarm zone western Saudi Arabia using potential field methods J Asian Earth Sci 169:79–92. https://doi.org/10.1016/j.jseaes.2018.07.024
Sastri VV, Bhandari LL, Raju ATR, Datta AK (1971) Tectonic framework and subsurface stratigraphy of the Ganga basin. Geolog Soc India 12(3):222–233
Stevens VL, Avouac JP (2015) Inter-seismic coupling on the main Himalayan thrust. Geophys Res Lett 42(14):5828–5837. https://doi.org/10.1002/2015GL064845
Shrestha BR, Rai RK, Marasini S (2020) Review of flood hazards studies in Nepal. Geograph Base 7:24–32
Surya RA, Abdullah F, Darisma D, Yanis M, Ismail N (2019) Sedimentation process in Kuala Gigieng Coast, Aceh Besar based on magnetic and gravity surveys. In: IOP Conference series: earth and environmental science, vol 348, No 1. IOP Publishing, p 012040
Tiwari VM, Rao MV, Mishra DC, Singh B (2006) Crustal structure across Sikkim, NE Himalaya from new gravity and magnetic data. Earth Planet Sci Lett 247(1–2):61–69
Tiwari VM, Rajasekhar RP, Mishra DC (2009) Gravity anomaly, lithospheric structure and seismicity of Western Himalayan Syntaxis. J Seismol 13:363–370. https://doi.org/10.1007/s10950-008-9102-6

Yanis M, Marwan (2019) The potential use of satellite gravity data for oil prospecting in Tanimbar Basin Eastern Indonesia. In: IOP Conference Series: Earth and Environmental Science 364(1):012032. https://doi.org/10.1088/1755-1315/364/1/012032

Yanis M, Faisal A, Yenny A, Muzakir Z, Abubakar M, Nazli I (2020) Continuity of great sumatran fault in the marine area revealed by 3D inversion of gravity data. J Teknologi 83(1):145–155

Yanis M, Simanjuntak AV, Abdullah F, Marwan M, Ghani AA (2023) Application of seismicity and gravity observation-based filtering model for mapping a fault structure in Weh Island, Indonesia. Iraqi Geolog J 260–274. https://doi.org/10.46717/igj.56.2A.20ms-2023-7-29

Chapter 9
GLOF Modeling to Assess the Higher Himalayan Floods: A Case Study from Tsho Rolpa Glacial Lake, Nepal

Shibangee Datta

Abstract Glacial Lake Outburst Floods (GLOFs) are becoming a major hazard in the Himalayas due to enhanced glacial retreat and lake expansion. The study of this paper is on Tsho Rolpa Glacial Lake in Nepal, one of the most potentially dangerous glacial lakes of the Himalayas. The study focused on simulating flood conditions through HEC-RAS 2D hydrodynamic modeling and remote sensing data from three years: 2015, 2020, and 2024. The spatio-temporal analysis verified the continuous growth of glacier size, width, and length because of the melting of the glacier. Symmetrical change patterns in cross-sectional profiles confirmed consistent broadening and deepening of the lake. hydrodynamic consequences—flood depth, velocity, and hydrograph—had increased magnitudes with time, and prominent downstream effects were observed. Flood susceptibility mapping supported the outcome of modeling to ascertain highly vulnerable zones in settlements such as Chankhu, Sikpasor Khani, and Jhyaku. GIS and remote sensing, combined with HEC-RAS, delineated a general picture of evolving GLOF hazards, enabling effective risk assessment and planning for disaster mitigation.

Keywords Remote sensing · HEC-RAS 2D · Flood susceptibility · Glacial lake expansion · Hazard mapping

9.1 Introduction

The Himalayas, called the "Third Pole," possess one of the most extensive depots of snow and ice in the world outside the polar regions. Among its features are the thousands of glaciers that feed the headwaters of the major South Asian rivers. The area, most notably the Higher Himalayas, has been experiencing a rapid withdrawal of glaciers during the last few decades due to the rise in global temperature. This has led to the formation and rapid expansion of glacial lakes, which are otherwise dammed by

S. Datta (✉)
Symbiosis International (Deemed University), Symbiosis Institute of Geoinformatics, Pune, India
e-mail: dattashibangee@gmail.com

M. Biswas et al. (eds.), *Himalayan Flood Hazards*, Advances in Geographical and Environmental Sciences, https://doi.org/10.1007/978-981-92-0702-2_9

weakly consolidated moraines. While these lakes are important indicators of climate change and act as high-altitude water storage systems, they do not pose severe natural hazards in the form of Glacial Lake Outburst Floods (GLOFs).

A GLOF is a sudden outburst of water from a glacial lake, usually due to the failure of its moraine dam. Failure is due to many natural reasons, like ice or rock avalanches, earthquakes, and most importantly, intense precipitation. They are characterized by steep peak discharges, rapid onset, and immense destructive energy, especially in mountainous terrain of high relief. In the Higher Himalayas, where the landscape is marked by elevation, glacial occurrence, and snowmelt, the GLOFs are not only increasing in numbers but also in uncertainty. All research conducted by ICIMOD (2011), Kattelmann (2003), and UNEP (2001) has emphasized that the frequency of GLOFs is increasing, coupled with a clear association with climate change.

Although the GLOFs are typically initiated in the Higher Himalayas, their worst and most devastating impacts often happen in the Middle Himalayas—a physiographic zone acting as a transition zone between the high mountains and low grounds. The Middle Himalaya zone of Nepal is particularly relevant here because it has high densities of human habitation, agricultural lands, hydropower schemes, and critical infrastructure. The deep gorges of these rivers in this region convey floodwaters from the higher mountains with much velocity, increasing their destructive force. It is because of this that the Middle Himalayas are most affected by floods, even though they are not located in the source area. This renders the Middle Himalaya geospatially and socioeconomically important in GLOF hazard planning and modeling for mitigation in the Nepalese context.

The Nepal Himalaya, in particular, is already well known to be one of the globe's hotspots for GLOF risks. With over 1400 glacial lakes spread across the area, and over 20 potentially hazardous, the risk of GLOFs to Nepal is among the highest in the world (ICIMOD 2011). The combination of glacial retreat, terrain in mountains, monsoon-driven precipitation, seismic instability, and population density creates a unique package of challenges. According to the Department of Hydrology and Meteorology (DHM), the hazard risk caused by GLOFs in Nepal is not only an environmental issue but a national issue due to its implications on disaster management, water resources, and infrastructure development.

Among the numerous glacial lakes in Nepal, Tsho Rolpa is considered one of the highest ranked in hazard potential. Located in Rolwaling Valley, Dolakha District, at around 4580 m' elevation, Tsho Rolpa is among the country's largest and most risky glacial lakes. It is formed due to the retreat of Trakarding Glacier, and the lake has significantly expanded in size over the past decades. Historical satellite imagery and topographic studies have shown that its surface area increased from as low as 0.23 km^2 during the 1950s to more than 1.6 km^2 during the early 2000s, making it the top priority on the list for glacial lake monitoring in Nepal. Its growth, combined with its steep topography and large catchment area, makes Tsho Rolpa a perfect subject for GLOF risk modeling.

The historical significance of Tsho Rolpa as a GLOF risk is well established. The level of the lake rose to a critically high point in 1997, and fear of an imminent outburst existed. In response, the Nepalese government, with the help of foreign aid organizations, undertook a massive intervention by constructing a controlled spillway to lower the water level artificially by 3 m. This intervention is overall believed to have prevented a catastrophic event on a massive scale. Nevertheless, despite these mitigating efforts, Tsho Rolpa remains a dangerous lake. Verifications by ICIMOD, UNDP, and DHM confirm that the moraine dam remains unstable, and the lake continues to grow due to ongoing glacier melting.

Precipitation, such as heavy monsoon rain, is also an important contributor to increasing the risk of GLOFs, in addition to glacier melting. The Tsho Rolpa catchment is located in a region with large summer rainfall. Heavy precipitation has the potential to result in sudden inflow, increasing the water level in hours or days. It may lead to greater hydrostatic pressure on the moraine dam or even overtopping. Heavy rain can also trigger landslides or icefalls onto the lake, pushing water out and creating wave surges, which can result in dam failure. Studies by Rana et al. (2020) have documented that precipitation-triggered GLOFs are now being recognized as a major threat in Nepal, especially in future climates where events of extreme rainfall are expected to increase.

These few factors—glacier retreat, rainfall, historical threat, and downstream vulnerability—render Tsho Rolpa a vital site for scientific research and for hazard modeling. The presence of satellite data, historical data, and topography provides an effective background for the application of methods like HEC-RAS 2D modeling and GIS-based mapping of flood hazard. Through the simulation of various outburst scenarios, scientists are able to forecast potential flood extents, peak discharges, and impacted areas. Such modeling is essential for planning early warning systems, evacuation procedures, and infrastructure construction in vulnerable downstream areas, especially along the Tama Koshi River Basin.

In short, glacial lakes in the Higher Himalayas, especially in Nepal, are both a cause of climate change and a vital environmental hazard. Tsho Rolpa, the biggest and most perilous glacial lake in Nepal, is a suitable example case for estimating GLOF risk in the Himalayas. The combination of glacier melt, monsoon precipitation, and downstream-exposed populations makes the lake an ideal candidate for simulation-based risk assessment. Studying Tsho Rolpa by means of hydrodynamic and geospatial modeling contributes not only to increasing the scientific understanding of GLOFs but also plays an important role in informing practical planning for disaster risk reduction across the Himalayas.

This whole discourse is focused on the Higher Himalayas, and glacial recession, intense monsoonal precipitation, and precarious moraine-dammed lakes such as Tsho Rolpa are all producing dangerous flood threats there. The mechanisms of GLOFs have their origin in this physiographic zone and thus become the focus for model-building and understanding high-altitude flood hazards in Nepal.

Glacial Lake Outburst Floods (GLOFs) are emerging as a powerful hydrometeorological hazard in the Himalayas, and especially in Nepal, due to accelerated glacier melting due to climate change. Rising frequency and intensity of GLOFs pose gigantic threats to lives, livelihood, and infrastructure of downstream societies. Among the risk-prone glacial lakes of Nepal that have been identified, Tsho Rolpa Glacial Lake has attracted significant attention from scientists, organizations, and policymakers. Shrestha and Shrestha (2019) refer to the high risk associated with the rapid growth of glacial lakes, especially in high downstream inhabited catchments and hydropower plants. Being among the biggest and most well-researched glacial lakes, Tsho Rolpa has been categorized as highly hazardous since the 1990s based on its failure potential as a moraine dam.

Hydrodynamic modeling has been a fundamental component of the simulation of GLOF events. Kayastha and Maskey (2024) employed the HEC-RAS model coupled with the NWS-BREACH module to simulate outburst floods under varied breach conditions in Tsho Rolpa. The research gave useful data on flood propagation characteristics, peak discharge estimation, and spatio-temporal flood depth and velocity patterns. Likewise, Shrestha and Nakagawa (2014) suggested using integrated modeling with terrain and lake volume data to predict possible flood extent and peak discharge and stressed the need for the application of high-resolution topographic input for credible simulation. Chen et al. (2021) used multi-scenario modeling for the same lake and produced a rich hydrodynamic dataset usable in hazard mapping and evacuation planning. Their research captures the possibility of integrating simulated depth and velocity layers with land use and infrastructure data for use in vulnerability mapping.

Remote sensing and GIS technologies also became more prominent in GLOF hazard studies. International Centre for Integrated Mountain Development (ICIMOD) in 2019 used the combination of high-resolution satellite imagery and in-field observations to trace lake expansion and determine indicators of moraine instability at Tsho Rolpa. Notwithstanding previous risk reduction measures like controlled draining of the lake water in 2000—the lake still shows indicators of expansion and possible hazard, which justifies continued monitoring. Shrestha and Gurung (2013) illustrated the potential of GIS-based hazard zonation based on satellite-derived inputs to effectively present flood-risk areas downstream and be relevant to early warning system planning and infrastructure design. Worni et al. (2013) reiterated, comparing different methods of modeling and recommending HEC-RAS and BREACH-type models due to their applicability and ability to model complex alpine terrain.

Glacier lake dynamics, or volume increase and morphological development, are the essence of GLOF hazard. Fujita et al. (2013) explained how small increases in the capacity of the lake, in combination with unstable moraine walls, can significantly advance the outburst threat. Sustained observation is also promoted by Richardson and Reynolds (2000), who initially classified Tsho Rolpa as one of the most hazardous Himalayan lakes. They emphasized the need for immediate concentration on mitigation well before the current lake extent. Peters and Bolch (2009) monitored long-term glacier retreat using Landsat time series and observed a distinct association between glacial thinning, enhanced meltwater input, and lake growth in the Rolwaling Valley. Shrestha et al. (2013) further validated these results through flume experiments for erosion-induced failure of moraine, which illustrated how fine-grained material and steep downstream slopes can create fast dam breach propagation.

It is further essential to know the overall regional context. Gurung et al. (2011) suggested a regional monitoring scheme for all Nepalese glacial lakes with high priority to Tsho Rolpa on account of its volume, moraine status, and downstream risk. Yao and Kattel (2022) linked the GLOF hazards with larger-scale cryospheric processes like monsoon variability and more intense precipitation capable of increasing the overtopping or hydrostatic failure hazard. Although numerous studies have simulated GLOFs for Imja, Tam Pokhari, and Lower Barun lakes, the methodology and results are strongly transferable to Tsho Rolpa, as topographies and climates are similar. For example, Osti and Egashira (2009) simulated the Tam Pokhari GLOF and showed that HEC-RAS simulations with different breach parameters yielded realistic flood predictions and inundation maps, which is a methodology transferable to Tsho Rolpa.

No attention can be left to institutional and historical interventions. ICIMOD and the Government of Nepal (2002) have even made structural interventions at Tsho Rolpa, such as a spillway construction, reducing the water level by a couple of three meters. Mool et al. (2001) had previously marked the lake as one among the top 20 most hazardous glacial lakes in Nepal, which induced multi-stakeholder intervention. However, satellite monitoring demonstrates that lake volume and area have continued to increase after intervention, supporting the need for redesigned modeling and adaptive mitigation measures. Alean and Hambrey contributed glaciological information on moraine formation, ice-core instability, and permafrost degradation—issues relevant to the interpretation of probable sources of failure in lakes such as Tsho Rolpa.

Evidence of necessity is also provided by recent investigations. Sattar et al. (2021) effectively used HEC-RAS and remote sensing information to model outburst events for South Lhonak Lake, eastern Nepal, with particular delineation of flood hazard areas and infrastructure susceptibility. The 2023 South Lhonak GLOF, studied by Choudhury et al. (2023), affirms the necessity of early simulation and the validity of simulation-based hazard maps. ICIMOD (2020) applied UAV and satellite imagery to validate the continued growth of Tsho Rolpa, supporting the high-risk designation of the lake. In addition, studies throughout the wider Himalaya and Tibetan Plateau, for example, Wang et al. of Galong Co and Liu et al. (2022) of JiwenCo, demonstrate the efficiency and utility of HEC-RAS-based flood modeling, particularly when integrated with InSAR and multi-source DEMs.

Generally, the literature-rich evidence refers to reintroducing hydrodynamic modeling, remote sensing analysis, ground truthing, and socio-economic vulnerability mapping into evaluating GLOF hazard mapping for lakes such as Tsho Rolpa. Not only is it crucial from an academic perspective, but also a hidden foundation for the planning of disaster risk reduction in the Himalayas.

From the above discussion, we can say that Glacial Lake Outburst Floods (GLOFs) are now a major climate-related hazard in high mountains, especially in the Himalayas. Globally, researchers have observed an increase in the frequency and intensity of GLOFs due to increased glacier retreat, subsequently triggered by rising temperatures. This has led to the rapid expansion of glacial lakes, endangering downstream human settlements and infrastructure with an intensifying impact. Scientific studies such as those by Allen et al. (2016) and Yao and Kattel (2022) have highlighted the means through which climate variability has restructured glacial hydrology, such that sudden lake breaches are more probable. In the context of Nepal, this threat is particularly severe because of the rugged terrain, glacial coverage, and dense human settlements along river valleys. Furthermore, Shrestha and Shrestha (2019) stressed that critical infrastructure, such as hydropower plants and roads, is increasingly vulnerable to GLOFs, and they emphasized the importance of integrating hazard assessment and mitigation into the planning process, especially for lakes like Tsho Rolpa.

From the modeling perspective, several researchers have formulated approaches for simulating potential outburst scenarios. Worni et al. (2013) analyzed previously developed GLOF simulation models and found that integration of breach models with hydrodynamic simulators such as HEC-RAS provides reliable flood extent and flow dynamics in steep mountain regions. Rounce et al. (2017) pushed the envelope by using probabilistic methods to represent uncertainties in breach depth, lake volume, and terrain characteristics. These are very powerful methods to improve the precision of flood forecasts, yet most studies are still generic at the regional scale and tend to omit site-specific modeling at high resolution. In the example of Tsho Rolpa, Shrestha and Nakagawa (2014) proposed a modeling approach that integrates terrain information with breach scenarios to quantify potential flood discharges and the areas of influence. Similarly, Kayastha and Maskey (2024) used models like NWS-BREACH and HEC-RAS to simulate the future GLOF events of Tsho Rolpa under climate change. But all these analyses concentrated on hydrodynamics only and did not examine the real spatial effects on settlements, infrastructure, and land use.

An attempt has been made to map and monitor Tsho Rolpa through remote sensing and GIS. For instance, Shrestha and Gurung (2013) applied satellite imagery to produce hazard maps of the Rolwaling Valley, identifying risk areas related to the lake. Chen et al. (2021) presented a detailed dataset of flood outputs through simulating different breach scenarios at Tsho Rolpa, which may be useful for risk assessments. Nevertheless, these works did not incorporate model outputs in full into GIS-based impact mapping. Similarly, although Fujita et al. (2013) and Peters and Bolch (2009) tracked lake growth according to long-term satellite time series, they did not model subsequent flood behavior or quantify impacts to human exposure. ICIMOD (2019) combined field surveys with satellite imagery to reassess the Tsho Rolpa risk, noting that even though mitigation has been undertaken, the lake remains a problem.

Yet, they also did not perform detailed flood modeling or vulnerability analysis. Other important contributions, like those by Mool et al. (2001), created inventories of glacial lakes and identified Tsho Rolpa as among the most dangerous, but lacked real-time modeling or impact-based outputs.

While significant research exists on GLOFs at both global and regional scales, and some specific studies have been conducted on Tsho Rolpa, a clear research gap persists. Most studies either focused solely on modeling flood dynamics without spatial impact analysis or created hazard maps without modeling real-time GLOF behavior. Very few works have integrated 2D hydrodynamic modeling with updated remote sensing data and GIS-based vulnerability assessments for this particular lake. As climate change accelerates glacier melt and lake expansion, and as new infrastructure continues to develop downstream, it becomes increasingly urgent to combine these methods to evaluate the full risk scenario. This integrated modeling and mapping gap has spurred this current study.

Thus, I have decided to research the Tsho Rolpa glacial lake in Nepal, renowned for its high rate of spatial growth and extreme GLOF hazard risk. The research is intended to contribute to substantial research gaps by integrating lake area evolution dynamics analysis and hydrodynamic flood modeling. By incorporating current multi-temporal satellite images, high-resolution digital elevation models, and land use data sets, I evaluate how the lake has changed over time and how such growth can impact the probability and magnitude of an outburst, if it were to occur. This requires applying the HEC-RAS 2D model alongside GIS tools in order to replicate the flood behavior of a GLOF event and determine resulting downstream hazard areas, impacted communities, and infrastructure. With this combined methodology, the study aims to better understand high-altitude Himalayan glacial lake behavior and flood propagation, ultimately leading to disaster preparedness, early warning systems, and resilient infrastructure planning in GLOF-vulnerable areas such as the Middle Himalayas.

While Glacial Lake Outburst Floods (GLOFs) have been well-researched in the Himalayas, the definite and essential gap lies in the convergence of multi-temporal remote sensing data with sophisticated hydrodynamic modeling methods for site-specific analysis of GLOFs. In Tsho Rolpa, although numerous investigations have evaluated its risk level, recorded lake growth, or modeled breach scenarios, very few merged lake areas change dynamics and 2D hydrodynamic modeling with GIS-based hazard mapping in an integrated high-resolution approach.

Most current literature either concentrates on breach modeling and hydrodynamic simulation (e.g., via NWS-BREACH or 1D models) or on remote sensing-based lake monitoring, with no integration of the two methods to measure real-time downstream effects. In addition, numerous previous studies have not applied newer high-resolution DEMs, multi-temporal satellite images, or included land use and infrastructure exposure in their flood modeling. This reduces the applicability of previous studies to disaster risk reduction and local planning.

Furthermore, the Middle Himalayan region, which is densely populated and most at risk from downstream flood impact, has been insufficiently covered in many current models. This spatial deficit underestimates actual vulnerability and impairs the establishment of site-specific mitigation measures.

Hence, the main research gap is the development of a complete integrated, dynamic, and spatially explicit model incorporating glacial lake evolution, GLOF simulation with HEC-RAS 2D, and exposure mapping with GIS for assessing the impacts. Filling the gap is crucial in advancing scientific knowledge, enhancing early warning systems, and enhancing infrastructure planning in the GLOF-vulnerable areas of Nepal. The Tsho Rolpa glacial lake was identified years ago as a high-priority location for GLOF risk observation in the Nepal Himalayas based on its ongoing growth and past hazard of outburst. Even after mitigation actions, including the development of a controlled spillway during 1997, the lake is still at risk of failure from glacier melt and heavy precipitation. Past research has provided a foundation for the comprehension of GLOF hazards, but does not include the incorporation of recent high-resolution DEMs, satellite images, and land cover data with sophisticated hydrodynamic modeling. Through the choice of Tsho Rolpa as a case study, this study looks to close that gap through a dual strategy: (1) assessing spatio-temporal alterations of the lake with geospatial analysis, and (2) application of HEC-RAS 2D modeling for simulating flood scenarios and delineating hazard areas. This combined methodology will allow for a greater understanding of the behavior, magnitude, and spatial extent of a hypothetical GLOF occurrence. The findings will aid local governments, NGOs, and disaster management authorities in formulating early warning systems, preparedness strategies, and infrastructure resilient to the hazards of GLOFs. Due to enhanced climate-induced vulnerability in the Himalayas, this research contributes scientifically and practically to sustainable mountain hazard management. The rapid increase in the growth and formation of glacial lakes in the Nepal Himalayas due to increased glacier retreat has resulted in mounting concerns for Glacial Lake Outburst Floods (GLOFs) having extreme potential effects on downstream villages, infrastructure, and ecosystems. Although numerous previous studies have appreciated the hazard potential of Nepal's largest and most hazardous glacial lake, Tsho Rolpa, existing studies have either focused on limited hydrodynamic modeling or performed generic hazard mapping without systematically integrating high-resolution spatial data with dynamic flood simulations. Moreover, the spatial growth of Tsho Rolpa over recent decades has never been comprehensively studied in relation to changing flood hazards. There exists a large research gap in integrating temporal lake dynamics with scenario-based GLOF modeling and impact-based exposure mapping. This hinders planners and policymakers from making sound data-driven decisions for minimizing disaster risk, particularly in the sensitive Middle Himalayan region, where a high human population and critical infrastructure are found in potential impact areas. In the absence of precise modeling of outburst situations and contemporaneous impact, risk mitigation measures are not sufficient, putting lives and livelihoods at recurrent threat.

The objective of this study is to examine the impacts potentially caused by a Glacial Lake Outburst Flood (GLOF) from Tsho Rolpa Glacial Lake in Nepal. By applying hydrodynamic modeling and geospatial techniques, the research attempts to simulate potential flood occurrences and assess the extent of the risk to downstream communities and infrastructure. The main objectives of the work are: to analyze the area change and dynamics of Tsho Rolpa Glacial Lake. This objective relates to analyzing the spatial development and morphological evolution of Tsho Rolpa Lake with multi-temporal satellite imagery and GIS techniques over the years. It seeks to explore how the retreat of glaciers has affected the evolution of the lake and its likely hazards. Secondly, the application of HEC-RAS 2D modeling for GLOF simulation and GIS-based hazard mapping.

9.2 Study Area

The research interest is Tsho Rolpa Glacial Lake, situated within the Rolwaling Valley of Dolakha District, in the province of Bagmati Province (Province No. 3) in eastern Nepal. It is situated within the Gaurishankar Conservation Area, a protected environment of the central Himalayas that is marked by high-mountain landscapes, snow-fed rivers, alpine flora and fauna, and active geological formations. Tsho Rolpa is located geographically at about 27.861° N latitude and 86.483° E longitude, with an elevation of about 4580 m above sea level (Fig. 9.1).

Tsho Rolpa was created through the receding of the Trakarding Glacier, which is part of the Himalayan cryosphere system. In the last several decades, as a consequence of global climate change, the glacier has been retreating abruptly, leading to the expansion of this proglacial lake. The lake has also become one of the biggest and most hazardous in Nepal, now occupying a surface area of about 1.65 km^2 with a water volume estimated to be over 90 million cubic meters. The lake is dammed by an unconsolidated moraine, which is geologically unstable and extremely prone to collapse by seismic activity, ice or rock avalanches, or excessive rainfall—all of which have the potential to cause a Glacial Lake Outburst Flood (GLOF).

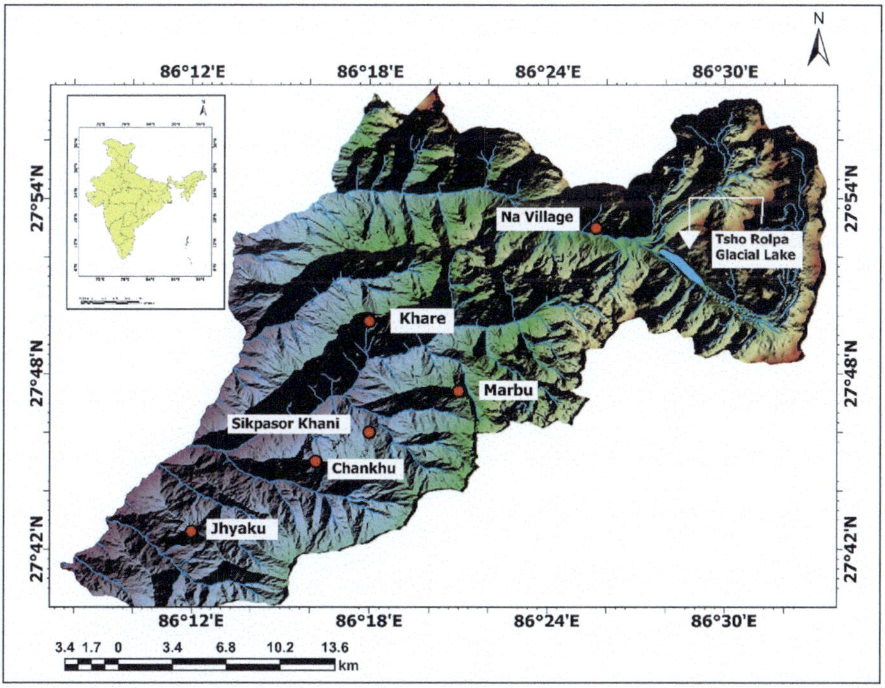

Fig. 9.1 Location map of the study area

Hydrologically, Tsho Rolpa drains to the Tsho Rolpa Khola, which drains downstream and joins the Tamakoshi River, one of the largest tributaries of the Koshi River system—a transboundary river basin that extends from Nepal to India and Bangladesh. The downstream section of the Tamakoshi valley is highly populated and has important infrastructure such as roads, footbridges, schools, and the Upper Tamakoshi Hydropower Project (UTKHEP)—Nepal's biggest hydropower development. An overnight release from Tsho Rolpa would have disastrous impacts on tens of thousands of people and billions of rupees' worth of investments in infrastructure and livelihood.

Topographically, the Rolwaling Valley is also narrow and steep and forms a natural channel for high-speed floodwaters in case of a GLOF. Na, Beding, and Dongang villages are located within the high-risk downstream area. The valley is also used as a trekking path to the Tashi Lapcha Pass, linking the Rolwaling area with the Khumbu Valley. Therefore, the region holds not only ecological but also economic and cultural value, and both local and foreign visitors are drawn there.

The 2010-established Gaurishankar Conservation Area covers the Tsho Rolpa area and contains rare Himalayan flora and fauna, such as snow leopards, red pandas, and a number of endangered birds. The conservation area is also an important habitat for yak herders and is culturally rich with Buddhist monasteries and traditional Sherpa villages scattered about the landscape. Because of its geographical remoteness, rugged terrain, and harsh climate, the area is poorly linked to infrastructure and lacks effective real-time hazard monitoring systems. The combination of rising lake volume, downstream vulnerability, and poor infrastructure for disaster preparedness renders Tsho Rolpa a priority site for scientific risk assessment and creation of early warning systems. This research thus tries to replicate the probable flood risk of a GLOF from Tsho Rolpa with digital elevation data and HEC-RAS 2D modeling protocols, combining remote sensing and GIS to create models for different breach assumptions.

9.3 Methodology

To conduct the study, the following methodology was discussed:

To complement the methodological setup, a series of reliable datasets was employed. Multi-temporal satellite imagery (Landsat 8 OLI-TIRS covering the periods 2015, 2020, and 2024) was acquired through downloading on the USGS Earth Explorer platform for hydrologic area change, glacial extent, and morphological changes assessments. The Digital Elevation Model (DEM) data, used to inform hydrodynamic modeling and terrain analysis, were obtained from Open Topography, which provides high-resolution elevation data. From this DEM, other layers like slope, aspect, drainage density, and elevation profiles were extracted for flood simulation and hazard mapping using HEC-RAS 2D. Administrative boundaries were received from the Survey of India, and LULC (Land Use Land Cover) datasets were also included to add spatial analysis. The datasets combined enabled in-depth flood modeling, vulnerability evaluation, and effective determination of high-risk zones in the area of the Tsho Rolpa Glacial Lake. Details in Table 9.1 (Fig. 9.2).

9.4 General Physiography of the Area

Tsho Rolpa Glacial Lake, found in the Rolwaling Valley of north-eastern Nepal in the Dolakha District, lies at an altitude of approximately 4580 m above sea level and is one of the biggest and most potentially dangerous glacial lakes of the Nepal Himalaya. Formed by the retreat of the Trakarding Glacier, the lake is bounded by active glacial processes, precipitous topography, and rugged terrain and is therefore highly responsive to climatic as well as geomorphological changes. The physiographic character of the region is defined by its high-altitude location, variable and steep topography, complex drainage system, and considerable change in climatic parameters such as temperature and precipitation from season to season.

To understand the natural environment and hazards of Tsho Rolpa, numerous physical parameters have been considered in this work. The slope direction factor represents how slope direction influences the rate of melting of the glacier from variances in solar radiation exposure. Drainage density represents the landscape support quality in the facilitation of runoff movement, which is crucial in glacial lake outburst cases. Patterns of elevation indicate topographic slopes controlling the direction of ice and water flow. Land use/land cover data help to assess the surface material—rock, ice, vegetation, or urban area—that controls water intake, flow, and sediment transport during runoff. Rainfall and temperature are basic climatic inputs that directly influence the melting of glaciers and the growth of lakes, especially during the summer monsoon season. Slope gradients play an important role in determining water flow velocity and erosive power, whereas stream ordering enables the ranking of the hierarchy of rivers and tributaries and depiction of the manner in which floodwater may be transported downstream.

This multiparameter physiographic assessment with support from GIS and remote sensing technologies serves as the basis for simulating potential GLOF events using hydrodynamic models such as HEC-RAS. It also assists hazard and vulnerability mapping in an effort to support early warning systems and disaster risk reduction planning for the downstream hazard-prone communities.

Aspect: The catchment area and Tsho Rolpa Glacial Lake orientation map illustrates the slope direction pattern, which is an important controlling parameter for sunshine exposure, snowmelt activity, vegetation growth, and overall hydrologic response. The map indicates a heterogeneous pattern of slopes in all cardinal and intercardinal directions. A predominance of the terrain is southeast-facing slopes (112.5–157.5) and south-facing slopes (157.5–202.5), represented by dark pink and light pink colors. The south-facing slopes typically receive more sun radiation, which has the effect of increased snow and ice melting, thereby causing increased runoff into the glacial lake.

Table 9.1 Description of the used datasets

Data sets	Description of layer	Source
Administrative Boundary of India (SHP)	Location map	Survey of India (SOI) https://surveyofindia.gov.in/
DEM (digital elevation model)	Location map	Open topography https://portal.opentopography.org/raster?opentopoID=OTSDEM.032021.4326.3
	Aspect	Open topography https://portal.opentopography.org/raster?opentopoID=OTSDEM.032021.4326.3
	Drainage density	Open topography https://portal.opentopography.org/raster?opentopoID=OTSDEM.032021.4326.3
	Elevation	Open topography https://portal.opentopography.org/raster?opentopoID=OTSDEM.032021.4326.3
	LULC (land use land cover)	Open topography https://portal.opentopography.org/raster?opentopoID=OTSDEM.032021.4326.3
	Rainfall	Open topography https://portal.opentopography.org/raster?opentopoID=OTSDEM.032021.4326.3
	Slope	Open topography https://portal.opentopography.org/raster?opentopoID=OTSDEM.032021.4326.3
	Stream ordering	Open topography https://portal.opentopography.org/raster?opentopoID=OTSDEM.032021.4326.3
	Temperature	Open topography https://portal.opentopography.org/raster?opentopoID=OTSDEM.032021.4326.3
Meteorological data	Rainfall in mm	WorldClim https://www.worldclim.org/data/worldclim21.html
	Temperature in degrees	WorldClim https://www.worldclim.org/data/worldclim21.html

(continued)

Table 9.1 (continued)

Data sets	Description of layer	Source
Landsat 8/9 OLI-TIRS (C2 L2)	Area changes of the lake	USGS earth explorer https://earthexplorer.usgs.gov/
Landsat 8/9 OLI-TIRS (C2 L2)	Symmetrical change of the lake	USGS earth explorer https://earthexplorer.usgs.gov/
River discharge data	GLOF hydrograph	GEOGIOWS hydroviewerV2.0 https://hydroviewer.geoglows.org/#=10.00&lat=18.00&zoom=3.00&definition =
	Maximum flood depth distribution	GEOGIOWS hydroviewerV2.0 https://hydroviewer.geoglows.org/#lon=10.00&lat=18.00&zoom=3.00&definition =
	Maximum flow velocity distribution	GEOGIOWS hydroviewerV2.0 https://hydroviewer.geoglows.org/#lon=10.00&lat=18.00&zoom=3.00&definition =
DEM (digital elevation model)	Flood susceptibility mapping	Open topography https://portal.opentopography.org/raster?opentopoID=OTSDEM.032021.4326.3

The northern aspects (0–22.5 and 337.5–360), in light blue and pale yellow, are not as prevalent and receive less solar exposure, possibly maintaining snow cover longer. The northeast (22.5–67.5) and northwest (292.5–337.5) slopes, indicated in purple and golden-yellow, are similarly extensive regions of the catchment and could also have intermediate melting regimes dependent on seasonal solar irradiation. The minor sections of flat terrain (color coded -1 in dark blue) occur in the vicinity of the lake region and valley bottoms, these being relatively more stable and less impacted by gravitational flow (Fig. 9.1).

The orientation trend of Khare and Mrabu settlements indicates varied aspects, suggesting microclimatic heterogeneity and varied exposure to environmental hazards. In general, the prevalence of south and southeast-facing slopes indicates a physiographic setting conducive to active glacial melting and concentration of runoff towards the Tsho Rolpa Lake. The spatial distribution of this aspect is therefore an important factor in determining the hydrological regime and potential glacial lake outburst flood (GLOF) hazards in the area (Fig. 9.3).

Drainage Density: The drainage density map of the Tsho Rolpa catchment is crucial in obtaining information about the concentration and distribution of stream channels in the region. Drainage density is a fundamental physiographic factor that impacts surface runoff, infiltration capacity, and flood hazards. The map indicates a spatial variation in drainage density ranging from 0.001 km/km^2 to more than 300 km/km^2, grouped into five different classes. Higher drainage densities are also prominently seen in the NE section in the area around Tsho Rolpa Glacial Lake, highlighted

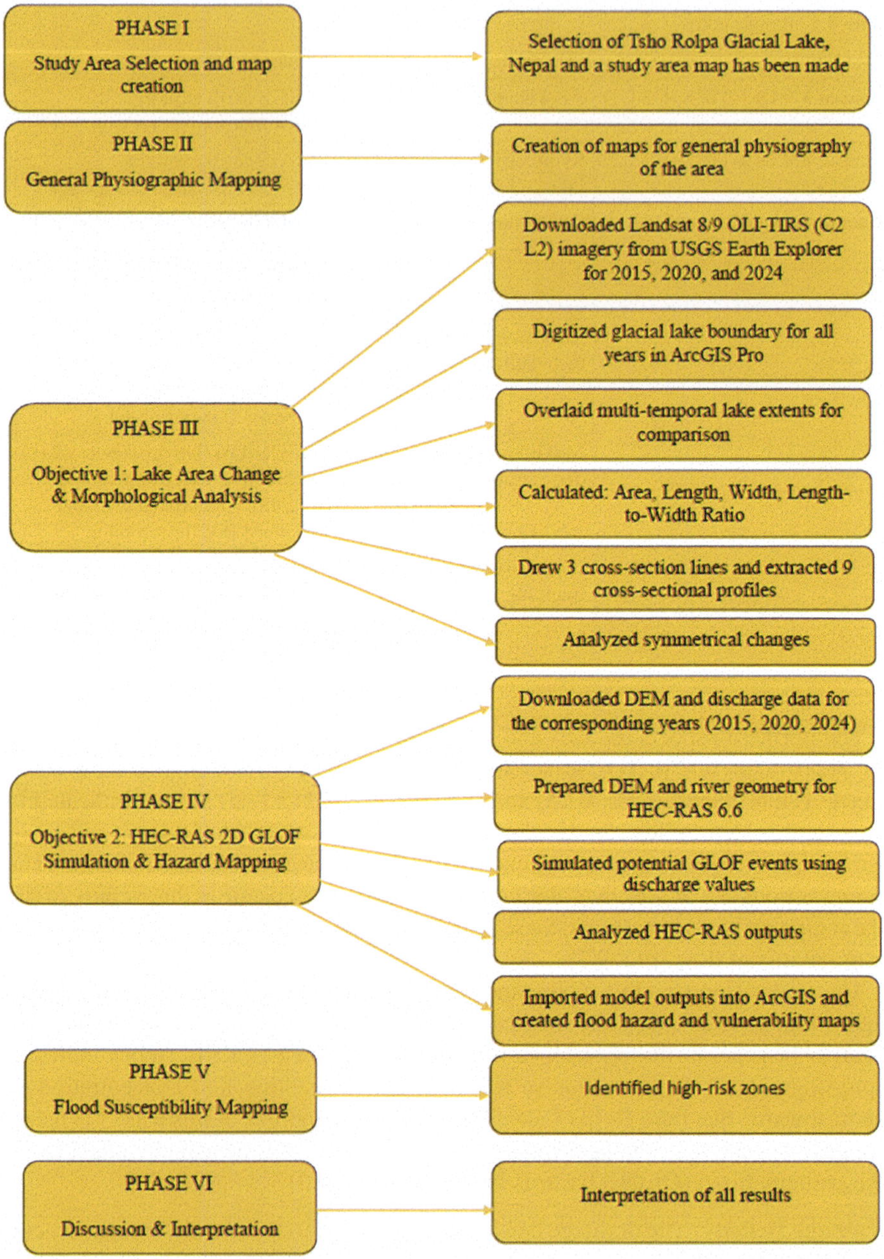

Fig. 9.2 Methodology shown in the flowchart

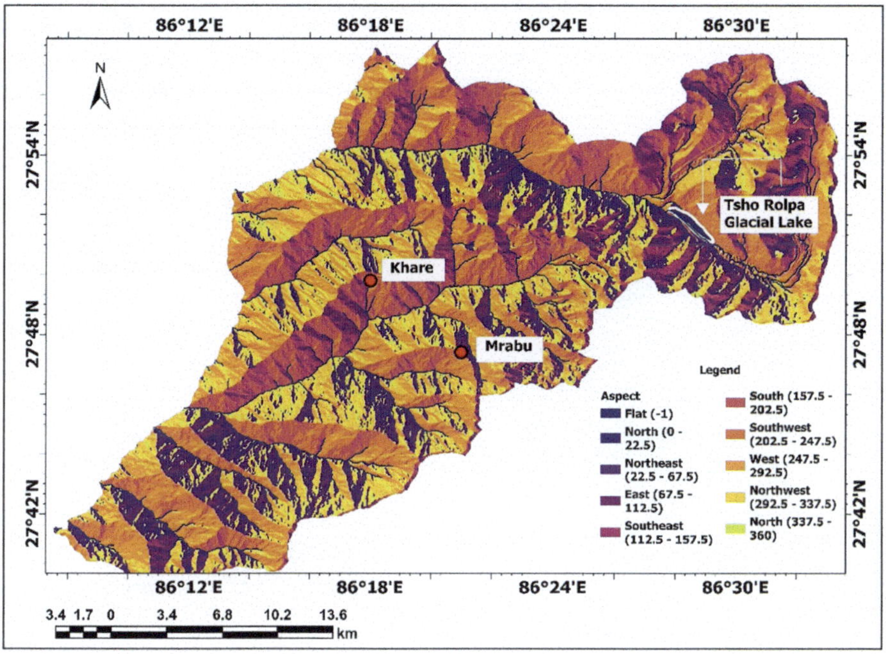

Fig. 9.3 Aspect map of the study area showing the slope direction

by dark violet colors (240.355–300.442 km/km^2). This reveals a highly dissected topography with a high density of stream channels, implying larger surface runoff propensity and higher sensitivity to erosion and flood hazards.

Conversely, the south and southwest regions of the basin, notably near Khare, have lower drainage densities (0.001–60.088 km/km^2) represented by light shades of violet, indicating rather smooth terrain and lower surface runoff. The Na Village region suggests a medium drainage density, reflecting its medium topographic complexity and hazard to high-rainfall or glacial melt occurrences. These drainage density changes reflect the conditions of underlying land cover, slope, and lithology, with lower permeability and steeper surfaces typically covering larger drainage densities (Fig. 9.4).

The density of streamlines converging close to the lake basin is also indicative that the Tsho Rolpa Glacial Lake acts as a drainage sink that receives high runoff from neighboring sub-basins. This hydrological context highlights the susceptibility of downstream settlements to GLOF (Glacial Lake Outburst Flood) incidents, as high drainage densities can enhance the propagation of flood waves. Therefore, the distribution of drainage density becomes a vital factor in the determination of both catchment hydrodynamic processes and the strategy for hazard assessments due to climate.

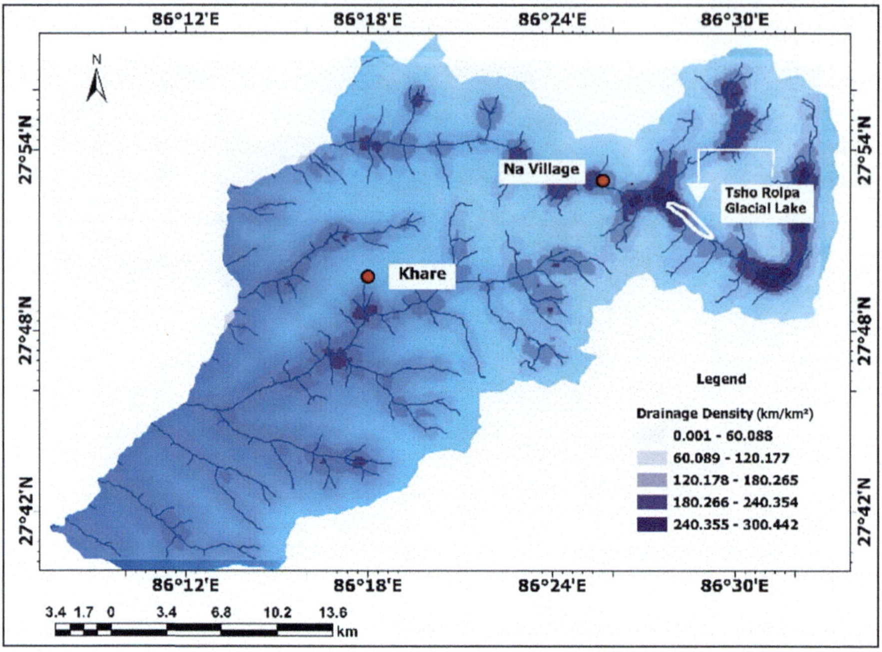

Fig. 9.4 Drainage density map

Elevation: The Tsho Rolpa catchment elevation contour map readily identifies the extremely variable and precipitous nature of the catchment area. From approximately 1000 m to over 6500 m above mean sea level, elevation varies with closely spaced contours hinting at steep gradients (Fig. 9.5).

The region to the east, where Tsho Rolpa Glacial Lake lies, has extremely high elevations over 6000 m, reflecting the high-altitude Himalayan range and white-capped summits. Such areas are painted in dark gray and dark brown tones, which symbolize low vegetation and permanent glacial cover. The middle region adjacent to Na Village falls in the 5000–5500 m elevation band, which represents an intermediate elevation zone that is still under alpine conditions.

Shifting toward the south-western portion, particularly in and around Sikpasor Khani, the elevation reduces dramatically to less than 2000 m, depicted in light blue and green shades, depicting fairly lower-altitude valleys and potentially more habitable areas. The pattern of contours reveals a largely northeast to southwest inclining terrain that is adding to the downstream flow of drainage towards susceptible settlements. The high relief in the upper catchment, defined by densely aggregated contours, has a strong effect on runoff speed and plays a central role in modeling the glacial lake outburst flood (GLOF). Hence, this relief profile is highly important to comprehend the vertical development of the basin, hydrological link, and geomorphic vulnerability of the area.

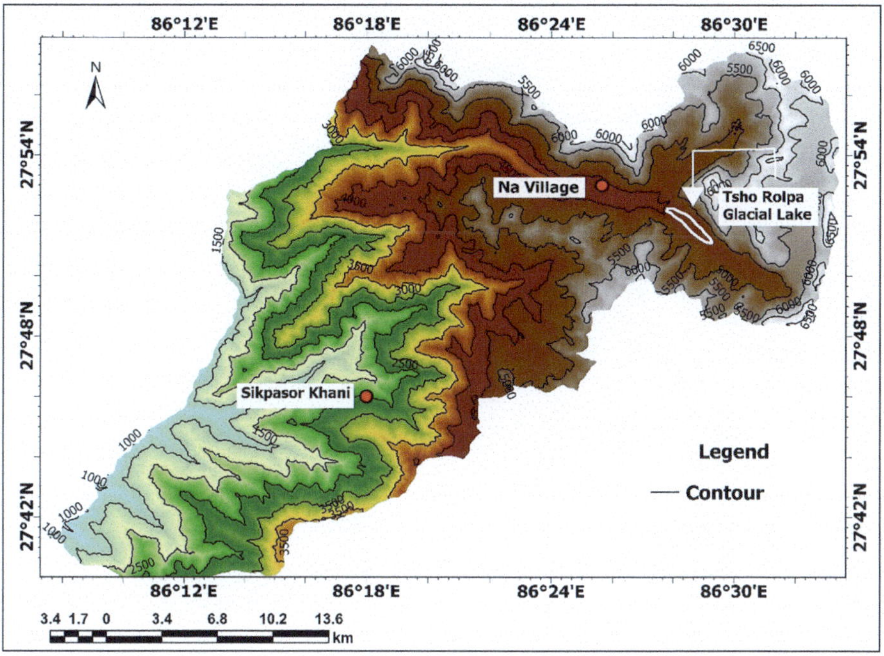

Fig. 9.5 Elevation map

Land use land cover (LULC): The Land Use Land Cover (LULC) map of the terrain around the Tsho Rolpa glacial lake is typical of the complex topography and biodiversity of the high Himalaya. The top-left corner of the map consists wholly of glacier and glacial meltwater landforms, with the Tsho Rolpa Glacial Lake as a dominant hydro-geomorphic feature. These regions point towards the origin of meltwater discharge and are extremely responsive to climatic fluctuations, especially glacial recession and melting due to rising temperatures.

Vegetation is primarily spread in the mid-altitudinal belts and within the river valleys, where the climatic and topographic conditions are superior. The green belt areas indicate the occurrence of alpine and sub-alpine vegetation, which is important in stabilizing the slopes and in minimizing erosion hazards. Hill sections (marked in brown) preponderate over the steeper and rougher topography, which are typically unsuitable for agriculture and habitation but make a major contribution to the region's overall hydrology and slope dynamics (Fig. 9.6).

Human habitations like Na Village, Khare, Marbu, and Jhyaku are found mainly in accessible areas close to water sources and on comparatively gentle terrain. Their proximity in space to the glacial lake and meltwater areas further exposes them to risk in the case of a glacial lake outburst flood (GLOF). The settlements rely predominantly on glacier streams for water resources and agriculture, and thus the identification of land cover types becomes crucial for hazard risk mitigation and sustainable development planning.

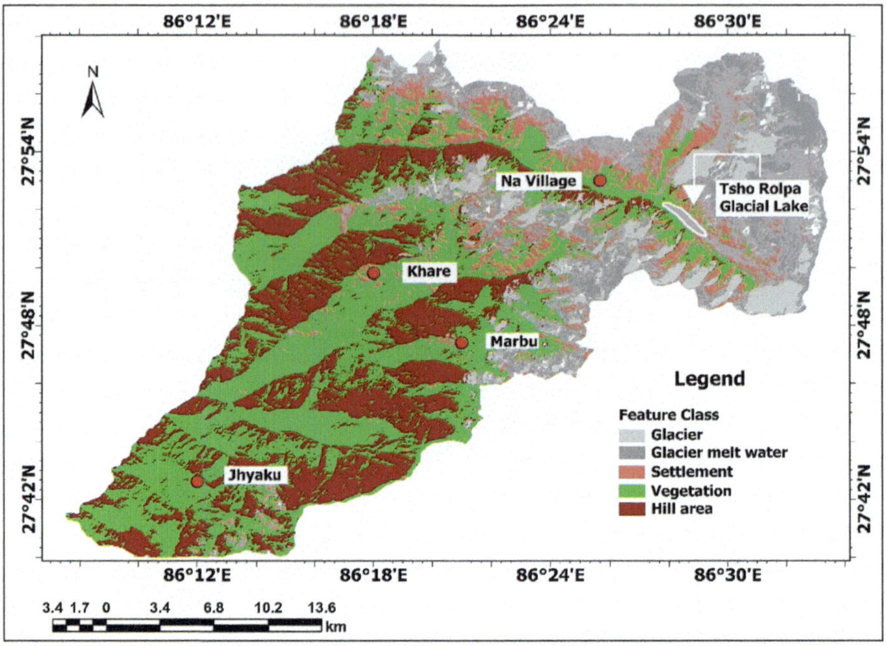

Fig. 9.6 Land use/land cover map of the study area

The spatial distribution of the LULC categories offers valuable information about the physiographic conditions of the study area and facilitates the recognition of areas of ecological importance as well as GLOF vulnerability. The maps are of prime importance in formulating adaptive land management policies, adaptation policy implementation towards climate change, and planning mitigation activities for the future.

Rainfall: The map of rainfall distribution in the Tsho Rolpa watershed shows high spatial heterogeneity in precipitation, divided into five classes with a variation from 60.495 mm to 552.48 mm. The heterogeneity is essential in the study of hydrological processes and flood vulnerability of the area.

The eastern and north-eastern regions, particularly the region around the Tsho Rolpa Glacial Lake, fall into the lowest rainfall classes, with a precipitation value ranging from 60.495 mm to 237.995 mm. They are the areas represented by pale blue colors that lie in higher elevations and are subjected to cold and dry climatic conditions more. Because of their location near glacial sources and snow-intensive precipitation, rainfall here is usually light, and snowfall could play a greater role in water storage than rain.

As we proceed towards the central and south-western parts of the watershed, there is an increased intensity of rainfall, with the maximum class of rainfall (475.306 mm to 552.48 mm) found in the south-western areas, especially around the settlements of Chankhu and Sikpasor Khani. These areas are at lower levels and are more directly under the monsoonal winds bearing moisture, hence receive greater orographic precipitation. The continuous progression from light blue to dark blue shades along the map shows a rainfall gradient from northeast to southwest.

This rain spatial distribution has straightforward consequences for hydrological modelization, the generation of runoff, and flood hazard mapping. The upper rainfall areas are at greater risk of surface runoff, saturation, and potential flash flooding, particularly when coupled with the contributions of meltwater from glacial catchments. Alternatively, low-rainfall areas around the glacial lake may receive less direct rainfall but remain significant since they pose a possible GLOF (Glacier Lake Outburst Flood) hazard.

Briefly, the rainfall pattern in the region of Tsho Rolpa is a controlling factor in flood exposure vulnerability downstream, erosion risk, and landslide risk, especially in inhabited areas. This map is a crucial layer for area-wide integrated hazard evaluation (Fig. 9.7).

Slope: The slope map of the Tsho Rolpa watershed indicates the change in steepness of terrain over the area, divided into five slopes ranging from 0.001 to 76.802°. This categorization is crucial for the determination of surface runoff behavior, landslide susceptibility, and GLOF risk transmission.

The gentle slope areas (0.001–15.36°), symbolized by dark green, occur largely in the lower south-western end of the basin, particularly surrounding Jhyaku and

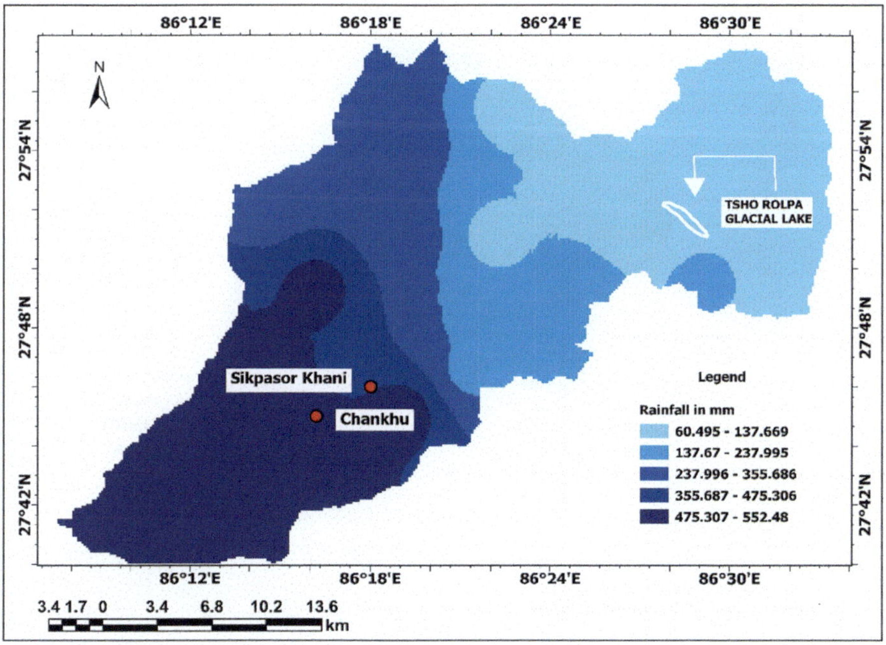

Fig. 9.7 Rainfall distribution map of the area

portions of the Chankhu areas. These gentle slopes are generally more favorable for habitation and farming, but also serve as floodplains where water can collect during periods of high discharge. These regions are, in turn, more prone to being flooded by inundation in case of a GLOF due to their flat topography and lower runoff speed.

Going north and east, particularly towards the middle and upper catchment areas, progressively increasing slope steepness is evident. Most of the central area is dominated by the moderate slope classes (15.361–46.081°), presented in light green to yellow. These middle-slope areas have a tendency for moderate erosion and enhanced surface runoff, particularly under intense precipitation or snowmelt, which is responsible for most of the drainage network development exhibited on the map.

The highest slopes (46.082–76.802°), colored orange to red, are found in the northeast part that encircles the Tsho Rolpa Glacial Lake. High-slope areas are represented by treacherous mountainous slopes on which glacial processes, rockfalls, and landslides occur more often. The steep slopes in this region promote quick downstream flow of glacial outburst floods, raising the hazard risk for locations further down the drainage route (Fig. 9.8).

Notably, the high slope along the lake and downstream gentle slopes set a topographic condition that favors high-velocity flow initiation coupled with water spreading in flatter areas, which needs to be taken into account while modeling GLOF

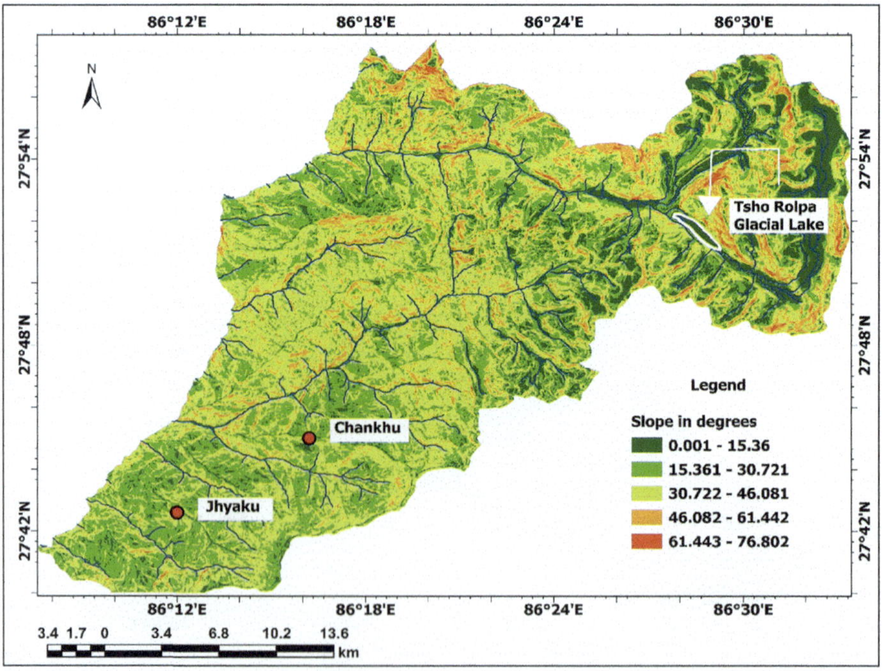

Fig. 9.8 Slope map of the study area

impacts. In addition, the slope pattern has a direct bearing on sediment transport, incision of channels, and the coverage of possible flood zones.

In summary, this slope map is an important input to hydrological modeling, flood hazard evaluation, and land use planning. It is evident from the map that although the top watershed is controlled by risky steep terrain, the lower reaches are more susceptible to overtopping, and thus it is important to devise measures of mitigation as such.

Stream ordering: The Tsho Rolpa watershed stream network has been ordered using the Strahler stream ordering system, which orders the drainage hierarchy based on tributary confluence. The streams have been assigned five orders from the first-order (blue) to the fifth-order (red), and each is colored in the map. First-order streams indicated in blue are the densest and are predominantly found in the uppermost part and the outermost margins of the basin. These are the youngest and smallest channels, normally developed from snowmelt, surface runoff, or seepage in high-altitude areas. They start near the Tsho Rolpa Glacial Lake and in the nearby steep slopes, suggesting active headwater development. These first-order streams channel their waters into the higher-order streams by aggregating water from localized catchments. Second- and third-order streams (purple and orange) are created where two first- or second-order streams meet. These are more pronounced in the middle section of the basin, between Tsho Rolpa and the lower regions such as Marbu. These are more voluminous, mean channel depth, and begin to show recognizable drainage channels. They indicate the prevailing tributaries that carry water and sediment from the upper glacial area to the middle and lower catchments. The fourth- and fifth-order streams (pink and red) are the basin's dominant trunk channels. These lie in the south-western sections of the watershed, downstream from Marbu and Jhyaku, where the landscape becomes wide, and the streams are well-interlinked. The fifth-order stream colored red is the main drainage route, which carries peak discharges from a high-rainfall or GLOF event. It most probably follows the principal river channel from the glacial source to the major river systems in the downstream direction. The hierarchical rise in stream order from east (Tsho Rolpa Lake) to southwest (Jhyaku) reflects natural flow accumulation and development of drainage. The pattern indicates a dendritic drainage system, characteristic of homogeneous terrain, in which the stream pattern is predominantly regulated by the underlying topography and not by lithology or faults. From the standpoint of a hazard assessment, the lower-order streams are more susceptible to flash flooding, particularly if there is a glacial lake outburst, while the higher-order streams play a critical role in defining the inundation path and velocity. The stream order map, thus, is critical in hydrological modeling, flood routing, and hazard zonation (Fig. 9.9).

In general, the stream order distribution reflects the hydrological maturity and drainage competence of the Tsho Rolpa watershed. The distribution provides a structural basis for further analysis of runoff production, sediment movement, and flood routing, especially in terms of GLOF hazard assessment.

Temperature: The temperature map of the Tsho Rolpa drainage has a definite spatial pattern, which is strongly regulated by altitude, topography, and exposure to solar

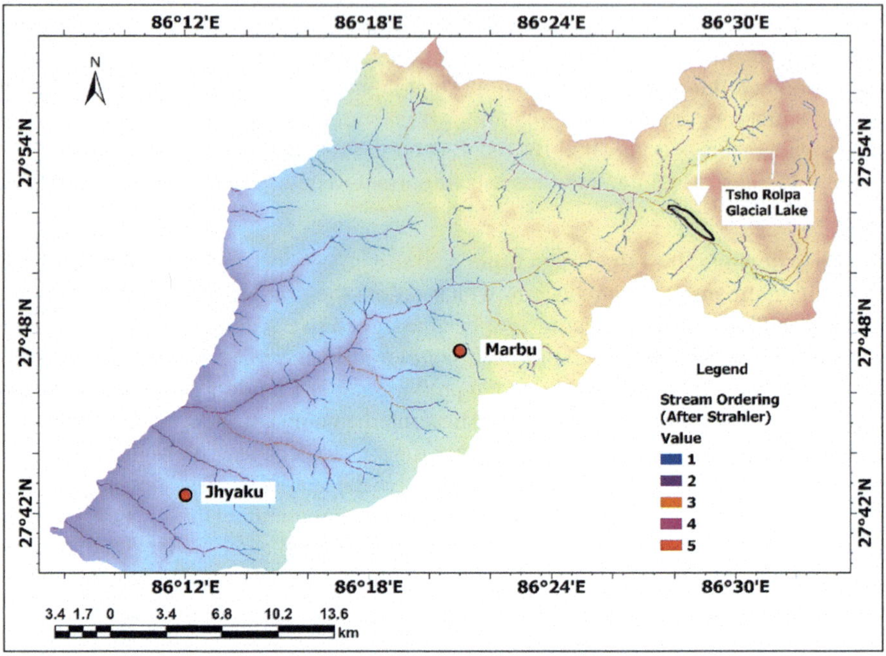

Fig. 9.9 Stream ordering map after the Strahler method

radiation. The range of the temperature values, interpolated from the dataset, ranges from 4.88 °C as the lowest value to 21.81 °C as the highest value, and can be classified into five definite temperature zones.

The north-eastern and eastern parts of the basin, including Tsho Rolpa Glacial Lake and Na Village, fall in the lowest range of temperature (4.876–8.129 °C). These areas are situated at higher altitudes of above 4500 m, where year-round snow cover and glacial influence result in lower heat conditions. The lower temperatures in these places are a key factor in glacial mass balance, stability of permafrost, and GLOF vulnerability (Fig. 9.10).

In the west and southwest directions, the temperature increases steadily. Marbu village lies in the intermediate zone (12.182–16.166 °C), which is the mid-elevation climatic belt. The transition zone possesses moderate temperatures with limited vegetation and seasonal melting of snow. Greater heat also means greater capacity for rain-fed hydrological inflows, which might contribute to downstream river runoff and flood water volume during hotter times.

The southern and south-western part of the basin exhibits the maximum temperature range (19.089–21.811 °C). They are situated at low levels, i.e., approximately 2000–2500 m, and have warmer climatic conditions year-round. The increased temperature raises evapotranspiration rates, agricultural productivity, and also leads

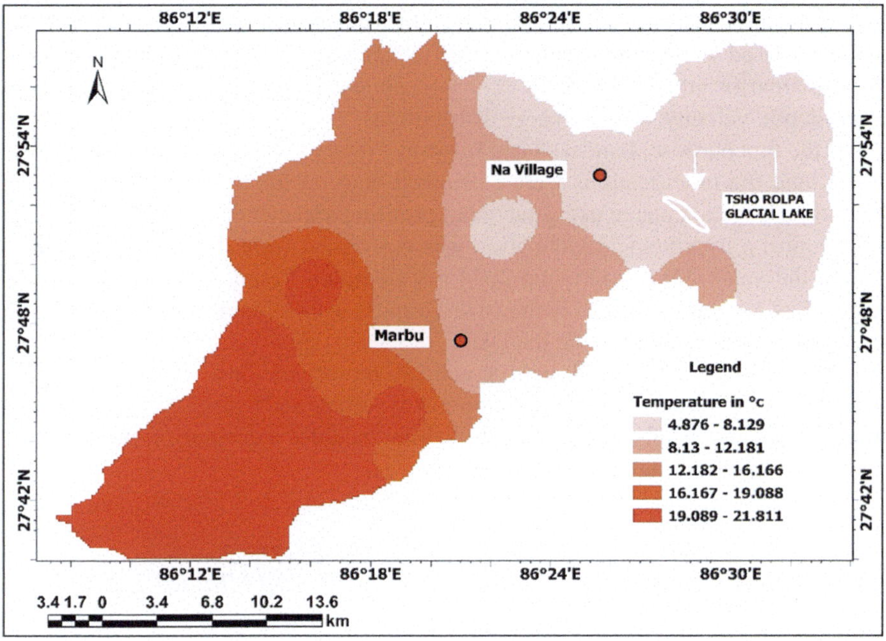

Fig. 9.10 Temperature distribution map

to rapid snowmelt and glacial retreat, which indirectly affects the hydrology and hazard profile of the basin.

The temperature gradient from northeast (Tsho Rolpa Lake) to southwest (lower basin) indicates altitudinal zonation of climate, a characteristic common in the mountain habitats of the Himalayas. This spatial differentiation is important for hydrological processes because glacial melting, the type of precipitation, and the generation of runoff are all temperature-dependent.

For the purpose of GLOF hazard modeling, the lower-temperature areas represent glacial accumulation and stability regions, while the higher-temperature areas are responsible for increased meltwater formation, which can enhance the volume and hazard of lake outburst events. The temperature map, therefore, represents an important layer for analysis of climate-hydrology interaction and facilitating the overall evaluation of flood vulnerability and hazard mapping in the Tsho Rolpa watershed.

9.5 Results

This section presents the outcomes of research on glacial lake evolution and hydrodynamic flood simulation of the Tsho Rolpa glacial lake catchment area. The results are organized in accordance with the two principal objectives of this research: (1) to investigate the temporal evolution of glacial lake coverage during years 2015, 2020, and 2024 based on satellite remote sensing techniques, and (2) to simulate potential GLOF scenarios under HEC-RAS 2D based on discharge input and terrain data to assess depth, velocity, and discharge distributions.

For the first purpose, Landsat 8 and Landsat 9 images were used to define the lake shores, and the modifications were calculated both on the basis of spatial growth and morphological change using sectional profiling and symmetry analysis. For the second aim, a hydrological simulation was performed from discharge data representing the years 2015, 2020, and 2024 key hydraulic results, such as peak flood depth, flow velocity, and discharge hydrographs, were obtained to evaluate flood propagation behavior following the lake.

Results are supplemented with graphical representations, maps, and cross-sectional profiles to show the spatio-temporal dynamics and hydrological response of the landscape. Each subsection of the following includes a qualitative interpretation of observed patterns and model outputs.

9.5.1 Spatio-Temporal Changes in Tsho Rolpa Glacial Lake

This figure consists of a sequence of images illustrating the advancing areal growth of the Tsho Rolpa Glacial Lake during the years 2015, 2020, and 2024 obtained from multi-temporal Landsat images. The year-by-year lake boundaries are also shown in another blue color—light blue for 2015, medium blue for 2020, and dark blue for 2024. The map also shows three cross-sectional transects (A–A1, B–B1, and C–C1) perpendicular along the longitudinal axis of the lake to visualize the symmetrical lake lateral growth over time.

9.5.2 General Expansion Pattern

The map clearly shows an even increase in the surface area of lakes during the last decade, which is primarily due to the increased ablation of the parent glacier resulting from higher air temperatures and mixed precipitation. Spatial pattern of growth is observed in the clear elongation towards the southeast, which reflects the vulnerability of the terminal moraine dam area (Fig. 9.11).

9.5.3 Upstream Zone (A–A1)

The northern end of the lake shows moderate lateral growth during 2015 and 2020, and trivial additional change from 2020 to 2024. The slow rate of growth in this area suggests that the head of the glacier is receding at a slower rate, or valley wall

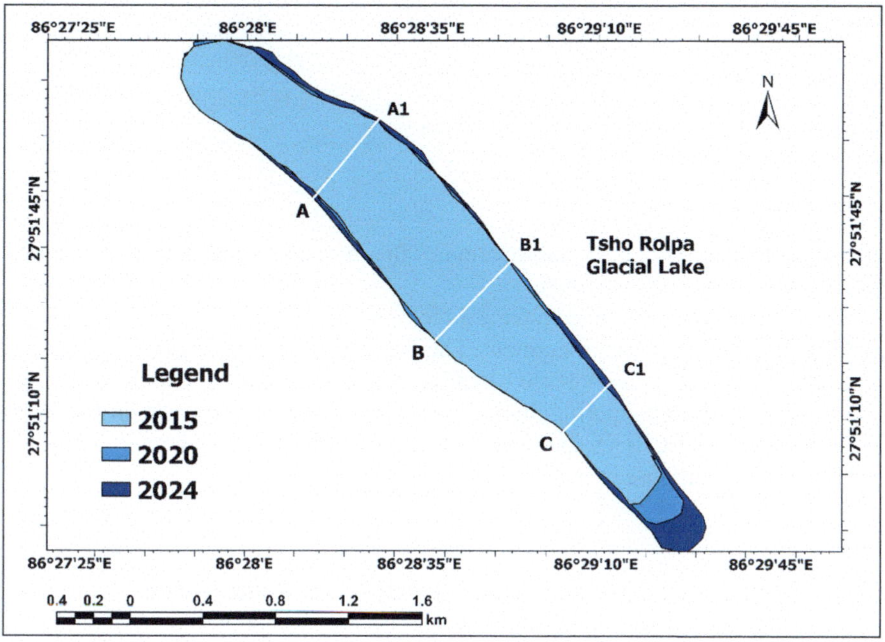

Fig. 9.11 Glacial lake area change and cross-section locations

confinement suppresses growth. This zone remains relatively stable with respect to hazard potential, but supplies meltwater consistently.

9.5.4 Middle Zone (B–B1)

The central section of the lake demonstrates symmetrical expansion on each eastern and western flank. The uniform broadening between years demonstrates balanced glacial retreat and testifies that this section of the lake is open to either side in a structural sense. This might also represent areas where ice-contact slopes or moraine ridges melt evenly, producing more uniform, predictable expansion.

9.5.5 Downstream Zone (C–C1)

The southern downstream region shows the most visible and striking changes, with the lake growing significantly between 2020 and 2024. In this region, marked by the moraine dam, there is intense outward growth in the direction of the southeast. This trend is alarming because the region is the potential GLOF breach location.

Table 9.2 Temporal variation in glacial lake dimensions

Year	Length (m)	Width (m)	Length/width ratio (L/W)	Area (km^2)
2015	3506.03	660.74	5.31	1.60
2020	3723.12	664.24	5.61	1.65
2024	3891.50	678.94	5.73	1.78

Any further lake growth or destabilization of the moraine in the region can directly increase flood hazard downstream.

The areal changes evidenced have a direct bearing on glacial lake outburst flood (GLOF) modeling and risk assessment. The constant expansion, especially towards the moraine dam, reflects increased hydrostatic pressure and potential instability. Such time-series morphometric change deserves further simulation with hydrodynamic modeling (HEC-RAS 2D) and justifies the importance of vulnerability and hazard zonation mapping in the downstream river basin (Table 9.2).

9.5.6 Temporal Cross-Sectional Analysis of Tsho Rolpa Glacial Lake

The cross-sectional profiles at the A–A1 transect, representing the upstream (northern) part of the Tsho Rolpa Glacial Lake, exhibit considerable morphological differences for the years 2015, 2020, and 2024. The 2015 profile presents a relatively narrow lake basin with a total width of approximately 644 m and an elevation of between 4550 and 4566 m.

The left flank, equaling the terminus of the glacier, is steep, while the right flank is more sloping, indicative of a shared proglacial lake shape confined by lateral moraines. By 2020, the lake had expanded to about 681 m—a 37-m increase. The elevation range also rises to 4574 m, which means that the sides' slopes have become steeper and the depth variation has increased. This represents progressive lateral erosion and ablation at the glacier's margin, which have contributed to widening and partial deepening of the lake basin. In 2024, the profile indicates an additional widening of lake width to 725 m, with an increased elevation range to 4587 m. The two sides of the slopes look steeper and more symmetric, in accordance with increased retreat of the surrounding moraines and possible destabilization resulting from ongoing glacial melting 81 m of width gain between 2015 and 2024 clearly reflects the increase of the glacial lake over the years, dominated by climatic forces of glacier retreat. A–A1 cross-section development in general signifies a progressive and cumulative development of the upstream lake basin, which is likely to become the source of future GLOF threats if the process continues.

B–B1 cross-sectional profiles indicate the center part of the Tsho Rolpa Glacial Lake and reflect incremental but subtle morphological changes between years. Along this transect in 2015, the cross-section of the lake has a combined width of 616 m between an elevation of 4550 m and 4564 m. The left slope, which is likely that of the glacial slope, has a maximum gradient of − 17.5%, whereas the right slope is considerably steeper, giving an asymmetrical lake shape. By 2020, the width just increases to 632 m with the elevation range increasing up to 4576 m. The steepness of the right bank continues to prevail with a maximum slope of − 14.3%, while the average slope on either side has increased slightly. This is an expression of lateral erosion and progradation of the lake towards the moraine-dammed right bank, perhaps by continuous melting and runoff. In 2024, the cross-section becomes a bit narrower at 619 m, but the elevation remains 4571 m. The pattern of the slope doesn't change much from that in 2020, but the intact steep right-side slope is evidence that the middle section of the lake equilibrated to a more mature morphology. While there was a slight narrowing in horizontal extent during 2020, the net deepening that can be seen in overall elevation gain/loss measurements is an indication of ongoing glacial retreat and sediment movement. Overall, the B–B1 cross-sectional profiles reveal an ongoing trend of widening and deepening of the central lake zone from 2015 to 2020, and a muted, yet next, stabilization in 2024, likely balanced by equilibrium being approximated within the central lake basin. These trends reinforce the role played by climatic and geomorphic processes over time.

Comparative cross-sectional profiles across the C–C1 transect of the Tsho Rolpa glacial lake in 2015, 2020, and 2024. The profiles indicate extensive morphological variation over the observed period. The cross-section of 2015 reveals approximately 348 m wide lake and elevation values ranging from 4550 to 4563 m, indicating moderate slope and depth features. By 2020, while the width slightly drops to 310 m, the profile is more even, with heights ranging between 4550 and 4557 m, indicating minimal deposition of sediments or unchanged water levels. By 2024, a major widening of the width to 369 m and the range of heights from 4550 to 4577 m is evident. This steepening and widening of the edges of the profile indicate greater glacial melt, resulting in more water accumulation and additional expansion of the lake laterally. The steeper slopes also seen in 2024 (max slope 62%) indicate greater erosion at the lake boundaries.

9.5.6.1 Symmetrical Expansion Pattern of Tsho Rolpa Glacial Lake

The comparison of periods 2015–2020 and 2020–2024 for Tsho Rolpa Glacial Lake shows a regular and symmetrical pattern of growth, especially along the east–west lake boundaries. For both time periods, the growth seems to be almost equal on both sides, indicating that the lake is expanding evenly and not expanding in one direction. The alterations are incremental and are not intermittent or sudden, but harmonize with the natural profile of the valley where the lake occurs.

Between the 2015–2020 interval, the lake experienced moderate growth, that is, observed at the southernmost point, which may be due to melting glaciers and ice mass melting in the snout vicinity. During the 2020–2024 interval, the expansion is stronger, with the expansion observable noticeably near the northernmost and the southernmost extremes. This shows a sustained rise in meltwater supply, which can be caused by increased temperature or increased glacial melting in the last years Tables 9.3 and 9.4 (Fig. 9.12).

Symmetrical nature of such changes is symptomatic of the fact that the lake basin itself is opening up in a structurally stable setting, wherein both lateral moraines are eroded over time. Also, the coincidence of lake boundaries newly formed with older ones is indicative of the fact that there is no topographic bar of considerable magnitude to hinder its extension, and the lake is advancing in a steady step-by-step manner along its axis. Such a symmetric development also suggests the prospect of future volume and risk potential increase, as wider lakes hold greater water capacity and apply more pressure on moraine dams.

Table 9.3 Symmetrical area change table between 2015 and 2020

Row	Lake feature (2015)	Lake feature (2020)	Changed area (km^2)	Interpretation
0	Present	Absent	0.03011	Lake is lost—existed in 2015 but not in 2020. Indicates lake shrinkage or seasonal water retreat
1	Absent	Present	0.078561	Lake area gained—newly inundated area in 2020 not present in 2015. Suggests lake expansion due to glacial melt or increased water input

Table 9.4 Symmetrical area change table between 2020 and 2024

Row	Lake feature (2015)	Lake feature (2020)	Changed area (km^2)	Interpretation
0	Present	Absent	0.029273	Lake is lost—existed in 2020 but not in 2024. Indicate water retreat or lake shrinkage
1	Absent	Present	0.162065	Lake is gained—newly inundated area in 2024, but not present in 2020. Suggests expansion due to glacial melt

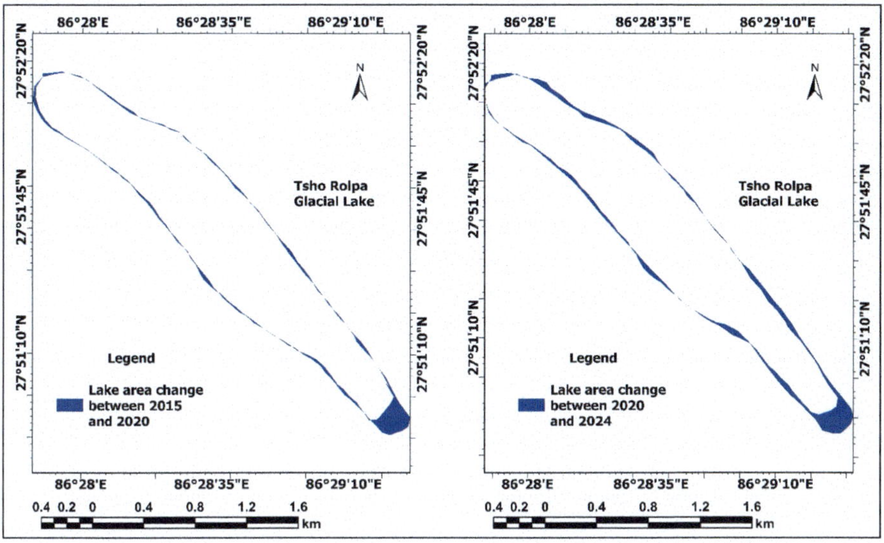

Fig. 9.12 Symmetrical expansion of glacial lake boundaries (2015–2020 and 2020–2024)

9.5.6.2 Flow Hydrograph Analysis of GLOF Events

It presents simulated flow hydrographs of the years 2015, 2020, and 2024 derived from HEC-RAS flood model results. Hydrographs are temporal variations in downstream discharge of flood at Tsho Rolpa Glacial Lake, and they illustrate intensity, duration, and peak flow pattern of each year's model flood event (Fig. 9.13).

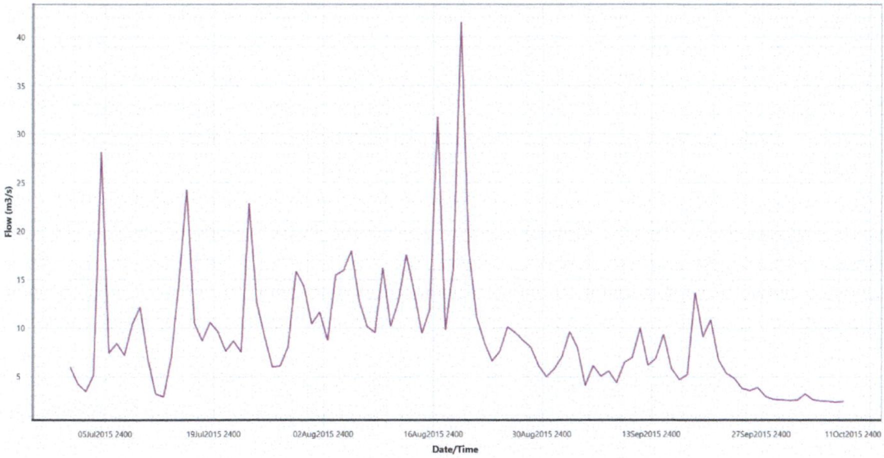

Fig. 9.13 Simulated GLOF hydrograph for the year 2015

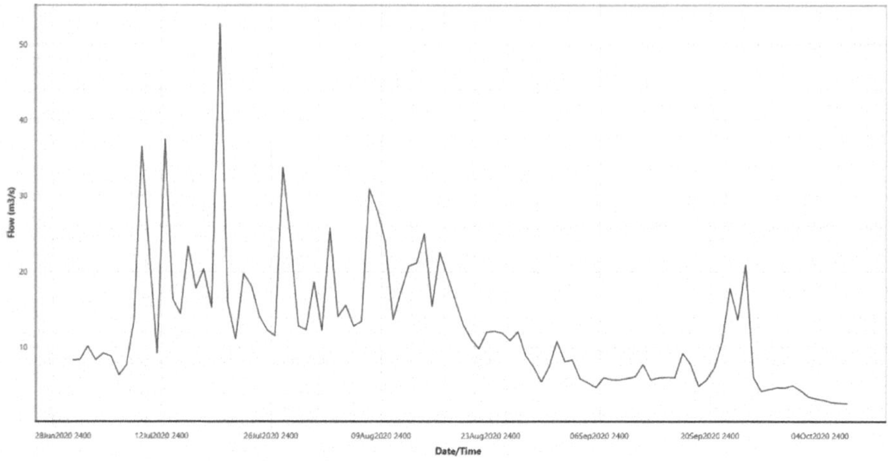

Fig. 9.14 Simulated GLOF hydrograph for the year 2020

During 2015, the hydrograph indicates several moderate to high-flow events, and discharge values are typically within the range of 4 m^3/s to 40 m^3/s. The flow peaks in mid-August, which is a large flood pulse that may be suspected to be related to glacial melt as well as monsoonal forcing. From the peak, the flow reduces gradually, and the minor oscillations persist up to September and early October (Fig. 9.14).

By 2020, the hydrograph reflects a rising and active pattern in the flood. The apex of the flow becomes more pointed and frequent, and discharge values exceed 50 m^3/s at some points. This reflects amplified flood activity, possibly due to increased volume in the lake, increased glacier melting, or weakening of the moraine with time. High-flow duration is also prolonged, reflecting increased flood duration (Fig. 9.15).

In 2024, the hydrograph once more has more energetic flood behavior with several peaks in the hydrograph and a final peak towards the end of the modeled period. The values of discharge continue to be higher than in 2015, frequently between 10 and 45 m^3/s, with greater variability and energy to flow. This reflects a growing hazard risk downstream as the lake and catchment react to continued climatic and geomorphic change.

Generally, the hydrograph plot exhibits a progressively deteriorating trend of flood frequency during the decade. The increasing value of peak discharge and extended high-flow durations reflects deteriorating risk of GLOF hazards to downstream society and infrastructure, requiring anticipatory hazard mapping and timely warning.

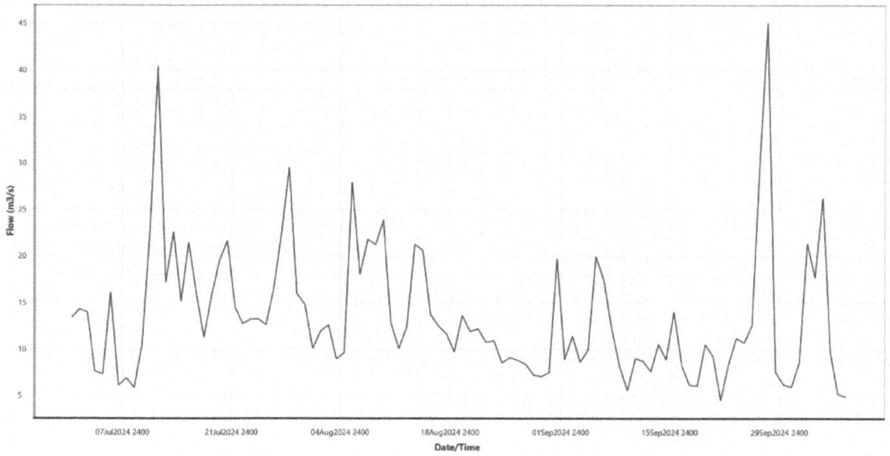

Fig. 9.15 Simulated GLOF hydrograph for the year 2024

9.5.7 Flood Depth Distribution from HEC-RAS Simulations

For assessing the potential future hazard generated by Tsho Rolpa's Glacial Lake Outburst Flood (GLOF), flood depth maps were modeled using HEC-RAS 2D for the years 2015, 2020, and 2024. The maps show spatial heterogeneity in the flood-water inundation within the downstream catchment, particularly in river corridors and habitation areas (Fig. 9.16).

In 2015, the simulated flood depth ranged from 0.003 m to a peak of 15 m, with the least flooding occurring in areas where the primary drainage streams slope downwards from the Tsho Rolpa Glacial Lake. Flood impact was minor in most sites, particularly in the uppermost basin, where the topography channels the initial wave efficiently. Villages such as Chankhu, Sikpasor Khani, and Jhyaku experienced rather shallow flooding. Places like Na Village and Marbu recorded moderate flooding, reflecting their sensitivity at the midstream.

By 2020, a significant increase in both the severity and extent of the flooding is expected. The area and depth of the flooding expanded, ranging from 0.007 to 15 m. A denser concentration of zones with high flood depth was evident, especially around the midstream regions. This is a 17–25% increment over 2015, which can be an outcome of the increase in lake volume and glacier retreat processes. This is a sign of increasing levels of severity of GLOF hazard, which may be induced by increasing amounts of the lake and changing glacial conditions.

By 2024, the flooding intensifies in magnitude, and intensification is also forecast by the model to increase further, with depths now extending more than 13 m in large downstream areas. The eastern villages, particularly Marbu and Khare, are also noted to be more impacted in 2024 as opposed to other years, possibly an increase and a downstream shift in flood-affected locations. This reflects an increase in high-risk flood-risk areas of 35–40% over 2015 and 15–20% over 2020. This indicates a clear

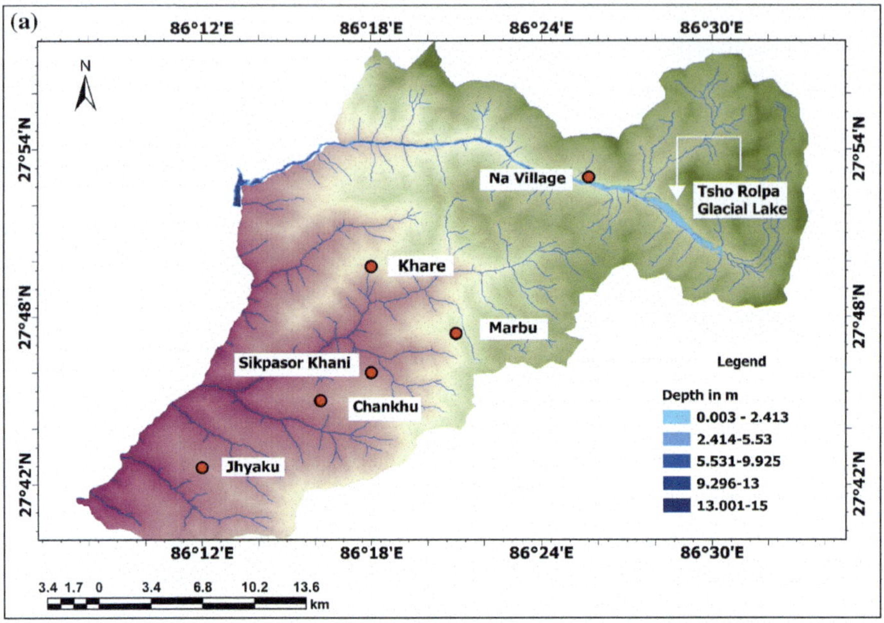

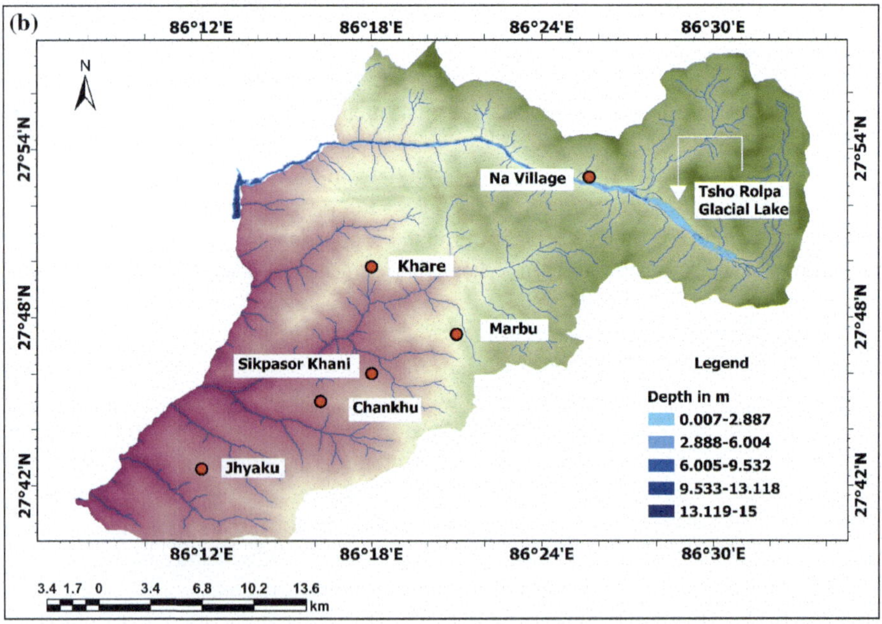

Fig. 9.16 **a** Maximum flood depth distribution—2015 GLOF scenario. **b** Maximum flood depth distribution—2020 GLOF scenario. **c** Maximum flood depth distribution—2024 GLOF scenario

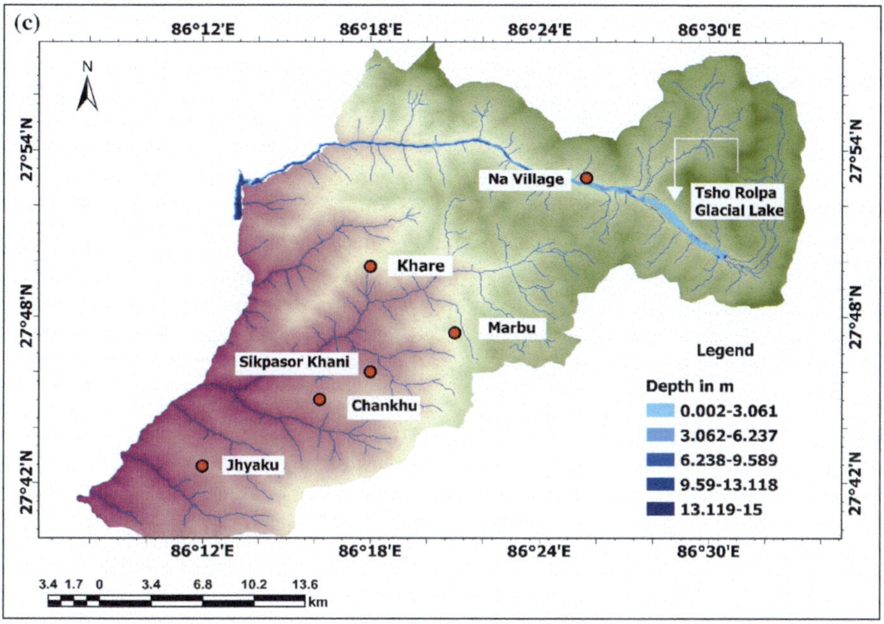

Fig. 9.16 (continued)

downstream shift of the hazard envelope, consistent with lake expansion patterns and changing glacial conditions. This is evidently apparent with documented lake expansion shifts and is consistent with the hypothesis of progressive enhancement of GLOF risk with time.

Generally, flood depth analysis demonstrates an increase in flood hazard between 2015 and 2024 at a constant rate, indicative of heightened downstream infrastructure and settlement susceptibility due to changing glacial dynamics.

9.5.8 Flow Velocity Patterns of Simulated GLOF Scenarios

It displays the simulated flood velocity distribution maps for 2015, 2020, and 2024 under HEC-RAS 2D simulation. Velocity is shown in meters per second (m/s), categorized into five, and illustrated spatially over the Tsho Rolpa downstream catchment.

During 2015, the flood velocity ranged from 0.005 to 15 m/s, with the highest velocity measured close to the outlet of the Tsho Rolpa Glacial Lake. The flow velocity drastically declines as the flood travels downstream, with lower velocities noted around Jhyaku, Sikpasor Khani, and Chankhu. This is evidence of a comparatively localized and contained impact area with high-velocity streams focused close to the lake outlet and Na Village. The existence of low to medium velocity zones downstream indicates restricted reach and momentum of floodwaters in 2015 (Fig. 9.17).

Up to 2020, the velocity distribution is more spread out in terms of high-velocity areas, but the peak value remains 15 m/s, with high-velocity areas above 7 m/s spread by a wide margin of 42% compared to 2015. Red and orange regions are more widespread, which is an increase in the areal coverage of flowing floodwaters. The downstream flow extends further towards Khare, Marbu, and Sikpasor Khani, indicating higher flow energy, most probably due to an increase in the discharge volume from the lake. Spatial extension in velocity is linked with increased depth footprint because deeper floodwaters with greater flows inundated areas that were not affected before. Several downstream villages that were formerly in areas of low velocity began to possess high to medium velocities, considerably adding to their flood hazard risk. The villages that were once under low-velocity zones now experience moderate to high flood velocities, raising flood hazard risk.

The velocity profile of the flood became more intensified and larger in 2024. Over 65% of the floodplain area showed velocities above 7.6 m/s, which was significantly greater compared to previous years. Na Village, Khare, Chankhu, and even Jhyaku had intense high-velocity flow, which is a sign of more flood energy and increased erosive capacity. This coincides very well with the rise in flood depth, as the majority of zones showed a depth of more than 5 m, indicating that the glacial lake outburst flood (GLOF) now has greater volume and momentum. The downstream extension of red and dark red velocity zones is a marker of an evolution from a local to a regional-scale hazard during the period. This development within the decade clearly shows that flood speed and, thereby, hazard potential are rising with the passage of time.

This decade-to-decade increase in velocity extent as well as depth intensity is tangible proof that flood energy is growing with the passage of time due to the volume increase in glacial lakes and augmented meltwater contribution. Such developments enhance the imperative for flood hazard zoning and structural mitigation in those high-risk downstream villages (Tables 9.5, 9.6 and 9.7).

9.5.9 Flood Susceptibility Mapping and Flow Dynamics Analysis

Flood susceptibility mapping in the vicinity of the Tsho Rolpa Glacial Lake makes an important contribution to knowledge of the possible downstream effects of a GLOF (Glacial Lake Outburst Flood) event. Spatial distribution of susceptibility zones from Very Low to Very High is not haphazard but is highly correlated with both topographic gradients and the hydrodynamic behavior under modeling, including flow velocity and direction.

As seen in the map, the south and southwest regions of the study site, that is, sites around Jhyaku, Chankhu, and Sikpasor Khani, are located in the "High" to "Very

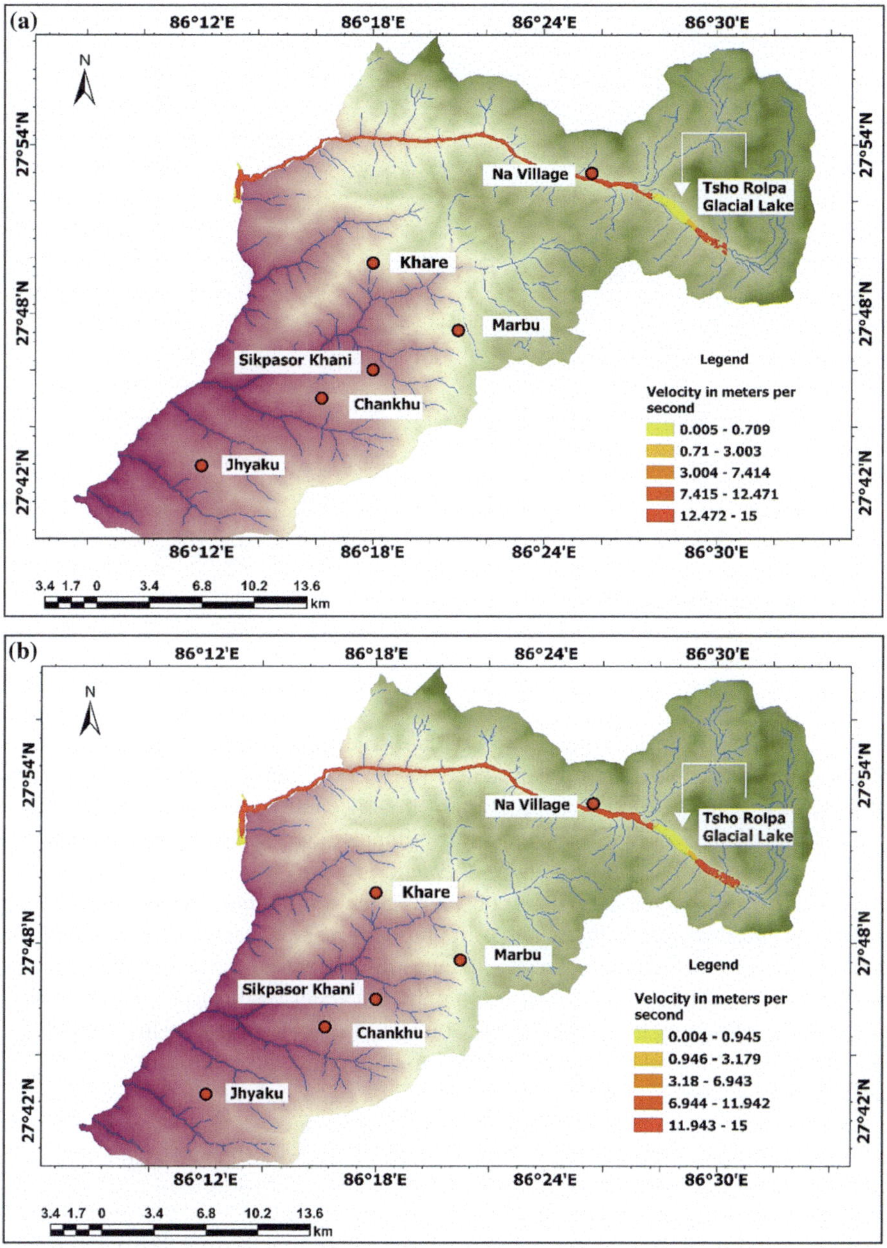

Fig. 9.17 **a** Maximum flow velocity distribution—2015 GLOF scenario. **b** Maximum flow velocity distribution—2024 GLOF scenario. **c** Maximum flow velocity distribution—2020 GLOF scenario

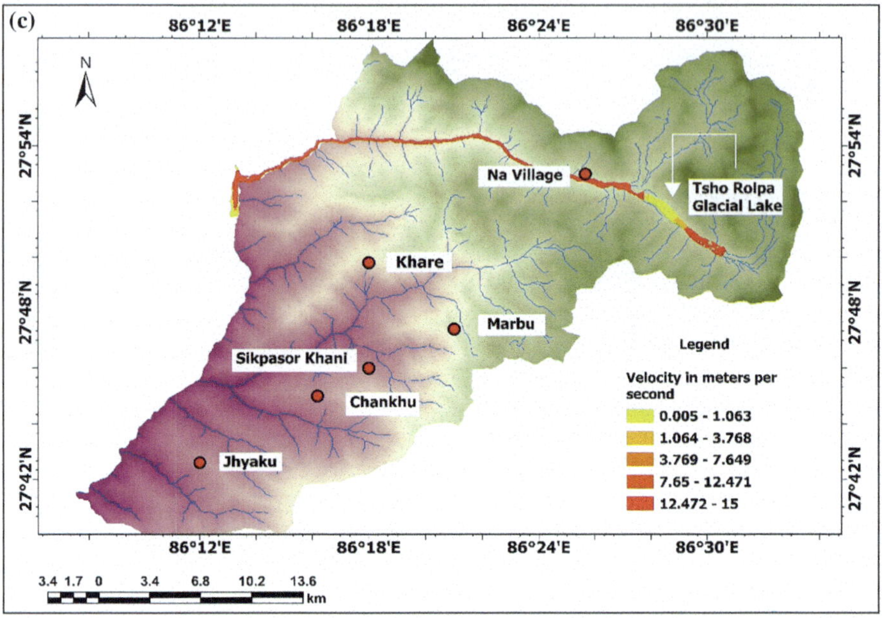

Fig. 9.17 (continued)

High" flood susceptibility zones. This spatial pattern also confirms the direction of flow and velocity maps derived from HEC-RAS 2D simulation, where the velocity vectors show that the dominant direction of flow is southwards from the lake. This trend supports the fact that floodwaters are funneled and speed down to the lower altitudes, where the landscape is more open and prone to flooding.

The verification is more evident considering the topographic gradient and slope of the channel steep gradients towards the higher reaches near the lake provide for higher velocities, which carry the floodwater downstream rapidly. When the water gets near flatter terrain in the lower catchment, the water begins to expand laterally, which leads to flooding in broader areas. Concentration of the velocity in the south (as shown on individual velocity maps) will be an important cause of high susceptibility ranking in this zone (Fig. 9.18).

In addition, this vulnerability mapping supports the model results and the past experience of flood-hit habitations. Existence of villages like Jhyaku, Chankhu, and Marbu in or around high-risk areas emphasizes the urgent necessity for early warning systems, evacuation planning, and infrastructure resilience.

Overall, the flood susceptibility map successfully validates the model result and interpretation of flow dynamics. The consistency between the susceptibility zones and high-velocity downstream channels supports the validity of the simulation and increases confidence in utilizing the output for hazard and vulnerability analysis (Table 9.8).

Table 9.5 Daily discharge data (m^3/s) from 01 July to 08 October 2015

Date	Discharge (m^3/s)	Date	Discharge (m^3/s)	Date	Discharge (m^3/s)
2015-07-01	6.01	2015-07-02	4.27	2015-07-03	3.45
2015-07-04	5.17	2015-07-05	28.13	2015-07-06	7.43
2015-07-07	8.45	2015-07-08	7.21	2015-07-09	10.38
2015-07-10	12.16	2015-07-11	6.81	2015-07-12	3.2
2015-07-13	2.9	2015-07-14	7.01	2015-07-15	14.81
2015-07-16	24.25	2015-07-17	10.5	2015-07-18	8.7
2015-07-19	10.58	2015-07-20	9.61	2015-07-21	7.62
2015-07-22	8.69	2015-07-23	7.51	2015-07-24	22.83
2015-07-25	12.6	2015-07-26	9.15	2015-07-27	6.01
2015-07-28	6.1	2015-07-29	8.05	2015-07-30	15.83
2015-07-31	14.32	2015-08-01	10.41	2015-08-02	11.64
2015-08-03	8.71	2015-08-04	15.45	2015-08-05	15.91
2015-08-06	17.96	2015-08-07	12.76	2015-08-08	10.15
2015-08-09	9.53	2015-08-10	16.16	2015-08-11	10.2
2015-08-12	12.77	2015-08-13	17.54	2015-08-14	13.75
2015-08-15	9.46	2015-08-16	11.88	2015-08-17	31.76
2015-08-18	9.79	2015-08-19	16.11	2015-08-20	41.51
2015-08-21	18.01	2015-08-22	11.18	2015-08-23	8.55
2015-08-24	6.61	2015-08-25	7.6	2015-08-26	10.11
2015-08-27	9.52	2015-08-28	8.74	2015-08-29	8.01
2015-08-30	6.11	2015-08-31	4.99	2015-09-01	5.8
2015-09-02	7.04	2015-09-03	9.62	2015-09-04	7.98
2015-09-05	4.09	2015-09-06	6.14	2015-09-07	5.03
2015-09-08	5.59	2015-09-09	4.36	2015-09-10	6.46
2015-09-11	6.97	2015-09-12	10	2015-09-13	6.18
2015-09-14	6.87	2015-09-15	9.35	2015-09-16	5.82
2015-09-17	4.71	2015-09-18	5.23	2015-09-19	13.68
2015-09-20	9.12	2015-09-21	10.88	2015–09-22	6.78
2015-09-23	5.38	2015-09-24	4.89	2015–09-25	3.83
2015-09-26	3.59	2015-09-27	3.9	2015–09-28	2.99
2015-09-29	2.69	2015-09-30	2.64	2015–10-01	2.59
2015-10-02	2.62	2015-10-03	3.24	2015–10-04	2.66
2015-10-05	2.53	2015-10-06	2.49	2015–10-07	2.4
2015-10-08	2.48				

Table 9.6 Daily discharge data (m^3/s) from 01 July to 08 October 2020

Date	Discharge (m^3/s)	Date	Discharge (m^3/s)	Date	Discharge (m^3/s)
2020-07-01	8.3	2020-07-02	8.43	2020-07-03	10.17
2020-07-04	8.35	2020-07-05	9.22	2020-07-06	8.82
2020-07-07	6.36	2020-07-08	7.69	2020-07-09	13.51
2020-07-10	36.52	2020-07-11	22.17	2020-07-12	9.2
2020-07-13	37.51	2020-07-14	16.42	2020-07-15	14.43
2020-07-16	23.37	2020-07-17	17.76	2020-07-18	20.37
2020-07-19	15.27	2020-07-20	52.73	2020-07-21	16.04
2020-07-22	11.08	2020-07-23	19.76	2020-07-24	18.1
2020-07-25	14.14	2020-07-26	12.27	2020-07-27	11.51
2020-07-28	33.79	2020-07-29	24.36	2020-07-30	12.83
2020-07-31	12.31	2020-08-01	18.73	2020-08-02	12.18
2020-08-03	25.77	2020-08-04	14.07	2020-08-05	15.61
2020-08-06	12.81	2020-08-07	13.48	2020-08-08	30.86
2020-08-09	28.03	2020-08-10	24.01	2020-08-11	13.73
2020-08-12	17.36	2020-08-13	20.74	2020-08-14	21.13
2020-08-15	25.02	2020-08-16	15.38	2020-08-17	22.55
2020-08-18	19.37	2020-08-19	16.22	2020-08-20	13.05
2020-08-21	11.11	2020-08-22	9.84	2020-08-23	12.03
2020-08-24	12.14	2020-08-25	11.88	2020-08-26	10.9
2020-08-27	12.08	2020-08-28	8.92	2020-08-29	7.45
2020-08-30	5.46	2020-08-31	7.43	2020-09-01	10.81
2020-09-02	8.14	2020-09-03	8.34	2020-09-04	5.87
2020-09-05	5.34	2020-09-06	4.74	2020-09-07	6.03
2020-09-08	5.74	2020-09-09	5.73	2020-09-10	5.92
2020-09-11	6.16	2020-09-12	7.77	2020-09-13	5.72
2020-09-14	6.02	2020-09-15	6.09	2020-09-16	6.05
2020-09-17	9.24	2020-09-18	7.74	2020-09-19	4.89
2020-09-20	5.74	2020-09-21	7.36	2020-09-22	10.7
2020-09-23	17.82	2020-09-24	13.62	2020-09-25	20.91
2020-09-26	6.04	2020-09-27	4.22	2020-09-28	4.47
2020-09-29	4.68	2020-09-30	4.67	2020-10-01	4.98
2020-10-02	4.34	2020-10-03	3.49	2020-10-04	3.25
2020-10-05	3.04	2020-10-06	2.76	2020-10-07	2.66
2020-10-08	2.63				

Table 9.7 Daily discharge data (m^3/s) from 01 July to 08 October 2024

Date	Discharge (m^3/s)	Date	Discharge (m^3/s)	Date	Discharge (m^3/s)
2024-07-01	13.34	2024-07-02	14.27	2024-07-03	13.93
2024-07-04	7.6	2024-07-05	7.3	2024-07-06	16.09
2024-07-07	6.09	2024-07-08	6.9	2024-07-09	5.83
2024-07-10	10.56	2024-07-11	22.93	2024-07-12	40.38
2024-07-13	17.17	2024-07-14	22.57	2024-07-15	15.13
2024-07-16	21.49	2024-07-17	16.01	2024-07-18	11.24
2024-07-19	15.84	2024-07-20	19.49	2024-07-21	21.66
2024-07-22	14.41	2024-07-23	12.71	2024-07-24	13.18
2024-07-25	13.2	2024-07-26	12.57	2024-07-27	16.62
2024-07-28	22.4	2024-07-29	29.53	2024-07-30	15.97
2024-07-31	14.83	2024-08-01	10.06	2024-08-02	11.94
2024-08-03	12.58	2024-08-04	8.93	2024-08-05	9.58
2024-08-06	27.91	2024-08-07	18.04	2024-08-08	21.78
2024-08-09	21.21	2024-08-10	23.85	2024-08-11	12.78
2024-08-12	10.04	2024-08-13	12.36	2024-08-14	21.24
2024-08-15	20.6	2024-08-16	13.64	2024-08-17	12.45
2024-08-18	11.62	2024-08-19	9.68	2024-08-20	13.61
2024-08-21	11.86	2024-08-22	12.16	2024-08-23	10.72
2024-08-24	10.87	2024-08-25	8.52	2024-08-26	9.07
2024-08-27	8.79	2024-08-28	8.35	2024-08-29	7.2
2024-08-30	7.07	2024-08-31	7.47	2024-09-01	19.78
2024-09-02	8.88	2024-09-03	11.39	2024-09-04	8.6
2024-09-05	9.88	2024-09-06	19.97	2024-09-07	17.41
2024-09-08	12.18	2024-09-09	8.12	2024-09-10	5.56
2024-09-11	8.99	2024-09-12	8.72	2024-09-13	7.57
2024-09-14	10.53	2024-09-15	8.85	2024-09-16	14
2024-09-17	8.31	2024-09-18	6.15	2024-09-19	6.07
2024-09-20	10.51	2024-09-21	9.19	2024-09-22	4.52
2024-09-23	8.42	2024-09-24	11.17	2024-09-25	10.67
2024-09-26	12.52	2024-09-27	30.03	2024-09-28	45.04
2024-09-29	7.55	2024-09-30	6.2	2024-10-01	5.92
2024-10-02	8.6	2024-10-03	21.37	2024-10-04	17.69
2024-10-05	26.24	2024-10-06	9.81	2024-10-07	5.19
2024-10-08	4.89				

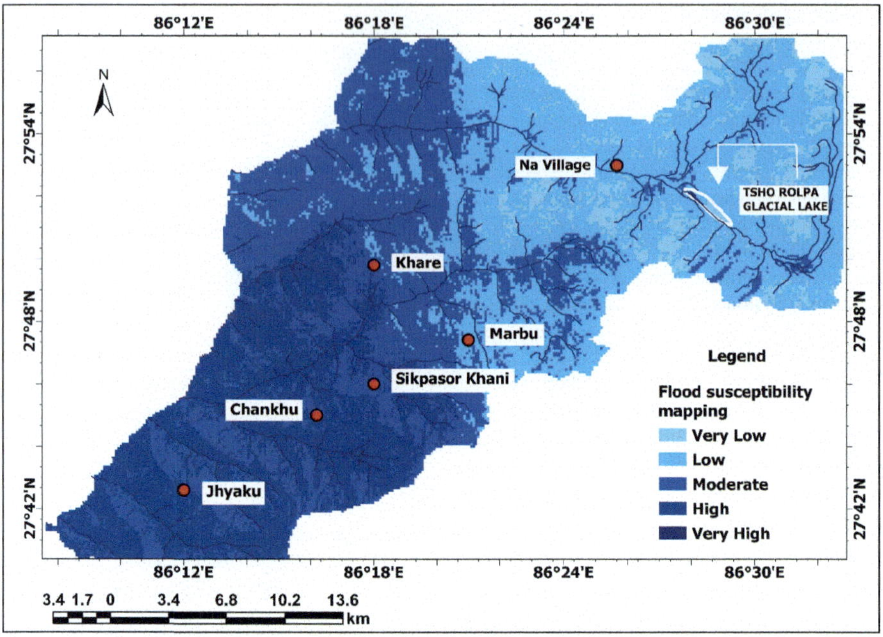

Fig. 9.18 Flood susceptibility map showing the different zones of flood magnitude

Table 9.8 Weight assigned to parameters in flood susceptibility mapping

S. No.	Parameters	Weight (%)
1	Aspect	18
2	Slope	12
3	Temperature	15
4	Rainfall	25
5	LULC	10
6	Drainage density	20
	Total	100

9.5.10 Major Findings

This current research pursued modeling of the Glacial Lake Outburst Flood (GLOF) event of Tsho Rolpa Glacial Lake through HEC-RAS 2D hydrodynamic modeling coupled with remote sensing and GIS-based analysis. The key findings are described as follows:

- **Spatio-Temporal Expansion of Tsho Rolpa Lake**:

The lake showed a successive increase in area from 1.60 km^2 in 2015 to 1.78 km^2 in 2024, reflecting a regular expansion due to intensified glacial melting.

- **Morphological Evolution in Cross-Sectional Profiles**:

Transect analysis along sections A–A1, B–B1, and C–C1 found symmetrical patterns of widening with time, indicative of stable lateral expansion under topographical constraint.

- **HEC-RAS 2D GLOF Simulation Results**:

Hydrograph analysis of major cross-sections disclosed peak discharges with sharp peaks, which manifested the characteristics of a sudden release pattern typical of GLOF behavior.

Maximum water depth exhibited downstream concentration, especially around high-risk settlements like Jhyaku and Chankhu. Maps of velocity revealed increased flow velocity in the lower catchment, confirming the downstream flood-prone nature of such areas.

- **Flood Susceptibility Mapping and Validation**:

A GIS-based flood susceptibility map classified the region into five classes. Villages like Jhyaku, Chankhu, and Sikpasor Khani came under "High" to "Very High" susceptibility zones. The spatial pattern also agreed quite well with HEC-RAS velocity and depth results, which supported the model's validity.

9.6 Concluding Remarks

The study is able to apply the integrated remote sensing, GIS, and HEC-RAS 2D modeling to estimate the probable effect of a Glacial Lake Outburst Flood (GLOF) in the area of Tsho Rolpa. Temporal extension of the lake, morphometry variation, and visualization of downstream flood hazard indicate the need for early warning and proactive planning in the Rolwaling Valley.

The similarity between the modeled flood output and the flood susceptibility map increases the accuracy of the results and provides an accessible reference for the calculation of hazards and decision-making. The method adopted in this study can be used as a template for future GLOF vulnerability analysis in the Himalayas.

Abbreviation

CSV	Comma-separated values
DEM	Digital elevation model
DN	Digital number
DSM	Digital surface model
DTM	Digital terrain model

GIS	Geographic information system
GLOF	Glacial lake outburst flood
HEC-RAS	Hydrologic Engineering Center—River Analysis System
ICIMOD	International Centre for Integrated Mountain Development
LULC	Land use land cover
RAS Mapper	River analysis system mapper
RGB	Red green blue
RS	Remote sensing
SHP	Shapefile
USGS	United States Geological Survey
UTM	Universal Transverse Mercator
WGS 84	World Geodetic System 1984

References

Allen SK, Linsbauer A, Randhawa SS, Huggel C, Rana P (2016) Glacial lake outburst flood risk in Himachal Pradesh, India: an integrative and anticipatory approach considering current and future threats. Nat Hazards 84(3):1741–1763

Chen F, Zhang M, Tian B, Li Z (2021) Hydrodynamic simulation and hazard assessment of glacial lake outburst floods in the Himalayas. Water 13(5):620

Choudhury BU, Das A, Ngachan SV (2023) Climate-induced changes in Himalayan glacier systems and associated flood hazards. Environ Dev Sustain 25(4):3152–3174

Fujita K, Sakai A, Nuimura T, Yamaguchi S, Sharma RR (2013) Recent changes in Imja Glacier, Nepal Himalaya, revealed by in situ surveys and glacier flow dynamics. Cryosphere 7(2):679–696

Government of Nepal (2002) Disaster management policy and action plan. Ministry of Home Affairs, Kathmandu

Gurung DR, Bajracharya SR, Shrestha AB (2011) Status of glacial lakes and associated hazards in Nepal Himalaya. ICIMOD, Kathmandu

ICIMOD (2011) Glacial lakes and glacial lake outburst floods in Nepal. International Centre for Integrated Mountain Development

ICIMOD (2019) The status of glacial lakes in the Hindu Kush Himalaya. International Centre for Integrated Mountain Development, Kathmandu

Kattelmann R (2003) Glacial lake outburst floods in the Nepal Himalaya: a manageable hazard? Nat Hazards 28(1):145–154

Kayastha RB, Maskey S (2024) Hydrodynamic simulation of glacial lake outburst flood and downstream hazard assessment in the Tsho Rolpa region, Nepal Himalaya. Nat Hazards 121(2):1567–1589

Liu J, Wang Y, Li X, Chen F (2022) Numerical modeling of glacial lake outburst floods and downstream impacts in the Himalayas. Nat Hazards 113(1):89–108

Mool PK, Bajracharya SR, Joshi SP (2001) Inventory of glacial lakes and identification of potential glacial lake outburst floods affected by global warming in the mountains of Himalayan region. ICIMOD

Osti R, Egashira S (2009) Hydrodynamic characteristics of the Tam Pokhari glacial lake outburst flood in the Mt Everest region, Nepal. Hydrol Process 23(20):2943–2955

Peters J, Bolch T (2009) Glacier lake changes in the Mt Everest region, Nepal Himalaya, during the last decades. Global Planet Change 66(1–2):1–13

Rana N, Shrestha AB, Gurung DR (2020) Monitoring of glacial lakes and associated flood hazards in Nepal Himalaya using remote sensing techniques. J Mt Sci 17(4):845–860
Richardson SD, Reynolds JM (2000) An overview of glacial hazards in the Himalayas. Quat Int 65–66:31–47
Rounce DR, Byers AC, Byers EA, McKinney DC (2017) Brief communication: observations of a glacier lake outburst flood from Lhotse Glacier, Everest area, Nepal. Cryosphere 11(1):443–449
Sattar A, Goswami A, Kulkarni AV (2021) Glacier retreat and glacial lake expansion in the Himalayas under climate change. Environ Monit Assess 193(5):286
Shrestha D, Gurung DR (2013) Glacial lake outburst flood hazards in Nepal Himalaya. J Nepal Geol Soc 45:75–84
Shrestha AB, Bajracharya SR, Mool PK (2013) Glacial lake inventory and outburst flood monitoring in the Nepal Himalaya. ICIMOD, Kathmandu
Shrestha M, Nakagawa H (2014) Assessment of outburst flood and debris flow caused by breaching of glacial lakes in the Nepal Himalaya. Nat Hazards 72(2):425–444
Shrestha AB, Shrestha S (2019) Glacial lake outburst flood risk assessment in the Hindu Kush Himalaya. International Centre for Integrated Mountain Development (ICIMOD), Kathmandu
Worni R, Huggel C, Stoffel M (2013) Glacial lakes in the Indian Himalayas—from an area-wide glacial lake inventory to on-site and modeling-based risk assessment of critical glacial lakes. Sci Total Environ 468–469:S71–S84
UNEP (2001) Nepal: state of the environment report. United Nations Environment Programme, Kathmandu
Yao T, Kattel DB (2022) Climate change impacts on glaciers and glacial lakes in High Mountain Asia. Sci Total Environ 822:153491

Chapter 10
Assessment of Flood Susceptibility and Effects on Urban Areas: A Case Study of Part of Mahananda River Basin

Sreya Banerjee

Abstract The paper is concerned with flood susceptibility mapping and inundation modeling and is limited to the northern sector of Mahananda river basin which falls partly in the Darjeeling and Jalpaiguri district of West Bengal, India. Comparing Analytical Hierarchy Process (AHP) and HEC-RAS modeling, the research assesses flood risk and depth in areas of hydrological sensitivity situated on slopes in the northern regions and plains in the southern regions. From the derived GIS maps of the influence factors including elevation, slope, LULC, soil, rainfall, and drainage density, multi-criteria decision analysis was conducted in order to produce a map showing flood susceptibility. In flood depth assessment, HEC-RAS was used in order to combine cross section of rivers, flow, and altitude data in order to estimate flood extent and flood depth and velocity along the river at different time of occurrence. Given below findings suggest that the western Jalpaiguri is more exposed to flood risks when compared to the eastern Darjeeling area primarily because of its flat topography. The GIS and hydraulic modeling analysis also reveals moderate to high risk along the river slope particularly where slopes are steep then gradual. The proposed integrated approach may improve the understanding of flood risks, contribute insights into efficient policies regulating urban growth across the floodplains and assist climate change adaptation for the advantage of the local government and other stakeholders aiming for protection of this environmentally sensitive area.

Keywords Flood susceptibility mapping · Hec-Ras modeling

10.1 Introduction

Flooding ranks among the most disastrous natural disasters with an impact on both; rural and urban settings all over the globe. Such areas and the degree of exposure and risk must be identified to minimize the social loss and damage to property (Rana

S. Banerjee (✉)
M.Sc. Symbiosis Institute of Geoinformatics, Symbiosis International (Deemed University), Pune, India

M. Biswas et al. (eds.), *Himalayan Flood Hazards*, Advances in Geographical and Environmental Sciences, https://doi.org/10.1007/978-981-92-0702-2_10

et al. 2023; Mandal et al. 2023). Over the recent years, GIS and remote sensing have been most widely used techniques in generating flood hazard maps for informing flood risks management and planning. These technologies have been used in some of the research studies to model flood risks and estimate social risk. For instance, Mandal et al. (2023) considered the effects of flood risks on social vulnerability of Malda district in West Bengal, India, through employing geospatial information system and MCDA. In the same way, Bhuyan et al. (2024) utilized machine learning algorithms in their context to evaluate the landscape ecological vulnerability to river-bank erosion in floodplains of the Brahmaputra River, Assam. Hydrological and hydraulic modeling mainly are used in flood hazard modeling, including the Hydrologic Engineering Centers River Analysis System (HEC-RAS) model, which is used to floodplain mapping successfully (Kim et al. 2020; Romali 2018). When HEC-RAS is combined with GIS, flood inundation areas are able to be modeled and the risks posed by floods better comprehended. This study by Peker et al. (2024) explains that flood hazard mapping through the application of GIS integrated with HEC-RAS offers useful information for disasters.

Some other works have been done on the urban flood exposure, where the effects of flood are compounded by factors like population density, poor infrastructural development and poor disaster preparedness. For instance, Sanyal and Lu (2003) have used GIS along with Remote Sensing technique for flood zoning in Gangetic West Bengal, India, and further focused on the effects of land use and infrastructure on flood risk. In the same regard, Roy et al. (2021) also showed the effectiveness of Geographical Information System (GIS) and Multi-Criteria Decision Analysis (MCDA) to estimate the level of risks in waterlogging in the city of Kolkata in India.

Other current works also investigate the use of decision-making tools like AHP in the unemployment evaluation of flood risks. Desalegn and Mulu (2021) applied GIS and HEC-RAS to simulate flood inundation in Upper Abbay Basin, Ethiopia, and Mitra et al. (2022) used AHP to model flood and assess flood susceptibility to geospatial factors where they are situated in Uttar Dinajpur district of India.

In sum, these works demonstrate that new geospatial tools and decision support systems are increasingly essential to flood risk and mitigation assessment and decision-making by authorities to minimize loss of life and economic losses from floods (Lee et al. 2023; Vashist and Singh 2023).

10.2 Study Area

The Himalayan districts of Eastern Darjeeling and Western Jalpaiguri, located in the northeastern part of West Bengal, are highly vulnerable to flooding. East Darjeeling, covering 1500–1600 km^2, lies between 26°31'N to 27°13'N and 88°01′E to 88°53′E, while West Jalpaiguri, covering 1100–1700 km^2, lies between 26°16'N to 27°0'N and 88°4′E to 89°53′E. The Mahananda River, originating from the Darjeeling hills and flowing through both districts, experiences significant flooding during the monsoon, especially in the plains of Jalpaiguri and the hilly regions of Darjeeling (Ghosh

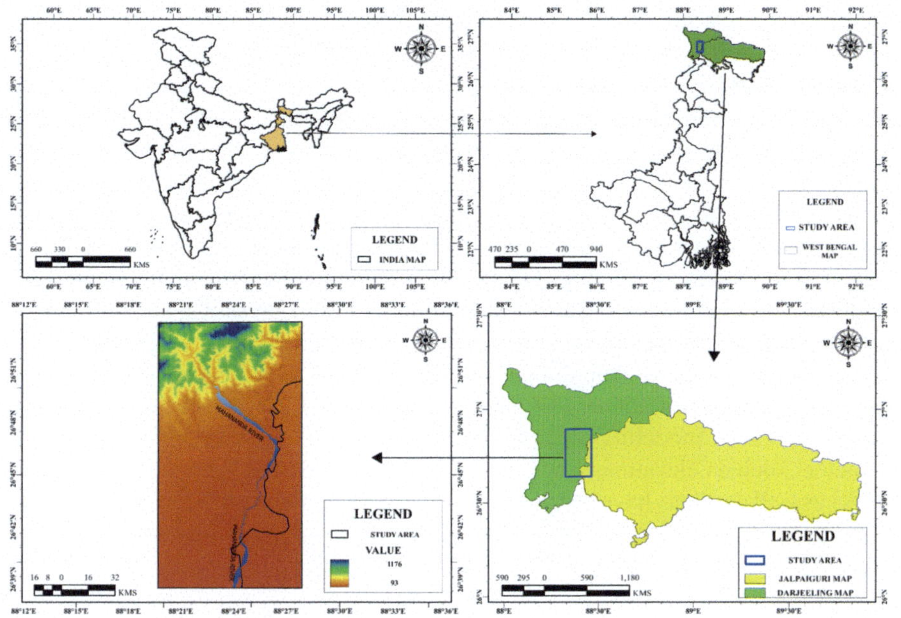

Fig. 10.1 Location of the study area (Darjeeling and Jalpaiguri District)

2014). The region's steep slopes in Darjeeling cause rapid water accumulation, while Jalpaiguri's flatter terrain leads to prolonged flooding (Mukherjee et al. 2018). Heavy monsoon rainfall, averaging 2500–3500 mm annually in both districts, exacerbates flooding, with East Darjeeling experiencing intense, high-duration rainfall (Sarkar and Maiti 2017). Urbanization, deforestation, and agricultural changes have altered land use and contributed to increased flood vulnerability (Roy et al. 2020). The 1968, 1993, 1998, 2000, 2015, 2017, and 2019 floods highlight the region's susceptibility (Das 2021). Additionally, the districts' population, which has steadily increased, faces heightened risks, especially among low-income ethnic groups, such as tea garden workers, who reside in flood-prone areas (Chakraborty and Saha 2016). This makes the region an essential area for flood risk study, given its diverse ecological landscapes and changing climatic conditions (Sharma 2019) (Fig. 10.1).

10.3 Materials and Methods

The evaluation of flood risks was positively informed by the use of several datasets that originated from various sources. Dem data were obtained from Copernicus DEM from Open Topography interface with setting the area of interest using the AOI and downloading the NASA SRTM DEM with 30 m resolution. The shape file of the Mahananda River was generated from the Google Earth engine code editor of Chow

(2005). Landsat-8 and Sentinel (2019 data and 30 m resolution) data downloaded from USGS earth explorer for NDVI, NDWI, and LULC map preparation. The images used in this study were obtained in June 2024 and mosaicked and clipped to the study area. Road data were collected from OSM and rainfall data obtained from the CRU in NetCDF format. Shore parallel lithosome, which forms shoreline of the Kathmandu Basin, and detailed geomorphological map and soil map were obtained from the Bhukosh Portal. The daily discharge data (2022–2024) were collected from the Irrigation and Waterways Department using the River station Hill Cart Road. The n values for the Manning roughness coefficients of channel and overbank areas were obtained from hydraulic reference tables and other best sources (Chow 2005), high-resolution imagery, and field evaluation. From using DEM and Google Earth Pro, slope changes and flood plain dimensions that are vital in HEC-RAS analysis were obtained. For the river channel geometry, a new shapefile was developed to allow realistic hydraulic modeling while not being computationally expensive. Therefore, 17 factors such as elevation, slope, LULC, drainage density, lithology, NDVI, and NDWI as well as others have been reclassified and weighted with the Analytical Hierarchy Process (AHP) for evaluating flood susceptibility successfully with parameter set of 17 eliciting varied values (Fig. 10.2).

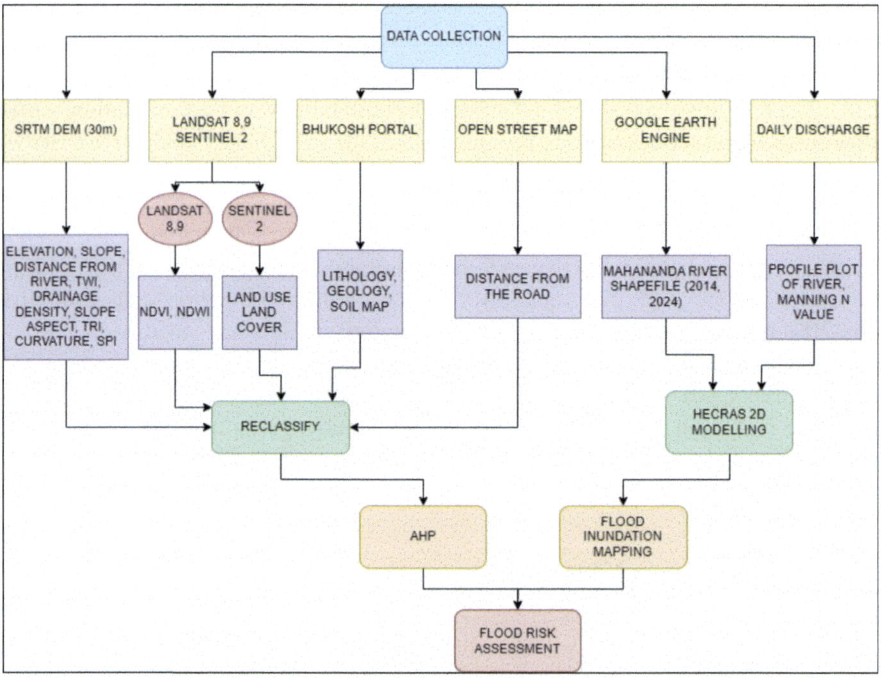

Fig. 10.2 Adopted methodology for the study

10.3.1 Construction of the Flood Susceptibility Map

These changes of parameters were done to have all the variables on equal scale for flood susceptibility analysis. Parameters were divided into nine classes showing the level of susceptibility from high to low, along with corresponding values (1–11) depending on flood risk. For this work, the 'Reclassify' tool in ArcGIS Pro was utilized. This step made enhancing the structural consistency of data interpretation particularly significant, especially when using an Analytical Hierarchical Process (AHP) model that has been established by Saaty in 1980 and, prior to this, studied by Myers and Alpert in 1968. AHP is applicable with sustainable development when undergoing environmental and risk assessment (Choudhury et al. 2022). The AHP model incorporated 17 flood-influencing parameters: Landform, Altitude, Land use/Land Cover, Drainage Network, Drainage Density, Distance to River, Topographic Wetness Index Vegetation Index, Water Index, Lithology, Distance from Road, Distance from Hospitals, Viewshed, Curvature, Topographic Roughness Index, Stream Transport Index and Soil. A pairwise comparison matrix was developed to get the relative importance of each parameter, which is arrived at by Saaty scale (Saaty 1980) and accompanied by the judgmental information. To minimize variability, the diagonal of the matrix was set to 1 (self-comparison), reciprocal values were chosen to make the matrix positive definite.

To verify reliability, the Consistency Index (CI) and Consistency Ratio (CR) were calculated using:

$$\text{C.I} = \frac{\lambda \max - n}{n - 1} \tag{10.1}$$

and

$$\text{C.R} = \frac{\text{C.I}}{\text{R.I}} \tag{10.2}$$

where λmax is the largest eigenvalue, n is the number of parameters (17 in this case), and R.I. is the Randomness Index. CR values ≤ 0.10 were considered acceptable; otherwise, recalibration was performed (Saaty 1980).

Post rebuilding, risk classes were then established with Very Low, Low, Moderate, High, and Very High risks using quantile classification or natural breaks. Since the protective factors were postulated to interact in a continuous fashion, sensitivity analysis was also performed by changing the weights of the parameters to examine shared vulnerabilities and assure the stability of the model. Cross-validation, past flood records, flood reports were conducted and approved flood vulnerability maps were complied with to make comparisons. For better understanding of the map, political boundaries, a legend, scale bar, and a north arrow were added on the finalized flood risk map. In this research, descriptive statistics such as the mean, median, and mode are used in the analysis of the spatial patterns of the risk classes with an aim of developing policies on flood control. Overlaying of thematic maps of individual parameters with the administrative boundaries was therefore done to enable the stakeholders and policy makers make some decisions. This comprehensive methodology highlights the integration of AHP, GIS tools, and sensitivity analysis for effective flood risk assessment in the Mahananda River Basin, demonstrating its utility for decision-makers in disaster management and planning (Choudhury et al. 2022; Saaty 1980).

10.3.2 Construction of Hec-Ras Model

In the HEC-RAS flood modeling, the preparation phase entails collection of topographic and hydrologic data, definition of area of interest (river centerline, banks and floodplain limits, and definition of transverse profiles across the floodplain to represent channel alteration). The roughness coefficients described above are given based on land use and vegetation density. Existing hydraulic structures such as bridges and culverts are considered in the study, and the upstream and downstream limits of the area are determined. Therefore, simulation parameters and analysis types are defined for a well031-structured river system model. During the processing phase, the preprocessed geometries are integrated with hydrologic parameters such as discharge data from gauge stations or from hydrological models (Rimba et al. 2017; Ouma and Tateishi 2014; Salazar-Briones et al. 2020). These flow rates act as steady or unsteady flow inputs, primarily when a runoff hydrograph is developed alongside rainfall data. The model calculates hydraulic parameters that includes energy equations for steady flow or St Venant's equations for unsteady flow, to determine water surface profiles, velocities, and hydraulic properties with allowances for losses and hydraulic structures. Forecasting and mapping of floods can, therefore, be made accurately. Such graphical outputs as water surface profiles, flow distribution, and velocities are typical examples of results obtained through post-progression. Flood inundation maps are created which show the comparative flood extent and depth at a given flood stage or for a specified number of years from a given flood event and the map data can be exported to GIS for additional analysis. Hydraulic variables such as flow velocity and shear stress are computed to look at erosion susceptibility and sediment conveyance. Such outputs as rating curves are used in flood frequency estimation control measures and flood risk management (Kumar and Jha 2023; Ullah and Zhang 2020; Ghosh et al. 2023).

Geometry of the Upper Reach of Mahananda River, draining across eastern part of Darjeeling district and western part of Jalpaiguri district, shows that the geomorphic system is quite complex and possesses appreciable relief. This reach mainly follows a northward trend with a few southeast loops. Transverse profiles, ranging from S-218 to S-24831, document the river morphology and floodplain; higher density samples the floodplain where geomorphology varies most sharply, especially in the steeper north. These cross sections, taken perpendicular to flow, as flow, are critical for hydraulic modeling, especially in the vicinity of sharp bends or confluences of tributaries.

This condition is due to the geographical position of the river, starts from high grounds of the northern region and empowering grounds of the southern region, the flow of water and flood incidents are likely to differ from each other. Location and separation distances: They are not constant; these are closer in mountainous country as they represent sharp changes in relief features than in the plains. This is beneficial for achieving a detailed depiction of the shape of the river while staying computationally inexpensive. Cross-sectional data elaborated in the floodplain include potential overbank flows and go beyond the immediate channel width to cover the entire floodplain area for hydraulic modeling. Sequentially numbered from downstream the cross sections described the morphological structure of the stream though providing accurate models of the hydraulic characteristics, flow, and flood potential (Manoharan and Ganapathy 2023; Taromideh et al. 2022; Korkmaz 2022).

An important realistic input of HEC-RAS, n Manning, is the resistance factors for the flow due to surface irregularities, vegetation, and channel characteristics. Here, the characteristic n values vary between 0.012 and 0.018 for plain concrete culverts and 0.1 or more for rough, vegetated flood plains. N values therefore require proper selection since they affect water surface levels, velocities, and flood depths. Standard tables, photographic comparisons, or field measurements are applied commonly, and n values are refined to match simulations to field conditions. Spatial variances in the Manning's n along the channel increases the capabilities of simulating floods.

Model Calibration:

HEC-RAS calibration involves processes of refining, among others, of the roughness coefficient (Manning's n), expansion/contraction coefficients, flow area so as to make the model results agree with historical data such as, water levels, discharges and inundation maps. Calibration is repeated, and it reduces error by using statistic like RMSE or NSE while at the same time it checks for model ability to respond to different flow condition in order to prevent over fitting. This process improves the credibility of models in flood forecasting and hydraulic design (Müller et al. 2011; Sarkar and Mondal 2019; Rana and Routray 2018).

Model Validation:

The model's ability to forecast was tested on a different three-year flood dataset by comparing simulated discharges against those obtained from the gauging stations without any calibration. NSE and RMSE, key metrics of model validation, reflected performance, making particularism of the correct timing of peak flows, and the magnitude in different terrains. Validation showed that the model provided robust estimates for flood risk and outlined key improvements that can further strengthen the model's utility in flood risk management in the Mahananda River basin.

10.4 Results and Discussions

10.4.1 Altitudinal Factors

The elevation map of Darjeeling and Jalpaiguri districts reveals significant topographical variation. The highest elevations (946.66–1176 m) are concentrated in the northeastern part, while the lowest (93.001–143.965 m) are in the south. The steep gradient from Darjeeling to Jalpaiguri influences hydrological processes, particularly flood dynamics. Darjeeling, part of the Lesser Himalayas, features steep hills and deep valleys, while Jalpaiguri consists of predominantly flat terrain with gentle slopes. The Mahananda River and its tributaries exhibit rapid flow in steep areas and slower movement in flatter regions, affecting flood patterns. Submerged floodplains in Jalpaiguri indicate high flood susceptibility, impacting land use and infrastructure planning.

The slope map highlights varying inclinations, with the steepest slopes (44.593–71.068°) in the north-west and the gentlest (0.001–2.508°) in the southwest and east. Darjeeling's steep slopes accelerate runoff, increasing flood risk and soil erosion, leading to sedimentation in downstream rivers. Jalpaiguri's gentle slopes allow higher infiltration but can lead to prolonged flooding due to soil saturation. Slope variations dictate land-use patterns, influencing flood vulnerability in agricultural and urban areas.

Slope aspect plays a crucial role in flood risk assessment. Slopes facing north and west (337.5–360°) are more prone to landslides due to increased moisture retention and weaker soil structures, while flat areas (−1°) indicate stable terrains. Steep slopes (>60°) increase susceptibility to landslides, especially under specific soil and land-use conditions. These geomorphological factors are critical for flood risk mapping and sustainable land management (Fig. 10.3).

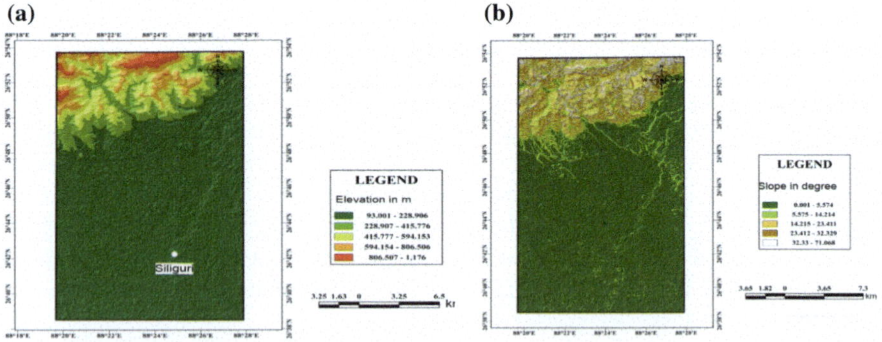

Fig. 10.3 **a** Elevation map, **b** slope map

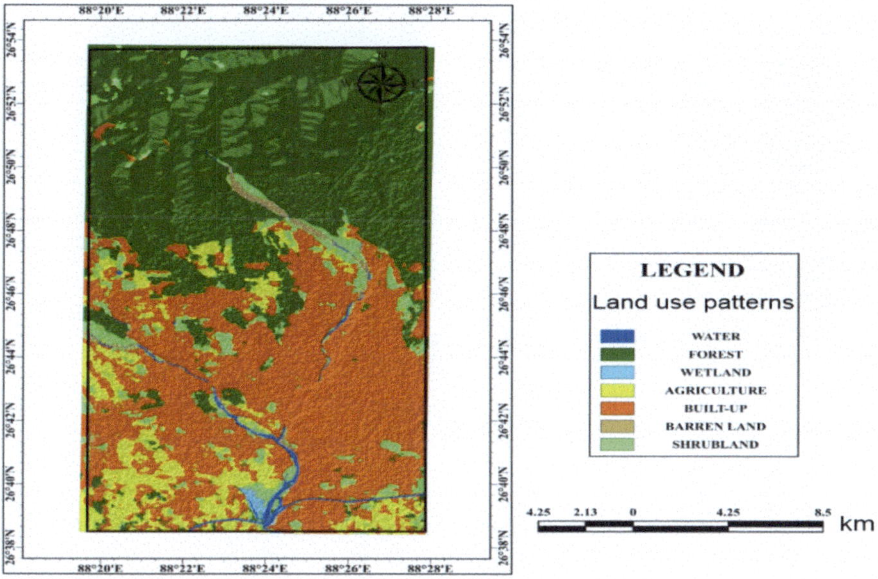

Fig. 10.4 Land use land cover map

10.4.2 Land Use Land Cover

As seen on the LULC map produced from SVM technique, large chunk areas of vegetation are seen in the Northern region that acts as a water reservoir and lessens the chances of surface flow and flash flood occurrences. As for central and eastern built-up regions, low permeability is further enhanced, aggravating the problems of runoff and floods. Irrigated land quantity is largest in the southwestern zone, and additional patches in the other zones; practices and cropping patterns affect runoff and water holding. Housing in areas with high stream density is influenced by seasonal floods and overflows from the river during showers. Some of the districts are more flooded than others owing to the water bodies mainly located in the southern part of the country (Fig. 10.4).

10.4.3 Spectral Indices

The Normalized Difference Vegetation Index (NDVI) for Darjeeling and Jalpaiguri ranges from 0.552 (highest) to −0.017 (lowest), reflecting variations in vegetation cover and health. NDVI serves as a key indicator in flood risk assessment, as higher values in northwestern regions indicate dense vegetation, which mitigates flood risk by enhancing infiltration and reducing runoff. Conversely, lower NDVI

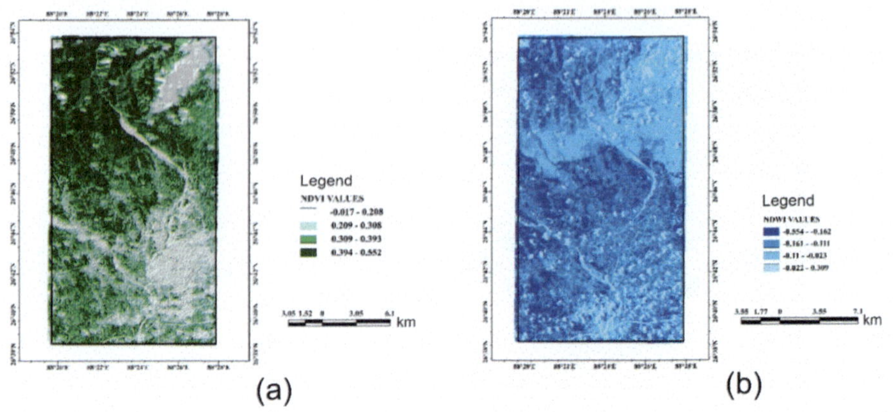

Fig. 10.5 **a** NDVI map of the study area and **b** NDWI map values

values in northeastern and southeastern areas signify sparse vegetation, contributing to increased flood susceptibility due to reduced water retention.

The Normalized Difference Water Index (NDWI), ranging from 0.309 to −0.554, highlights the spatial distribution of water bodies and moisture content. High NDWI values in central and southwestern areas correspond to saturated soils, rivers, and wetlands, increasing flood risk due to higher runoff volumes. In contrast, low NDWI values in northeastern areas indicate drier conditions with reduced flood potential. These indices provide critical insights for flood risk management by assessing vegetation health and water accumulation dynamics (Fig. 10.5).

10.4.4 Accessibility and Connectivity Parameters

The Distance from River map for Darjeeling and Jalpaiguri highlights how topography influences water flow, with values ranging from 0.001–0.002 (near rivers) to 0.017–0.021 (farther away). Areas closer to rivers (darker blue) are more flood-prone due to their location within floodplains. However, flood susceptibility is also influenced by elevation, land use, and tributary networks, meaning even distant low-lying areas can experience flooding.

Similarly, the Distance to Roads map, ranging from 0.001–0.001 (near) to 0.034–0.04 (far), illustrates accessibility variations. Flood-prone areas farther from roads (darker green) in the north and north-east face greater challenges in evacuation and disaster response. Conversely, regions near roads (lighter shades) in the central and southeastern parts benefit from improved accessibility, though factors like topography and drainage also affect flood risks. These spatial analyses aid in flood risk assessment and disaster preparedness.

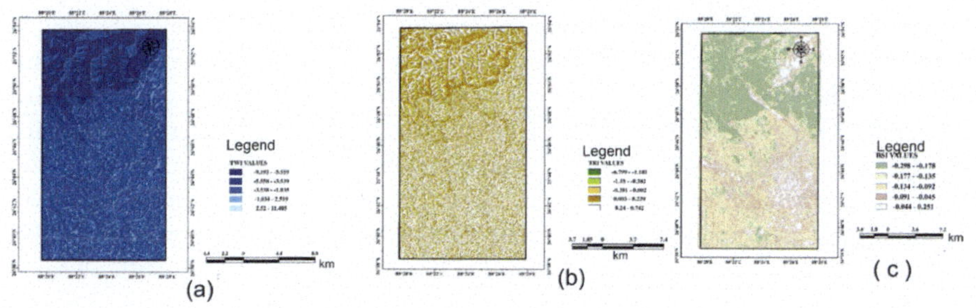

Fig. 10.6 **a** TWI map, **b** TRI map, **c** BSI map

10.4.5 The Index Parameters

The Topographic Wetness Index (TWI) map of Darjeeling and Jalpaiguri (range: −9.193–11.405) highlights terrain wetness variations. Higher TWI values indicate low-lying areas prone to water accumulation and flooding, particularly in the southern and southeastern regions, Conversely, steep slopes with lower TWI values facilitate runoff, reducing flood susceptibility.

The Topographic Roughness Index (TRI) map (range: −6.799–0.742) depicts surface complexity. Higher TRI values, concentrated in the north and north-west, correspond to rugged terrain with rapid runoff and lower flood risk. Conversely, flatter areas in Jalpaiguri exhibit higher flood susceptibility due to water accumulation.

The Stream Power Index (SPI) map (range: 0.001–0.334) highlights variations in erosive potential. Higher SPI values in northern Darjeeling indicate strong water flow and sediment transport, increasing erosion risk. In contrast, lower SPI values in western Jalpaiguri suggest reduced stream power but higher flood vulnerability due to water stagnation in floodplains.

The Bare Soil Index (BSI) map (range: −0.298–0.251) reveals spatial variations in soil exposure. Higher BSI values in western Jalpaiguri and eastern Darjeeling correlate with increased susceptibility to erosion and surface runoff, exacerbating flood risk. Forested areas exhibit lower BSI values, indicating better water retention and lower flood susceptibility. Temporal analysis suggests rising soil exposure in western Jalpaiguri due to land-use changes, highlighting the need for conservation measures (Fig. 10.6).

10.4.6 Regional Parameters

The geological composition of Darjeeling and Jalpaiguri districts is diverse, with five major formations: Quaternary (dominant in the study area), Proterozoic (north-west), Mesoproterozoic (north-west strip), Permian (north and north-west), and Pliocene-Pleistocene (north and north-east). The Proterozoic and Mesoproterozoic formations, comprising crystalline rocks like granite and basalt, have low permeability, leading to high runoff and increased flood risk downstream. Permian formations exhibit multi-layered permeability variations, while Pliocene-Pleistocene deposits, composed of sedimentary materials, influence localized flood behavior. The Quaternary deposits, largely covering the region, consist of young, unconsolidated sediments that, despite high permeability, saturate quickly, and intensifying flood risks (Mallick et al. 2023).

Soil distribution impacts flood susceptibility. Humic Acrisols (north-west) have low infiltration, increasing surface runoff. Dystric Regosols (north-east), weakly developed and shallow, drain quickly but are erosion-prone. Humic Cambisols (east) retain moisture, reducing flood impact. Eutric Fluvisols, widespread in low-lying Jalpaiguri, form in alluvial deposits and are highly flood-prone due to poor drainage. This distribution highlights the need for soil-specific flood management strategies (Sharma et al. 2018; Timbadiya et al. 2011; Darabi et al. 2019).

The lithology of the region plays a crucial role in flood dynamics. Quartzarenite is the most resistant, limiting erosion, while weathered sand, silt, and clay contribute to sediment deposition, altering river flow and increasing flood hazards downstream. Low-permeability rocks (quartzite, gneisses) promote surface runoff, whereas weathered granite, with higher infiltration, mitigates flood risks. Understanding these geological, soil, and lithological factors is essential for effective flood resilience and land-use planning in the Mahananda River Basin (Shrestha 2022; Ogras and Onen 2020).

10.4.7 Drainage Density and Curvature

The drainage density map of Darjeeling and Jalpaiguri districts reveals variations in stream network distribution, influencing flood susceptibility. High drainage density (401.802–446.446) occurs in limited northern and southern regions with well-developed stream networks, increasing flood risks. In contrast, areas with low drainage density (0.001–44.645) experience slower runoff, but prolonged rainfall may still lead to localized flooding.

Curvature analysis highlights diverse topographic patterns, with high positive values (249,406,304,946.699–693,360,001,024) in the north and north-west, indicating convex slopes (ridges) that facilitate runoff and reduce flood risk. The transition from hills to plains exhibits mixed curvature, necessitating targeted flood mitigation strategies (Shahiriparsa et al. 2013; Saini and Kaushik 2012; Isma'il and Saanyol 2013).

10.4.8 Flood Susceptibility Map

Flood susceptibility in Darjeeling and Jalpaiguri districts is primarily influenced by elevation (15%), distance to river (13%), and slope (11%). The Mahananda River and its tributaries significantly impact flood risk, with low-lying areas in southern and central regions exhibiting high susceptibility due to lower elevation, flat terrain, and proximity to water bodies. Conversely, the elevated, steeper slopes of eastern Darjeeling show reduced flood risk due to rapid runoff and resistant geological formations (Swain et al. 2020; Dekaraja and Mahanta 2021; Ghosh et al. 2019).

Intermediate vulnerability zones, situated between high-risk floodplains and low-risk uplands, require adaptive management due to land-use changes and climate variability. The spatial distribution of flood risk highlights a strong negative correlation between elevation, slope, and flood-prone areas, with Quaternary deposits and Eutric Fluvisols dominating high-susceptible zone zones. Agricultural and urban expansion in low-lying areas further exacerbates flood vulnerability.

The flood susceptibility map serves as a crucial tool for flood risk mitigation, urban planning, and land-use management. Future refinements should integrate historical flood data, local knowledge, and field verification, along with climate change projections and hydrological analyses. This study provides a foundation for improved regional resilience, disaster preparedness, and sustainable development in the Mahananda River basin (Fig. 10.7) (Tables 10.1 and 10.2).

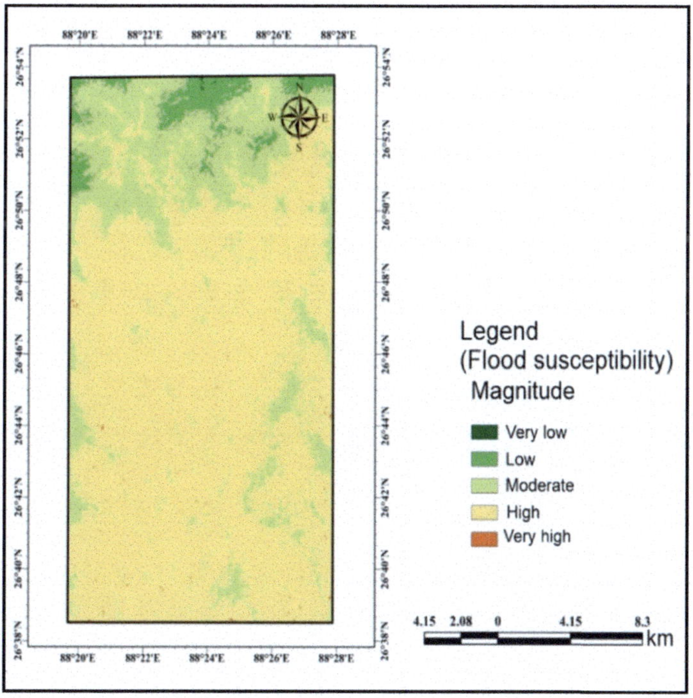

Fig. 10.7 Flood susceptibility map showing the five zones of flood risk area

Table 10.1 Pairwise comparison matrix for flood susceptibility mapping

Criterion	Elevation	Distance from river	Slope	TWI	Drainage density	LULC	Soil	NDWI	SPI	Geology	Lithology	Curvature	TRI	Aspect	NDVI	Distance to road	Bare soil index
Elevation	1	2	3	4	5	6	7	8	9	9	9	9	9	9	9	9	9
Distance from river	1/2	1	2	3	4	5	6	7	8	9	9	9	9	9	9	9	9
Slope	1/3	1/2	1	2	3	4	5	6	7	8	9	9	9	9	9	9	9
TWI	1/4	1/3	1/2	1	2	3	4	5	6	7	8	9	9	9	9	9	9
Drainage density	1/5	1/4	1/3	1/2	1	2	3	4	5	6	7	8	9	9	9	9	9
LULC	1/6	1/5	1/4	1/3	1/2	1	2	3	4	5	6	7	8	9	9	9	9
Soil	1/7	1/6	1/5	1/4	1/3	1/2	1	2	3	4	5	6	7	8	9	9	9
NDWI	1/8	1/7	1/6	1/5	1/4	1/3	1/2	1	2	3	4	5	6	7	8	9	9
SPI	1/9	1/8	1/7	1/6	1/5	1/4	1/3	1/2	1	2	3	4	5	6	7	8	9
Geology	1/9	1/9	1/8	1/7	1/6	1/5	1/4	1/3	1/2	1	2	3	4	5	6	7	8
Lithology	1/9	1/9	1/9	1/8	1/7	1/6	1/5	1/4	1/3	1/2	1	2	3	4	5	6	7
Curvature	1/9	1/9	1/9	1/9	1/8	1/7	1/6	1/5	1/4	1/3	1/2	1	2	3	4	5	6
TRI	1/9	1/9	1/9	1/9	1/9	1/8	1/7	1/6	1/5	1/4	1/3	1/2	1	2	3	4	5
Aspect	1/9	1/9	1/9	1/9	1/9	1/9	1/8	1/7	1/6	1/5	1/4	1/3	1/2	1	2	3	4
NDVI	1/9	1/9	1/9	1/9	1/9	1/9	1/9	1/8	1/7	1/6	1/5	1/4	1/3	1/2	1	2	3
Distance to road	1/9	1/9	1/9	1/9	1/9	1/9	1/9	1/9	1/8	1/7	1/6	1/5	1/4	1/3	1/2	1	2
Bare soil index	1/9	1/9	1/9	1/9	1/9	1/9	1/9	1/9	1/9	1/8	1/7	1/6	1/5	1/4	1/3	1/2	1

Table 10.2 Weightage of each criteria for flood susceptibility mapping

Sl. No.	Parameters	Weightage (IN %)
1.	Elevation	15
2.	Distance to river	13
3.	Slope	11
4.	TWI	9
5.	Drainage density	8
6.	LULC	7
7.	Soil	6
8.	NDWI	5
9.	SPI	5
10.	Geology	4
11.	Lithology	4
12.	Curvature	3
13.	TRI	3
14.	Aspect	2
15.	NDVI	2
16.	Distance to road	2
17.	Bare soil index	1

10.4.9 Manning's n Value

Manning's n value diagrams illustrate roughness coefficients across the Mahananda River's upper reach, with sectional profiles classified into the left overbank (LOB), main channel, and right overbank (ROB). The main channel exhibits stable roughness values (~0.03–0.04), while the overbanks show greater variability (~0.05–0.08), likely due to vegetation, land use, and topographic features. These coefficients are crucial for accurate hydraulic modeling and flood risk assessment (Kute 2014; Paul and Sarkar 2022; Mehebub et al. 2015).

10.4.10 HEC-RAS 2D Model

The HEC-RAS model for the Mahananda River has been post processed with the help of version 6. 4. 1 (June 2023), devoted to the examination of the steady flow simulation results in the UR particularly at the cross section 672. The model calculated the supercritical profile for the D26 scenario and produced GIS data for the "ALOSDEM" flood inundation plan. Other steps involved included computation of edge line, generation of cross section interpolation surface and writing event condition data. Geometry, flow, and plan calculations took 3 s of the simulation time and offered a detailed set of the water surface profile results. No maps were created directly for this run, but data which has been worked on forms the foundation of flood inundation mapping, computation of water surface profiles, flood hazard risk analysis along the Upper Reach of Mahananda River. The post processing phase of this research takes the raw simulation data and converts them into useful information that can be used in flood risk assessment and management in the study area concerning the flood potentials and their effects (Figs. 10.8, 10.9, and 10.10).

Narrowing of Mahananda River Boundaries

The 2014 map of the Mahananda River depicts a broader, more meandering course, suggesting a larger floodplain with fewer human interferences. In contrast, the 2024 map reveals significant narrowing, particularly in the middle and lower reaches, with

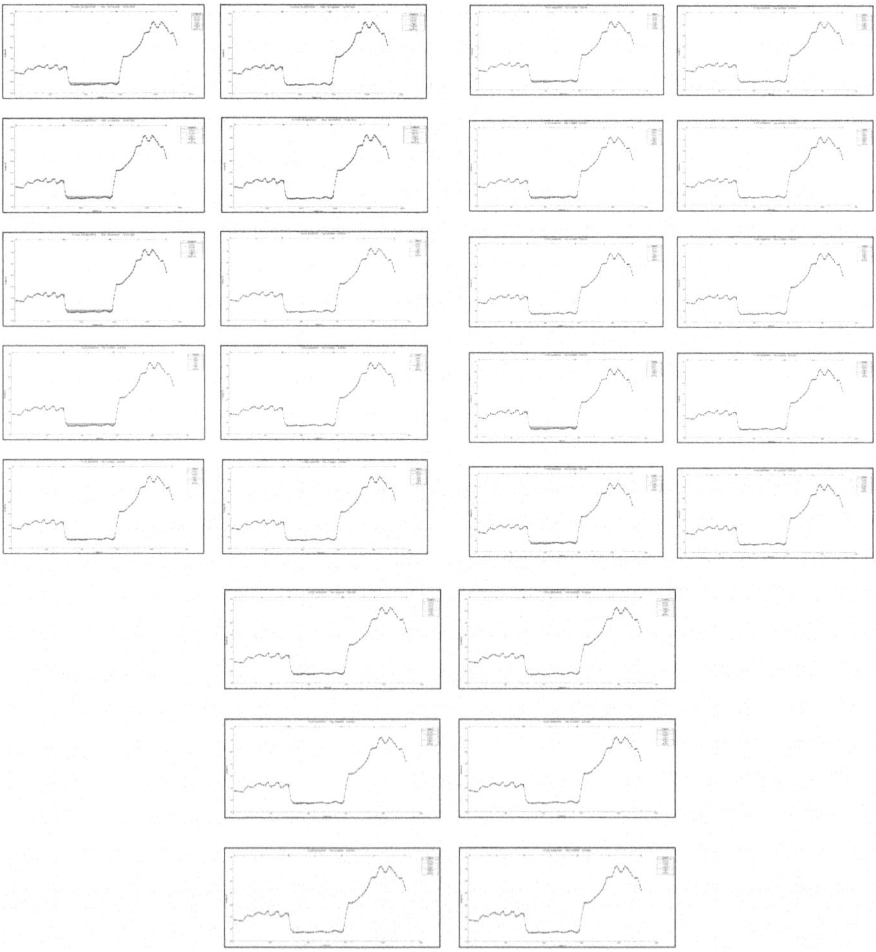

Fig. 10.8 Profile plot of the highest discharge

Fig. 10.9 Profile plot of the intermediate discharge basin

reduced sinuosity. This constriction is likely due to urbanization, land-use changes, and infrastructure development, such as bridges, which have restricted the river's natural width.

These alterations have critical implications for flood susceptibility. A narrower channel limits floodwater dispersion, increasing the risk of overbank flooding during extreme events. The loss of natural meanders and floodplains reduces the landscape's ability to mitigate floods, leading to faster and intensified flood pulses downstream (Fig. 10.11).

Fig. 10.10 Profile plot of lowest discharge basin

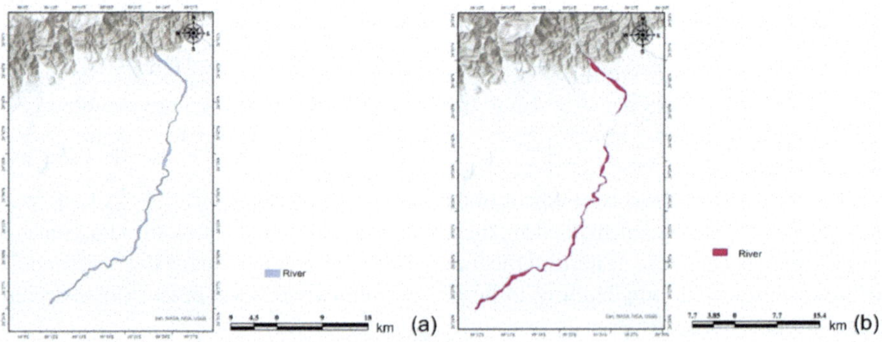

Fig. 10.11 Mahananda river in **a** 2014 and **b** 2024

The rapid transformation of the river over just a decade underscores the urgent need for effective watershed management and flood control strategies to mitigate risks and enhance resilience in the Mahananda River Basin.

10.5 Conclusion

This study evaluates flood vulnerability in the Mahananda River Basin, particularly in Darjeeling and Jalpaiguri districts, using AHP for flood susceptibility analysis and HEC-RAS for flood modeling. Western Jalpaiguri is most flood-prone due to its topography and land use. Factors such as drainage density, river proximity, and soil erosion were integrated into GIS-based flood risk assessments. The study highlights the necessity of improved drainage, early warning systems, and land-use regulations to mitigate flood risks (Abdrabo et al. 2023).

Additionally, it underscores the importance of incorporating climate change projections into flood risk management, given the expected increase in extreme precipitation events. The methodologies employed can be adapted to similar regions, providing valuable insights for flood resilience planning and disaster mitigation.

Declaration The authors declare that they have no known competing financial interests or personal relationships that could have appeared to influence the work reported in this paper.

References

Abdrabo KI, Kantoush SA, Esmaiel A, Saber M, Sumi T, Almamari M, Elboshy B, Ghoniem S (2023) An integrated indicator-based approach for constructing an urban flood vulnerability index as an urban decision-making tool using the PCA and AHP techniques: a case study of Alexandria, Egypt. Urban Clim 48:101426. https://doi.org/10.1016/j.uclim.2023.101426

Bhuyan N, Sajjad H, Saha TK, Roshani N, Sharma Y, Masroor M, Rahaman MH, Ahmed R (2024) Assessing landscape ecological vulnerability to riverbank erosion in the middle Brahmaputra floodplains of Assam, India using machine learning algorithms. Catena 234:107581. https://doi.org/10.1016/j.catena.2023.107581

Darabi H, Haghighi AT, Mohamadi MA, Rashidpour M, Ziegler AD, Hekmatzadeh AA, Kløve B (2019) Urban flood risk mapping using data-driven geospatial techniques for a flood-prone case area in Iran. Nord Hydrol 51(1):127–142. https://doi.org/10.2166/nh.2019.090

Dekaraja D, Mahanta R (2021) Riverbank erosion and migration inter-linkage: with special focus on Assam, India. Environ Syst Res 10(1). https://doi.org/10.1186/s40068-020-00214-0

Desalegn H, Mulu A (2021) Mapping flood inundation areas using GIS and HEC-RAS model at Fetam River, upper Abbay Basin, Ethiopia. Sci Afr 12:e00834. https://doi.org/10.1016/j.sciaf.2021.e00834

Ghosh A, Chatterjee U, Pal SC, Islam ARMT, Alam E, Islam MK (2023) Flood hazard mapping using GIS-based statistical model in vulnerable riparian regions of sub-tropical environment. Geocarto Int 38(1). https://doi.org/10.1080/10106049.2023.2285355

Ghosh NIC, Islam NMA, Mondal NS, Prodhan NS, Ghosh NPK (2019) Impacts of flood on ecological environment of Dwarka River basin (West Bengal, Jharkhand). https://doi.org/10.25215/1387453440.008

Isma'il M, Saanyol IO (2013) Application of remote sensing (RS) and geographic information systems (GIS) in flood vulnerability mapping: case study of river Kaduna. Int J Geomat Geosci 3(3):618–627. http://www.ipublishing.co.in/jggsvol1no12010/volthree/EIJGGS3151.pdf

Kim V, Tantanee S, Suparta W (2020) Gis-based flood Hazard mapping using Hec-Ras model: a case study of lower Mekong River, Cambodia. Geographia Technica 15(1):16–26. https://doi.org/10.21163/gt_2020.151.02

Korkmaz M (2022) The use of Hec-Ras Modeling in flood risk analysis. Eng Sci 17(4):54–66. https://doi.org/10.12739/nwsa.2022.17.4.1a0482

Kumar N, Jha R (2023) GIS-based flood risk mapping: the case study of Kosi River basin, Bihar, India. Eng Technol Appl Sci Res 13(1):9830–9836. https://doi.org/10.48084/etasr.5377

Kute S (2014) Flood Modeling of river Godavari using HEC-RAS. Int J Res Eng Technol 03(21):81–87. https://doi.org/10.15623/ijret.2014.0321017

Lee S, Choi Y, Ji J, Lee E, Yi S, Yi J (2023) Flood vulnerability assessment of an urban area: a case study in Seoul, South Korea. Water 15(11):1979. https://doi.org/10.3390/w15111979

Mallick RH, Bandyopadhyay J, Halder B (2023) Impact assessment of river bank erosion in the lower part of Mahanadi River using geospatial sciences. Sustain Horiz 8:100075. https://doi.org/10.1016/j.horiz.2023.100075

Mandal KK, Dharanirajan K, Meena M, Jaman T, Rana S (2023) Application of geospatial tools in the assessment of flood hazard impact on social vulnerability of Malda district West Bengal, India. Natural Hazards Research. https://doi.org/10.1016/j.nhres.2023.11.008

Manoharan SG, Ganapathy GP (2023) GIS based urban social vulnerability assessment for liquefaction susceptible areas: a case study for greater Chennai, India. Geoenviron Disasters 10(1). https://doi.org/10.1186/s40677-022-00230-5

Mehebub, S., Raihan, A., Nuhul, H., & Haroon, S. (2015). Assessing flood inundation extent and landscape vulnerability to flood using geospatial technology: a study of Malda district of West Bengal, India. Forum Geografic/Revista Forum Geografic, XIV(2), 156–163. doi:https://doi.org/10.5775/fg.2067-4635.2015.144.d

Mitra R, Saha P, Das J (2022) Assessment of the performance of GIS-based analytical hierarchical process (AHP) approach for flood modelling in Uttar Dinajpur district of West Bengal, India. Geomat, Nat Hazards Risk 13(1):2183–2226. https://doi.org/10.1080/19475705.2022.2112094

Müller A, Reiter J, Weiland U (2011) Assessment of urban vulnerability towards floods using an indicator-based approach—a case study for Santiago de Chile. Nat Hazards Earth Syst Sci 11(8):2107–2123. https://doi.org/10.5194/nhess-11-2107-2011

Ogras S, Onen F (2020) Flood analysis with HEC-RAS: a case study of Tigris River. Adv Civ Eng 2020:1–13. https://doi.org/10.1155/2020/6131982

Ouma YO, Tateishi R (2014) Urban flood vulnerability and risk mapping using integrated multi-parametric AHP and GIS: methodological overview and case study assessment. Water 6(6):1515–1545. https://doi.org/10.3390/w6061515

Paul P, Sarkar R (2022) Flood susceptibility assessment using geospatial multi-criteria framework: a study in lower Mahananda Basin, India. Research Square (Research Square). https://doi.org/10.21203/rs.3.rs-1472638/v1

Peker İB, Gülbaz S, Demir V, Orhan O, Beden N (2024) Integration of HEC-RAS and HEC-HMS with GIS in flood Modeling and flood Hazard mapping. Sustainability 16(3):1226. https://doi.org/10.3390/su16031226

Rana IA, Routray JK (2018) Multidimensional model for vulnerability assessment of urban flooding: an empirical study in Pakistan. Int J Disaster Risk Sci 9(3):359–375. https://doi.org/10.1007/s13753-018-0179-4

Rana SS, Habib SA, Sharifee MNH, Sultana N, Rahman SH (2023) Flood risk mapping of the flood-prone Rangpur division of Bangladesh using remote sensing and multi-criteria analysis. Nat Hazards Res. https://doi.org/10.1016/j.nhres.2023.09.012

Rimba A, Setiawati M, Sambah A, Miura F (2017) Physical flood vulnerability mapping applying geospatial techniques in Okazaki City, Aichi prefecture, Japan. Urban Sci 1(1):7. https://doi.org/10.3390/urbansci1010007

Romali NS (2018) Application of Hec-Ras and arc Gis for floodplain mapping in Segamat town, Malaysia. Int J GEOMATE 15(47) https://doi.org/10.21660/2018.47.3656

Roy S, Bose A, Singha N, Basak D, Chowdhury IR (2021) Urban waterlogging risk as an undervalued environmental challenge: an integrated MCDA-GIS based modeling approach. Environ Challenges 4:100194. https://doi.org/10.1016/j.envc.2021.100194

Saini SS, Kaushik S (2012) Risk and vulnerability assessment of flood hazard in part of Ghaggar Basin: a case study of Guhla block, Kaithal, Haryana, India. Int J Geomat Geosci 3(1):42–54. http://www.ipublishing.co.in/jggsvol1no12010/volthree/EIJGGS3104.pdf

Salazar-Briones C, Ruiz-Gibert JM, Lomelí-Banda MA, Mungaray-Moctezuma A (2020) An integrated urban flood vulnerability index for sustainable planning in arid zones of developing countries. Water 12(2):608. https://doi.org/10.3390/w12020608

Sanyal J, Lu XX (2003) Application of GIS in flood hazard mapping: a case study of Gangetic West Bengal, India. ResearchGate. https://www.researchgate.net/publication/238661989

Sarkar D, Mondal P (2019) Flood vulnerability mapping using frequency ratio (FR) model: a case study on Kulik river basin, indo-Bangladesh Barind region. Appl Water Sci 10(1). https://doi.org/10.1007/s13201-019-1102-x

Shahiriparsa A, Heydari M, Arumugan MARBMA (2013) Determination of flood zoning by HEC-RAS model (case study: Johor River—Kota Tinggi region). ResearchGate. https://www.researchgate.net/publication/294218878

Sharma SVSP, Roy PS, Chakravarthi V (2018) Assessment of social vulnerability to the impact of flood Hazard: a case study of Kopili River basin, Assam, India. In:˜the œInternational archives of the photogrammetry, remote sensing and spatial information sciences/international archives of the photogrammetry, remote sensing and spatial information sciences, XLII(5), 455–460. doi:https://doi.org/10.5194/isprs-archives-xlii-5-455-2018

Shrestha A (2022) Flood risk mapping of Kankai Basin: a case study of Shivasatakshi municipality and Kankai municipality. ResearchGate. https://www.researchgate.net/publication/363864092

Swain KC, Singha C, Nayak L (2020) Flood susceptibility mapping through the GIS-AHP technique using the cloud. ISPRS Int J Geo Inf 9(12):720. https://doi.org/10.3390/ijgi9120720

Taromideh F, Fazloula R, Choubin B, Emadi A, Berndtsson R (2022) Urban flood-risk assessment: integration of decision-making and machine learning. Sustainability 14(8):4483. https://doi.org/10.3390/su14084483

Timbadiya PV, Patel PL, Porey PD (2011) Calibration of HEC-RAS model on prediction of flood for lower Tapi River, India. J Water Resour Prot 03(11):805–811. https://doi.org/10.4236/jwarp.2011.311090

Ullah K, Zhang J (2020) GIS-based flood hazard mapping using relative frequency ratio method: a case study of Panjkora River basin, eastern Hindu Kush, Pakistan. PLoS One 15(3):e0229153. https://doi.org/10.1371/journal.pone.0229153

Vashist K, Singh KK (2023) HEC-RAS 2D modeling for flood inundation mapping: a case study of the Krishna River basin. Water Pract Technol 18(4):831–844. https://doi.org/10.2166/wpt.2023.048

Chapter 11
Assessment of River Channel Dynamics and Flood Propagation of the Lower Subansiri River

Pratiti Debnath, Sandipan Das, and Kaustav Baidya

Abstract The Lower Subansiri River, a significant northern tributary of the Brahmaputra, showcases considerable morphodynamic variability characterized by shifts in sinuosity, braiding intensity, curvature radius, and channel movement over the past forty years due to flood migration. This paper combines multi-temporal Landsat images (1986–2025), GIS-based morphometric analysis, and one-dimensional unsteady hydraulic modelling to evaluate long-term channel change and flood propagation between the Subansiri dam and Goriamari. Findings indicate periods of migration alternating between the left and right bank, with significant shifts (up to 8 km) observed from 1995 to 2005, a phase of partial stabilization from 2005 to 2015, followed by renewed local instability starting in 2015. The devastating flood of 2000 was a geomorphic turning point, as it caused massive avulsions, bar reworking, and straightening of the master channel alterations continuing until 2025. Land use/land cover analysis shows extensive conversion of natural vegetation to agriculture and built-up area expansion, especially after 2005, augmenting exposure to flood and erosion risks. The combined analyses of morphometry, hydrodynamics, and land use/land cover highlight the role of significant hydrological events, variations in sediment load, and human activities in driving changes in river channels. The significance of these findings is crucial for managing the floodplain in the Subansiri–Brahmaputra confluence area, as well as for hazard zoning and land use planning aimed at sustainable development.

Keywords Channel migration · Spatio-temporal scale · HEC-RAS flood modelling

P. Debnath · S. Das (✉)
Symbiosis Institute of Geoinformatics, Symbiosis International (Deemed University), Pune, India
e-mail: sandipan@sig.ac.in

K. Baidya
Department of Geography, Presidency University, Kolkata, India

M. Biswas et al. (eds.), *Himalayan Flood Hazards*, Advances in Geographical and Environmental Sciences, https://doi.org/10.1007/978-981-92-0702-2_11

11.1 Introduction

The Subansiri River, originating from the western slope of Mount Porom in the Tibetan Himalayas, is the largest tributary of the Brahmaputra River. Stretching over 520 km and covering the catchment of around 37,000 km^2, the river flows through the hilly region of Arunachal Pradesh before entering the alluvial plains of Assam, where it is controlled by the tectonically unstable terrain. The study area faces vulnerability to agriculture, habitation, and infrastructure, for such a significant study on morpho-dynamic behaviour and flood propagation dynamics. Regular channel shifting and flooding over an area with a 130 km stretch in the state of Assam. Although the river flows in a steady course along the mountainous area, on the other hand, in the plain region, it experiences heavy channel migration and erosion because of a rapid decrease in gradient. As per NHPC, the mean slope of the riverbed decreases from 130 cm/km at the foothills to a negligible 24 cm/km at Chaldhowaghat, resulting in an abrupt loss in flow energy and rise in sediment deposition. The riverbanks in this reach, made up of sand, gravel, and silt, are extremely susceptible to erosion. Variation in seasonal rainfall also affects the hydrology of the river, where monthly precipitation varies between 0.02 and 615.79 mm as per IMD data. Consequently, the months of January and July receive the lowest and highest rainfall. The current research is pertinent in this regard as it investigates the hydrological variability-sediment dynamics-channel migration relationship with the help of multi-temporal satellite data (1986–2025). Even though significant research has been conducted on the Brahmaputra and its major tributaries, the lower Subansiri has been somewhat overlooked, particularly regarding hydro-sediment relationships. This research addresses that shortcoming by examining how discharge patterns and geomorphic characteristics influence river instability and the occurrence of bank line avulsions. The results offer the potential of understanding the Himalayan foreland fluvial processes and provide insights of significance for erosion hazard evaluation, floodplain management, and land use planning in a region increasingly impacted by river-triggered hazards. In the study area, river channel dynamics are controlled by an intricate interaction among hydrology, sediment transport, bank material composition, and anthropogenic factors. Substantial work has been carried out in the last three decades across the world and in the region to comprehend the mechanisms of channel migration, including the focus on the Brahmaputra basin. The present work has provided a platform for studying the hydro-sedimentological behaviour that helps to form the floodplains in the lower Subansiri area of Assam.

There are the availability of previous research articles focusing on the Brahmaputra River's fluvial dynamics and its tributaries, such as the Subansiri River, with emphasis on bank erosion and channel migration. Among the notable, Coleman (1969) and Goswami (1985) theorized on Brahmaputra basin channel processes and sedimentation. Multi-decadal changes in channel alignment in the Subansiri were later documented by Goswami et al. (1999) and Kotoky et al. (2005) with strong erosion and depositional fluvial activities along its banks. Sarma and Phukan (2004,

2006) and Debnath et al. (2017) used remote sensing and GIS to analyse river-induced transformation of land in Assam and Tripura, validating extensive lateral migration based on hydrological variability. Hazarika et al. (2015) and Das et al. (2012) also made valuable contributions by linking channel shifts with land use land cover (LULC) transitions in Brahmaputra and Subansiri floodplains. Acharjee et al. (2022) enriched this by analyzing Subansiri's channel migration from 1973 to 2017 using Landsat imagery, Support Vector Machine (SVM), and Artificial Neural Network (ANN) classification techniques to assess LULC change. Their findings demonstrated considerable built-up area expansion and agricultural land loss as a direct consequence of river migration. But even with this growing research, a key gap still exists. Previous studies have mainly focused on morphological change and LULC results, paying little attention to the intrinsic hydro-sediment dynamics, sediment texture, riverbed slope, and discharge fluctuation that are responsible for these physical changes. This study aims to complete this lacuna by combining sedimentological and hydrological variables with spatial channel behaviour, thus contributing towards a more process-based explanation of river migration within the lower Subansiri basin. Research work like that of Coleman (1969) and Schumm and Lichty (1963) provided some of the early conceptual models of sedimentation and river migration, highlighting sediment load, variability in discharge, and channel gradient as dominant influences on river form. These ideas remain of fundamental interest to Himalayan tributaries, where seasonality in high rainfall and tectonics add further complexity to fluvial processes. In India, important work has been done by scientists working on the Brahmaputra River and its major tributaries. Goswami et al. (1999, 2007) and Sarma and Phukan (2004, 2006) presented historical channel migration and bank erosion trends in Assam from multi-temporal satellite images and topographic maps. These investigations showed extensive lateral bank shifting of rivers, particularly in the alluvial plains, under the force of high-discharge, loose sediment texture, and low channel slope. Although these studies provide a thorough explanation of morphological evolution, they mainly focused on spatial-temporal trends, with little regard to the behaviour of sediments and hydrology as jointly driving forces. Recent studies by Neog (2018) and Imsong et al. (2018) applied GIS-based analyses to delineate erosion and deposition areas and delineate areas of active channel instability. Their application of contemporary remote sensing technology to evaluate channel shift in other north Indian northeast rivers like Kulsi and Dibru that offer useful methodological analogues to Subansiri. Yet these studies were more concerned with the product of migration than with underlying hydrological and sedimentological processes. Hydro-sediment behaviour, one of the main themes of the present research, has been dealt with by scientists like Kotoky et al. (2005) and Couper (2003). Their research on sediment grain-size distribution, soil cohesion, and bank material composition has demonstrated how sediment texture is related to erosion susceptibility. These results are especially relevant to the lower reaches of Subansiri, where the river changes from a steep, narrow Himalayan profile to a wide, meandering alluvial plain with erosion-prone banks comprising sand, silt, and gravel. Rainfall and discharge variability are the crucial aspects of hydrological behaviour, and have also been examined. For instance, Lahiri and Sinha (2012) examined the influence of tectonics and rainfall intensity in modifying the morphodynamics of the Brahmaputra basin. In the same vein, Das et al. (2014) and Mani et al. (2003) correlated channel instability with intense monsoonal precipitation and overbank flow that is flood-driven, illustrating how peak discharge promotes bank erosion as well as redistribution of sediment. Such evidence provides a regional context that explains hydro-sediment dynamics within the Subansiri floodplain.

Further studies, like those of Borgohain et al. (2016) and Annayat et al. (2022), have examined land use/land cover (LULC) change linked with channel shifting. These investigations identify how river shifts affect agriculture, human habitation, and wetland habitats. The role of sediment dynamics as a factor influencing land use and land cover (LULC) change has, however, not been thoroughly investigated in these studies, highlighting a significant gap that this research addresses. In addition, Sharma et al. (2003, 2005) and Kotoky et al. (2003) analyzed the geomorphology of riverine islands such as Majuli and adjacent swamps, formed by long-term sedimentation and avulsion processes due to tributary dynamics—particularly in Subansiri. Although these studies recognized the significance of Subansiri in changing the course of Brahmaputra and its sediment load.

Famous writers, Florsheim and Mount (2008) and Lawler (1993) highlighted riverbank erosion as a natural and, at times, desirable phenomenon promoting ecological diversity and renewal of floodplains. In densely populated and agriculturally valuable areas such as Assam, though, uncontrolled channel migration may lead to land loss, displacement, and infrastructural danger, as illustrated by Debnath et al. (2016) and Sarthak et al. (2015) in settlement vulnerability and hazard mapping studies. In spite of this large corpus of work, there is still an evident gap in research in terms of collective studies that relate hydrological fluctuations, sediment properties, and long-term spatial channel behaviour, especially in such tributaries as the lower Subansiri. The literature has generally dealt with erosion, hydrology, and sedimentation as distinct phenomena, with no holistic, process-oriented study connecting these parameters. Leopold and Wolman (1957) postulated basic information about channel patterns—meandering, braided, and straight—and how sediment load, bank erodibility, and slope control morphology. Howard and Hemberger (1970) developed process-based modelling of meandering river channel migration and floodplain sedimentation, with a focus on point-bar deposition and cut-bank erosion cycles. Lawler (1993) integrated lateral bank migration measurement methods, highlighting systematic change detection as vital in trajectory-based research. Zolezzi et al. (2012). Verma et al. (2021) investigated the influence of spatial width changes and curvature in meandering rivers close to the braiding transition. They established that non-linear curvature-width interactions have a significant impact on bend stability and migration rates. Khosronejad et al. (2022) used high-resolution LES combined with sediment dynamics to compute equilibrium patterns in meandering rivers and demonstrate how curvature-to-width ratio affects point-bar development and scour at bend apexes. (2021) examined channel morphology and mid-line migration of the Ganga River (1975–2020) via GIS and ARIMA modelling—exhibiting temporal patterns of sinuosity and tortuosity, and providing strong predictive models ($R^2 \approx 0.9$) of future migration. Annayat and Sil (2020) identified the Barak River, estimating morphometric meander parameters and predicting channel shifts from Landsat-based analyses; it shows how meander radius, sinuosity, and channel curvature change over time. Ghosh and Mukhopadhyay (2021) studied the Dwarkeswar River in Eastern India, with a detailed examination of avulsions, bankline migration, and landform adjustments under monsoon forces, underscoring the transition of channel patterns from meandering to braided and vice versa.

Mukherjee (2022) studied the Ramganga River in the Ganga basin, providing quantitative measures on channel planform dynamics, bend evolution, and geomorphological changes over time, underscoring the connection between bed characteristics and meander development. Wahiduzzaman and Yeasmin (2022) reported a transition from meandering to braided morphology in the Ganges–Padma River through bracing indices and sinuosity ratio analysis between 1973 and 2016, posing further questions regarding threshold transitions in channel behaviour. Gran and Paola (2001) and Tal and Paola (2007) illustrated how riparian vegetation dynamics modulate the transformation from braided to single-thread meandering channels by stabilizing bank erosion and affecting sedimentation patterns. The objectives of the study are to compare and prioritize the main geomorphic and hydrological factors controlling bankline migration. Secondly, to identify avulsion hazard zones along the Subansiri-Narayanpur stretch using geospatial technology.

11.2 Study Area

The Subansiri River, a north-flowing tributary of the Brahmaputra, originates in the eastern Himalaya and flows into Arunachal Pradesh before entering Assam, where it runs as the lower Subansiri in the plains. The downstream parts of the river, spreading over Lakhimpur and Dhemaji districts of Assam, fall under the study area, roughly covering an area between 27° 00′ N to 27° 45′ N latitude and 93° 30′ E to 94° 15′ E longitude. Here, the river enters the plain after its mountainous course at Gerukamukh and develops a wide alluvial valley, typified by numerous anabranching channels, braid bars, and oscillating meander loops. The elevation is between approximately 85 m closer to the Himalayan foothills and 70 m closer to the Brahmaputra confluence, constituting a gentle slope that supports lateral channel migration and monsoon overbank flooding. Geomorphically, the basin falls in an active tectonic area of the eastern Himalayas, where steep upstream gradients, high relief, and regular seismicity cause strong sediment delivery and quick channel adjustments in the floodplain. The area has a humid subtropical climate with an annual precipitation of more than 2000 mm, which is mainly focused from May to September and causes high sediment flux as well as regular floods. The productive floodplains accommodate high-density agrarian settlements, yet repetitive bank erosion, channel avulsions, and seasonal flooding threaten the settlements and agricultural lands. This dynamic hydro-geomorphic environment makes the lower Subansiri a suitable place

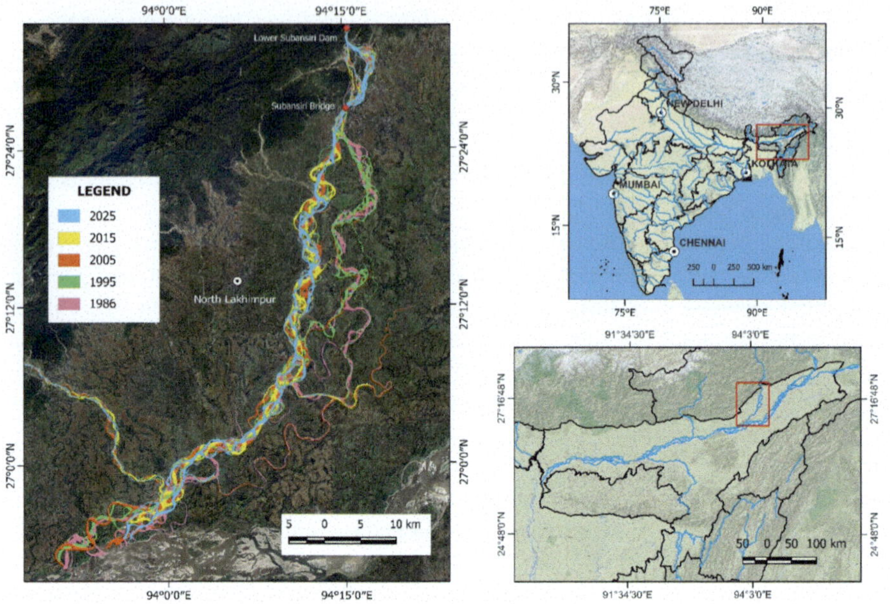

Fig. 11.1 Location map showing the extension of the basin of the Subansiri River

to study long-term channel evolution, sinuosity, and braiding pattern changes, and their influence on flood hazard in the Brahmaputra valley (Fig. 11.1).

11.3 Methodology

For the fulfilment of the objectives, the following methodology has been followed.

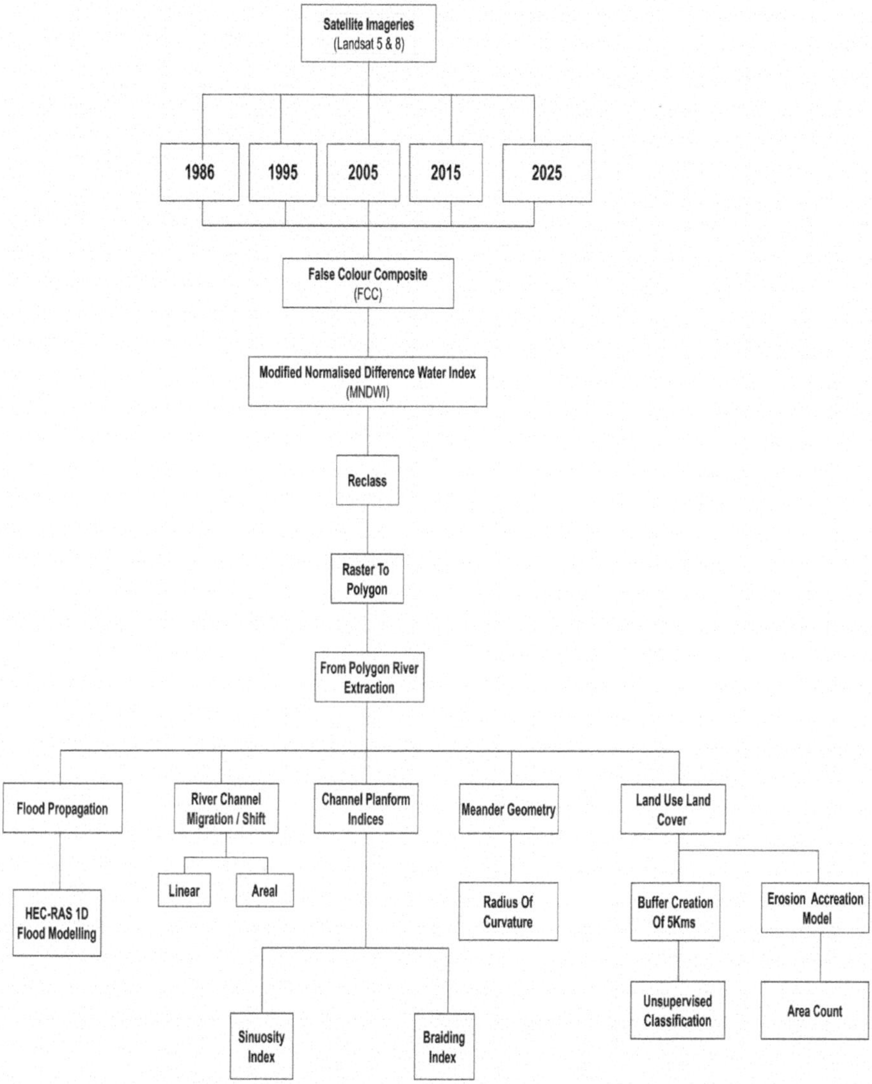

Historical channel avulsions have been well examined by workers utilizing topographical maps along with satellite images, together with GIS methods for analysis of river morphological development. In this research, Remote Sensing and GIS have been utilized for evaluating the channel dynamics of the Lower Subansiri River and its influence on surrounding land use. Multi-temporal Landsat 5 and Landsat 8 data from 1986, 1995, 2005, 2015, and 2025 were used, as these collections offer continuous multispectral coverage at 30 m spatial resolution and are easily accessed from the USGS Earth Explorer portal. The river section between the Subansiri dam site and the point where it joins the Brahmaputra was separated into three different reaches for analysis. Along these reaches, 1 km-spaced cross-sections were surveyed along the second reach and at 5 km intervals for the entire river course to record differences in channel geometry.

Major channel parameters, including channel width, lateral shift, sinuosity index, and erosion areas, were calculated from digitized riverbank lines (left and right banks) obtained from the satellite images (Table 11.1).

Cross-section-based analysis enabled the observation of channel widening trends, while reach-wise observation allowed the calculation of the sinuosity index and overall dynamics in the channel area. Erosion, deposition, and zones of historical migration were mapped using overlay techniques, reflecting spatio-temporal channel adjustments. The sinuosity index, a quantitative measure of meandering, was computed using the formula developed by Schumm (1963). It is expressed as follows:

Sinuosity index

$$S = \frac{\mathrm{OL}}{\mathrm{EL}} \tag{11.1}$$

where: S = sinuosity index, OL = observed (actual) length of a river, and EL = expected straight distance of a meandering river.

To measure the intensity of braiding of the Subansiri River, mid-channel and lateral bars were located and mapped from false colour composite (FCC) satellite images for the chosen years. The combined length of the bars (ΣL_b) was determined and compared to the combined reach length of the river (L_r) to calculate the Braiding Index (BI) based on the technique suggested by Brice (1964). It is expressed as:

$$\mathrm{BI} = \frac{2L_b}{\sum L_r} \tag{11.2}$$

where: $\sum L_b$ = total length of mid-channel bars within the reach, L_r = total length of the river reach.

Changes in the meandering course of the Subansiri River were analyzed for a period of five selected years, and quantitative measures were performed for every

Table 11.1 Characteristics of the reaches of the Subansiri

Parameters	Reach-1	Reach-2	Reach-3
Areal extent	Subansiri dam to Chaldhowa Ghat	Chaldhowa Ghat to Na-Ali Koibarta	No.2 Na-Ali Koibarta to the confluence of the Brahmaputra
Length	14.393	46.9	37.226
Sinuosity index	1.274	1.193	1.176
Morphological character	Sinuous and Anabranching	Straight to sinuous	Meandering to braided

period. The channel thalwegs were georeferenced with GIS software, and important parameters like radius of curvature and sinuosity were calculated with vector measuring tools based on the procedure described by Leopold and Langbein (1966) to analyze meander geometry and planform attributes. This method allowed a thorough evaluation of the temporal changes in the meandering trend of the river and channel behaviour.

The formula for meander geometry is expressed as:

$$L_{\mathrm{m}} \times \frac{\sqrt{k}}{13} \times \sqrt{k-1} \tag{11.3}$$

where: L_{m} = Meander length (the length of one complete meander loop along the thalweg), k = Number of meander loops in the selected river reach, 13 = Empirical constant used in meander geometry computation.

The 2000 flood event (June and August) in the Subansiri River was simulated using HEC-RAS 1D unsteady flow simulation. River geometry was prepared by digitizing center line, left and right-bank lines from topographic/DEM data, and defining cross-sections along the reach. June and August discharge hydrographs were entered as upstream boundary conditions, with a uniform Manning's roughness coefficient (n) value of 0.40 assigned to simulate channel and floodplain resistance. The model was run for both months, giving simulated water surface elevations and depths. The maximum flood depths were pulled out and mapped to give inundation zones for the event period. LU/LC maps of the 2000 flood event time were incorporated to evaluate the land use patterns that are affected by floods.

11.4 General Physiography

The Brahmaputra's largest northern tributary, the Subansiri River, takes origin in the Tibetan Plateau close to Mount Porom in China and descends through the mountainous region of Arunachal Pradesh's tough terrain before bursting into the Assam plains. Following its fall from the Himalayan foothills at Gerukamukh, the river flows into the rich alluvial floodplains of Lakhimpur and Dhemaji districts to create its lower reach, which forms the subject of this study. Its lower course of the Subansiri stretches roughly between 27° 00′ N to 27° 45′ N latitude and 93° 30′ E to 94° 15′ E longitude. Stretching from the dam location at Gerukamukh to its junction with the Brahmaputra, the length is approximately 130 km, with channel width significantly fluctuating due to the high dynamic braided and meandering nature of the river. When the river flows from the precipitous Himalayan slopes to the gradual Assam plains' slopes, the speed of the river drops significantly, promoting sedimentation, bar creation, and frequent bifurcations of channels. The alluvial floodplain around the reach is distinguished by natural levees, abandoned meander scars, oxbow lakes, and seasonal wetlands that attest to its dynamic fluvial history. The area has a humid subtropical monsoon climate with more than 2000 mm of annual rainfall, which fuels high seasonal discharge and frequent flooding. The unconsolidated alluvial deposits that make up this reach are rich but vulnerable to serious bank erosion that leads to continual lateral channel migration. Moreover, tectonic activity in the eastern Himalayan front controls the morphology of the river, causing abrupt avulsions and time-consistent channel adjustments (Fig. 11.2).

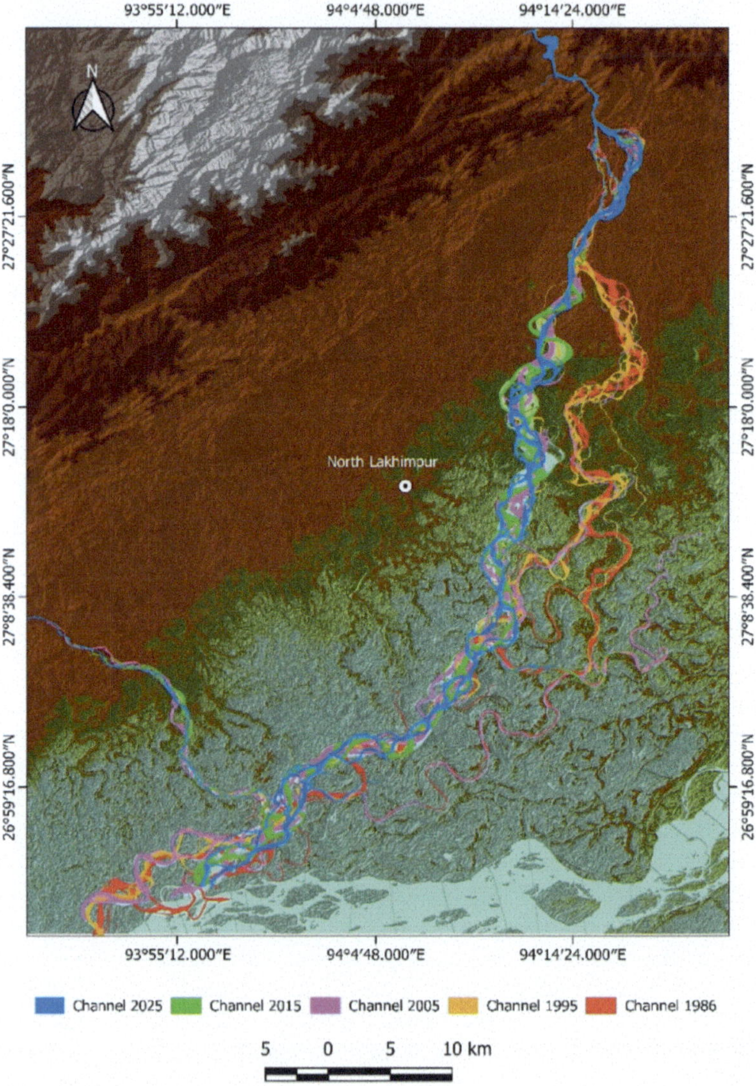

Fig. 11.2 Channel migration in different years

The gradient of the Subansiri River has a strong contrast between its upper mountainous sections and the lower floodplains of Assam. Starting from the high-relief Tibetan Plateau and flowing through the mountainous topography of Arunachal Pradesh, the river plunges sharply before spilling out into the Assam plains at Gerukamukh. In the lower Subansiri reach, extending some 130 km from Gerukamukh to where it enters the Brahmaputra, the mean channel slope decreases considerably, varying from around 0.7–1.2 m/km in the foothill zone to less than 0.5 m/km in the central floodplain sector. This sharp decrease in slope allows for large-scale sedimentation, braid bar formation, and lateral migration of the channel and is responsible for the dynamic nature of the river's morphology. The gentle gradient also promotes (Fig. 11.3) frequent overbank flooding during monsoon peaks, leading to the formation of levees, scroll bars, and oxbow lakes characteristic of this alluvial landscape. Such low-slope conditions in the lower reach are a key driver of the evolving sinuosity and braiding patterns that dominate the Subansiri's floodplain geomorphology (Fig. 11.4).

The aspect slope direction (Fig. 11.5) in the terrain, which is represented by the Subansiri River basin, signifies the gradation from the Himalayan foothills that are steep to the low-gradient Assam plains. The slopes, which face south and southwest primarily, steer the river's fall towards the Brahmaputra valley in its upper region, closer to Arunachal Pradesh. When the river enters the floodplains of the lower Subansiri in Assam, the aspect pattern becomes more mixed with east and west-facing slopes connected with meander bends, braid bars, and natural levees along channel margins. The aspect pattern varies and affects insolation, moisture distribution, and erosion–deposition processes on the floodplain. South-facing slopes within the foothill area get more solar radiation, enhancing snowmelt and runoff during pre-monsoon periods, whereas the lower-lying north-facing parts in the plains are more moisture-holding and favourable for sediment deposition and vegetation growth. Knowledge about the aspect pattern is thus crucial to understand channel migration trends, floodplain morphology, and land use dynamics within the lower Subansiri basin.

The drainage pattern of the Subansiri River is an integral component of the north Brahmaputra basin and exhibits a sub-dendritic to sub-parallel pattern in its headwaters, evolving into a more braided and anabranching network in the plains of Assam (Fig. 11.6). The river rises in the high-relief Tibetan Plateau and receives many tributaries from Arunachal Pradesh's rugged terrain in the form of the Ranganadi, Dikrong, Panyor, and Kamala rivers, before entering the alluvial floodplains of Lakhimpur and Dhemaji districts. In its lower section, covering about 130 km from Gerukamukh to the Brahmaputra confluence, drainage becomes highly distributary as a result of Himalayan sedimentation and decreasing channel slope. The basin has a high drainage intensity in the foothill area, which diminishes considerably in the floodplain area where broad channels, numerous bars, and abandoned channels predominate the topography. Seasonal monsoon runoff and intense sediment supply result in recurring channel avulsions and lateral migrations, generating a complex network of active and inactive channels, oxbow lakes, and floodplain wetlands. This dynamic

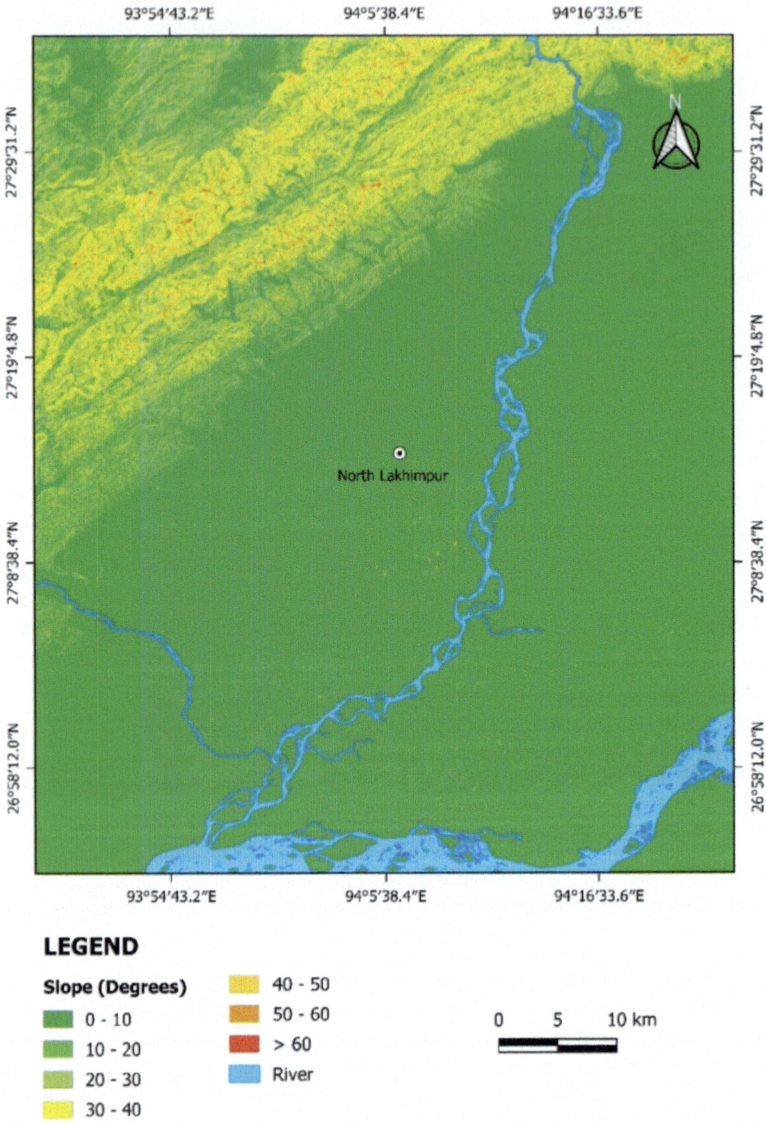

Fig. 11.3 Slope map of the study area

drainage pattern is of significant importance in determining the river's flood hazard potential and land use patterns in the lower Subansiri basin.

The shaded relief of the Subansiri River basin gives a vivid visualization of its topographic changes, emphasizing the gradient from rugged Himalayan foothills to the gently inclined floodplains of Assam. Extracted from high-resolution Digital Elevation Models (DEMs), the shaded relief clearly shows steep gradients and

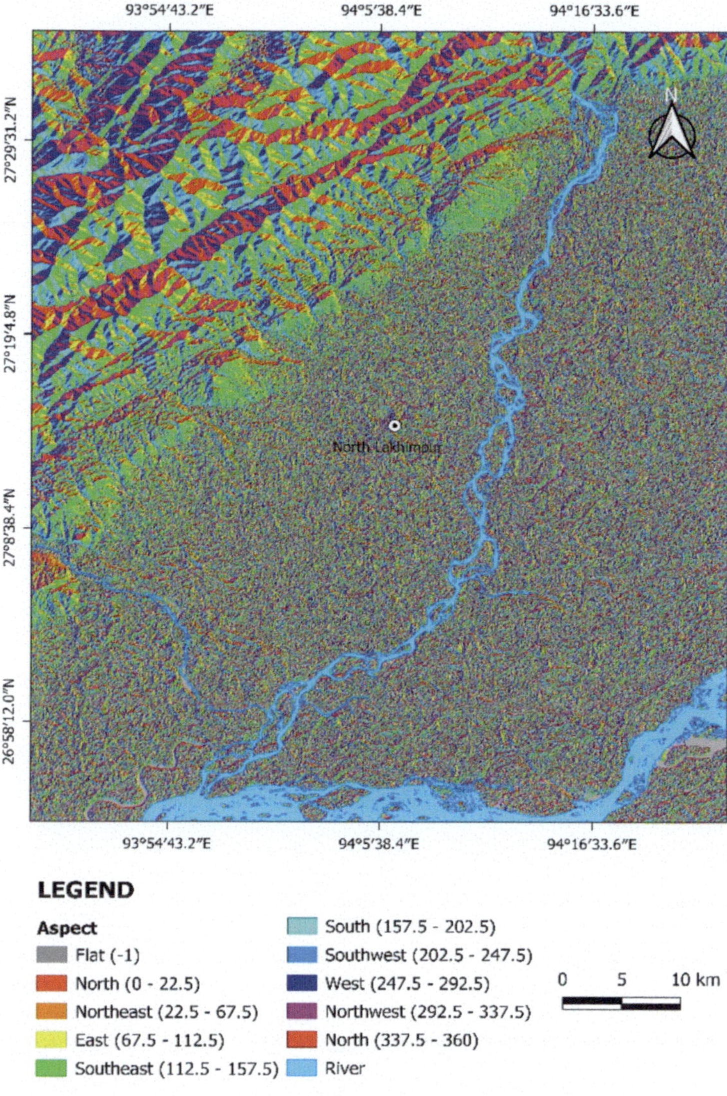

Fig. 11.4 Aspect slope direction map of the area

dissected terrain in the upper Arunachal reach, where elevations are above 1000 m, as compared to the broad alluvial plains of the lower Subansiri that range between 65 and 85 m above mean sea level. The relief map highlights features like valley incisions, natural levees, meander scars, and braid bars, and in so doing, provides information about both erosional and depositional processes. The upper catchment's steep slope allows high sediment delivery, whereas the lower reach's low relief encourages sediment deposition and repeated channel migration. This clear topographic

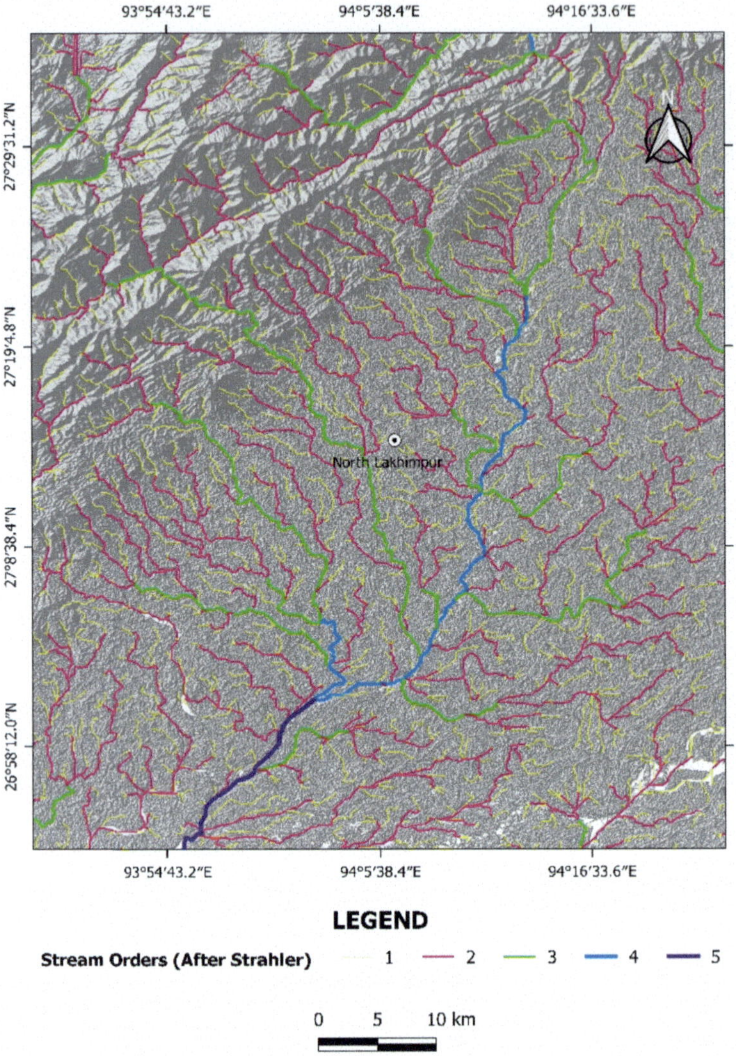

Fig. 11.5 Drainage map with stream ordering of the Subansiri River area

contrast found through river shading analysis is important to recognize in interpreting flood propagation, bank erosion pattern, and channel stability in the lower Subansiri floodplain.

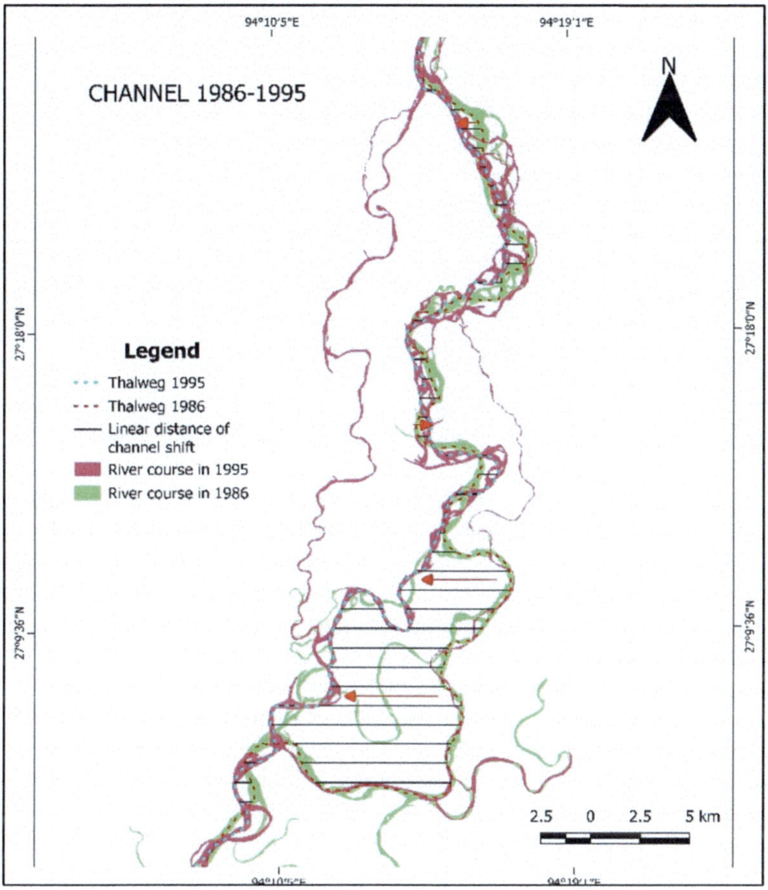

Fig. 11.6 Channel shift of 1986–1995

11.5 Results

11.5.1 Channel Migration/Shift

Lateral channel erosion is a well-marked geomorphic process along the Subansiri River that is a major factor in the progressive enlargement and migration of the channel through time. In the 1 km cross-section of Stretch-2 of the comparative years (1986–1995), (1995–2005), (2005–2015), (2015–2025), it has been observed that a drastic shift occurred between the years 1995–2005 with a highest value of 8.1683 km. The channel has shifted in the west direction. The 1986–2025 channel migration study of the Lower Subansiri River identifies marked spatio-temporal variability in lateral migration and planform change. In 1986–1995, the river underwent extensive left-bank migration, especially in the upstream and mid-reach parts (A1–A22), with maximum migrations of 6–8 km and marked meander loop extension. During 1995–2005, the trends of migration reversed to right-bank dominance, with mid-reaches (A19–A31) moving up to 4–6 km, coupled with cutoff meanders and avulsions. The 2005–2015 phase was one of relative stabilization, with movements limited to 0.5–2 km and balanced migration on both banks, with a reduction in thalweg curvature. Nevertheless, 2015–2025 saw renewed right-bank activity, particularly between A25–A33, with 2–3 km movements and a significant westward channel realignment and straightening close to the Brahmaputra confluence. These patterns indicate alternate bank erosion stages and dynamic morphodynamics shaped by hydrological and sediment processes. The latest phase (2015–2025) shows an upsurge in channel activity, with resurgent right-bank migration focused between stations A25–A33 and 2–3 km lateral shifts documented. The phase is characterized by westward channel realignment and gradual straightening close to the confluence with the Brahmaputra River, indicating a transition from meandering to semi-straight planforms in terms of morphodynamics. Confluence migration and pronounced thalweg adjustments observed in this period underscore the sensitivity of this segment to both hydrological variability and sediment redistribution processes (Figs. 11.7, 11.8, 11.9, 11.10, 11.11, 11.12, 11.13, 11.14, 11.15, 11.16, 11.17, 11.18, 11.19 and 11.20; Table 11.2).

For meander geometry, it has been observed that the sinuosity of Stretch 2 for both 1986 and 1995 is more than 1.5, which states it is meandering in nature, whereas in 2025, the sinuosity value remains constant between the 1.1–1.2 range. Its high sinuosity in 1986 and 1995 in Stretch 2 (Chaldhowa Ghat to Na-Ali Koibarta) is a result of its geomorphic as well as hydrological environment. The stretch falls in the middle floodplain zone where the river gradient falls precipitously after leaving the Himalayan foothills, forming environments conducive to lateral channel migration and development of meander loops. High sediment input from the upstream

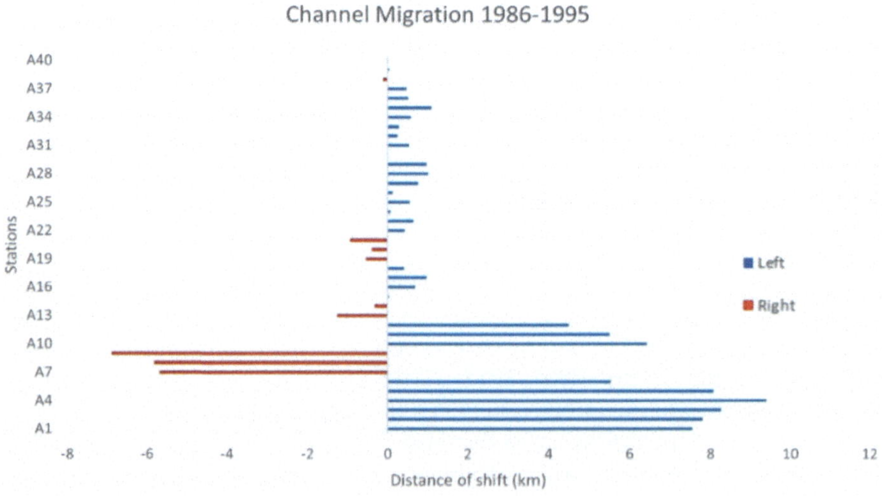

Fig. 11.7 Graph for channel shift of 1986–1995

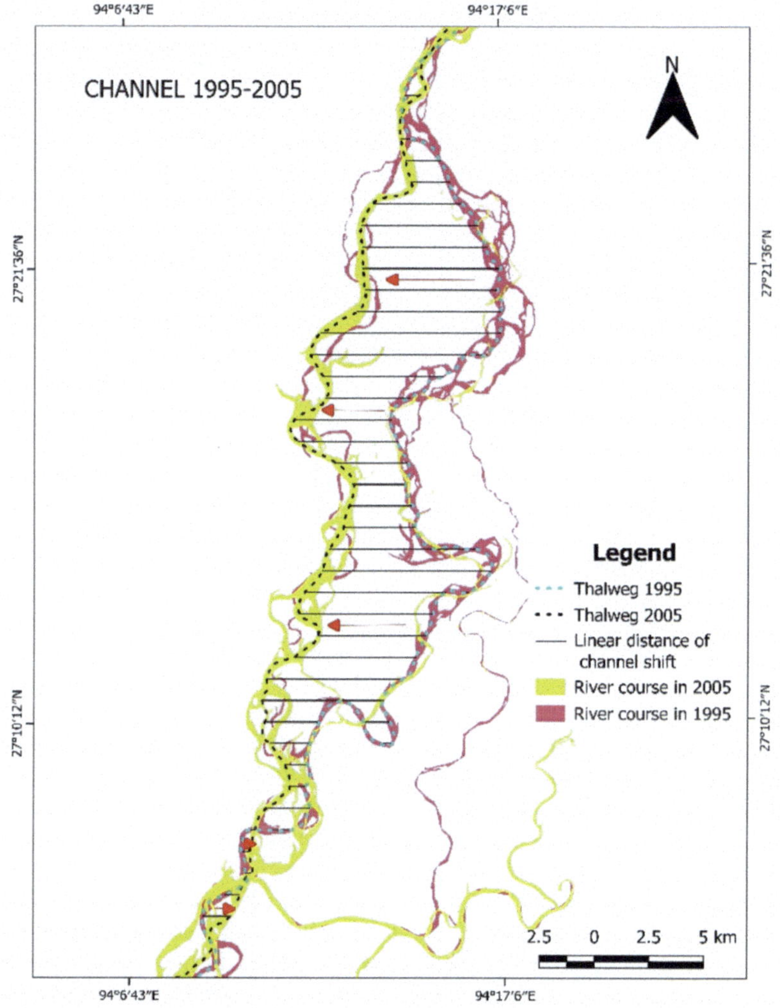

Fig. 11.8 Channel shift of 1995–2005

Himalayas during monsoons probably favoured bar development and flow deflection, enhancing channel curvature even further. Availability of loose alluvial soils and ample floodplain room enabled the river to swing unconditionally, leading to a greater development of bends. The occurrence of extreme floods during the late 1980s and early 1990s escalated bank erosion and channel migration, enhancing sinuosity in this reach (Figs. 11.21, 11.22, 11.23, 11.24 and 11.25; Tables 11.3 and 11.4).

For the braiding pattern in the Subansiri River, Stretches 1 and 3 are taken into account, as there is no braiding pattern found in Stretch 2. The differences in braiding index across the Subansiri River indicate pronounced spatio-temporal variations in

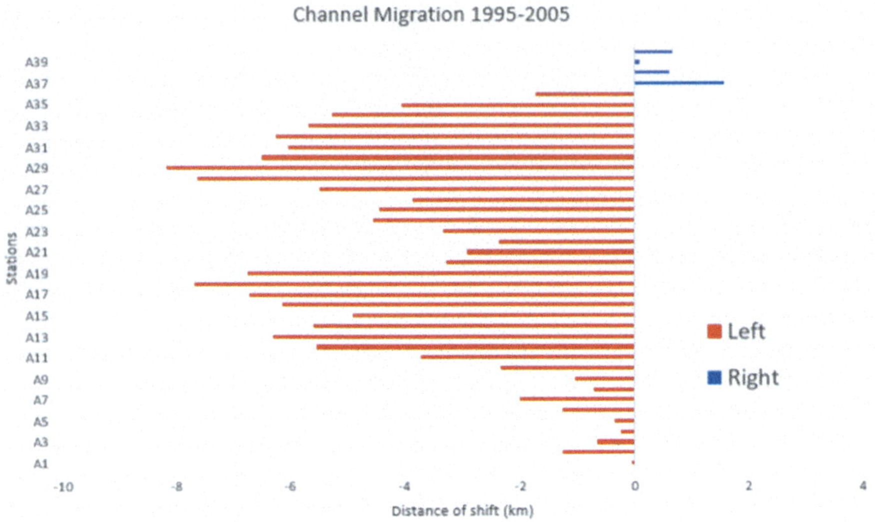

Fig. 11.9 Graph for channel shift 1995–2005

channel dynamics. In Stretch 3 (Na-Ali Koibarta to Brahmaputra confluence), the braiding index attained its maximum of 2025, reflecting widespread channel splitting and bar growth in the lower reach. This increase can be attributed to the vast sediment load brought in from upstream with the significantly decreased gradient in the vicinity of the Brahmaputra confluence, which encourages sediment deposition and the formation of several anabranches. Further encouragement of lateral channel dispersal is given by the broadening of the floodplain in this reach, augmenting braiding intensity over the last few years. Conversely, Stretch 1 (Lower Dam to Chaldhowa Ghat) displayed its peak braiding index in 1986, and this can be explained through the post-monsoon sediment bursts during the early study years as well as natural adjustment of the river after high-discharge events during the same years. The existence of unconsolidated alluvium within this upper reach also permitted channel instability, resulting in several mid-channel bars. Throughout the following decades, controlled discharges from upstream hydraulic facilities and progressive floodplain stabilization seem to have attenuated braiding intensity within this reach, while the downstream reaches went on to develop dynamically under dynamic sedimentation and floodplain processes (Figs. 11.25 and 11.26).

Comparison of Land Use/Land Cover (LULC) maps for 1986, 1995, 2005, 2015, and 2025 along the Subansiri River identifies extensive changes in the floodplain environment. During the decades, agricultural land has increased significantly, frequently substituting natural vegetation cover, which demonstrates a clear decrease, particularly after 2005. Built-up areas demonstrate consistent growth, with high growth rates in 2015 and 2025, indicative of fast settlement development and infrastructural extension along riverbanks; nevertheless, slight classification inconsistencies within these years overestimate built-up areas to some extent in specific segments.

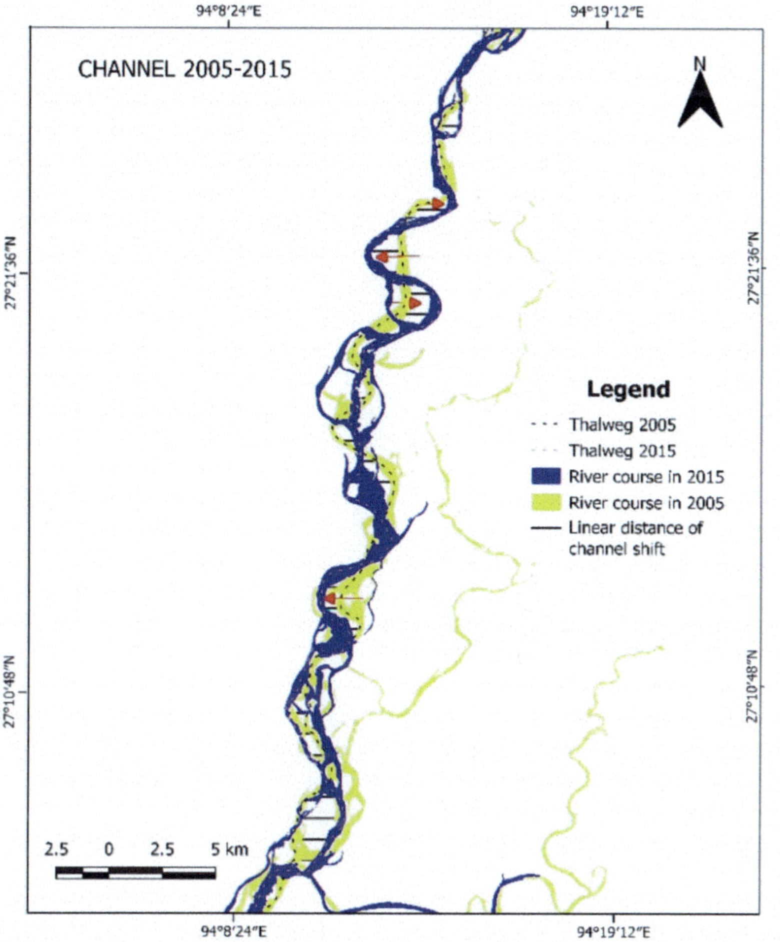

Fig. 11.10 Channel shift of 2005–2015

Concurrently, river sediments and arid land vary with channel migration, erosion, and sediment deposition during years of high floods. These mobile shifts directly affect livelihood activities—more agricultural conversion favours local agricultural economies, but declining vegetation and unstable river sediments increase exposure to flood risks, erosion loss, and displacement of communities. The development of built-up areas within active channels also highlights increasing risk exposure for settlements in this morphologically active floodplain (Fig. 11.27a–e).

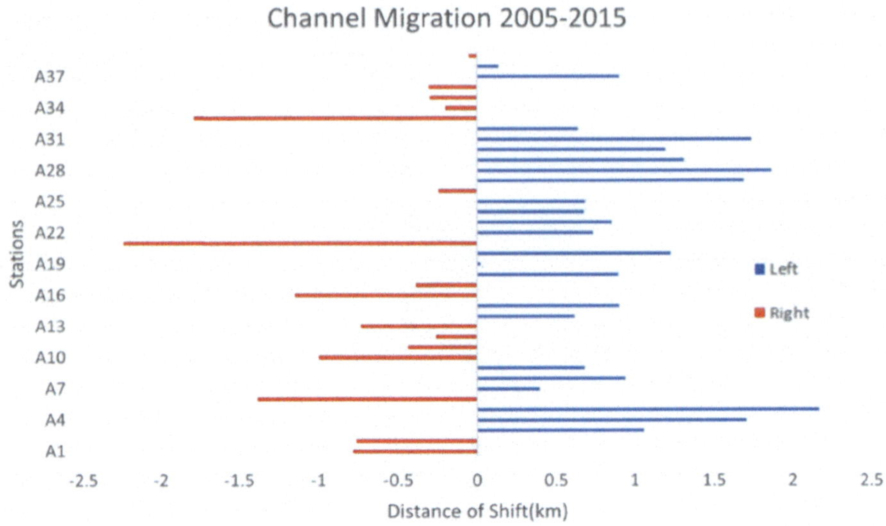

Fig. 11.11 Graph for channel shift of 2005–2015

11.5.2 Flood Hazard Modelling

One-dimensional unsteady hydraulic modelling method, combined with GIS-based river geometry, was used to simulate the flood hydrodynamics along the Subansiri River from the dam to Goriamari. The output of the June model showed a generalized flooding trend, with depths of flooding spanning from about 0.001 m on the outer floodplain boundaries to about 7.20 m at maximum in the channel and deep anabranch reaches.

The deepest depths were found in the upstream braided reaches and confluence regions, where there is high channel capacity and stream waters are channel, and the floodplain had moderate depths of 1–3 m. The August simulation displayed a broadly similar spatial inundation pattern but with a reduction in maximum depth to around 6.04 m, alongside a slight contraction in the inundated extent, particularly in the marginal floodplain areas. The downstream section approaching Goriamari experienced sustained inundation in both months, with depth variations reflecting local topography and channel alignment. Generally, the June situation reflected the maximum flood stage, whereas August exhibited a partially reduced wave of the flood, albeit still with significant inundation over the floodplain (Fig. 11.28a, b).

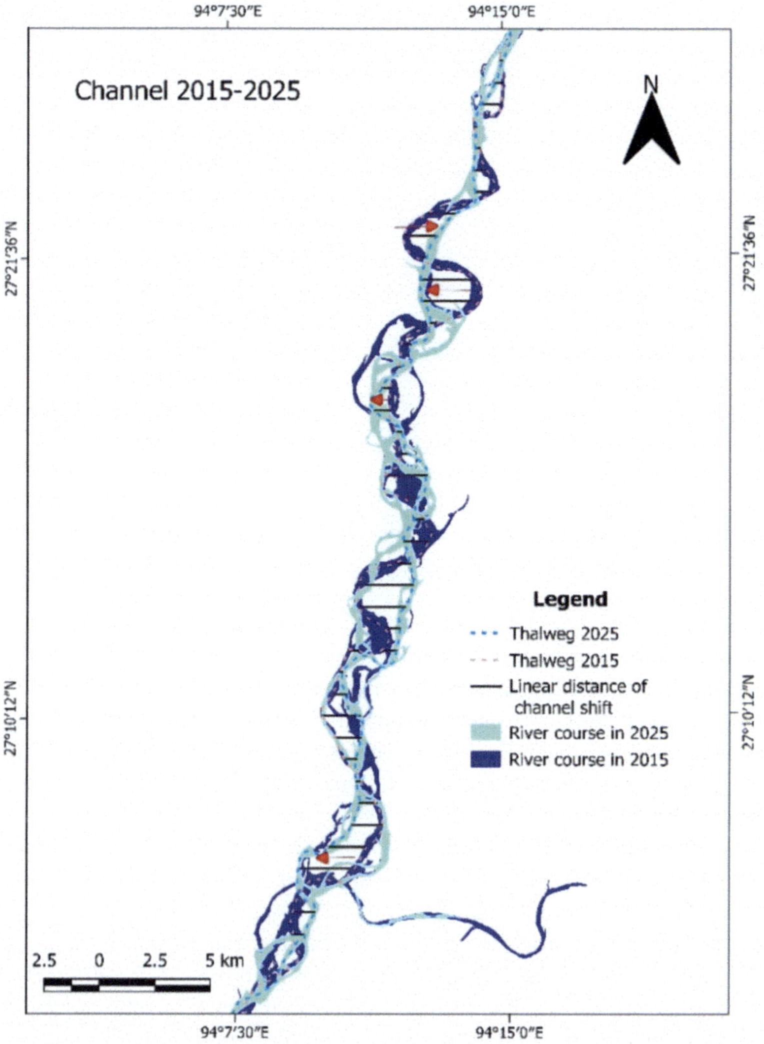

Fig. 11.12 Channel shift of 2015–2025

11.6 Discussion

The Subansiri River shows significant morphodynamic variability in its lower reaches, with measurable change occurring in radius of curvature, sinuosity, braiding intensity, channel migration, and the surrounding land use/land cover (LULC) between the years 1986 and 2025.

Meander geometry analysis shows that large-scale variations in radius of curvature and sinuosity occur between various temporal time frames. During 1986 and 1995,

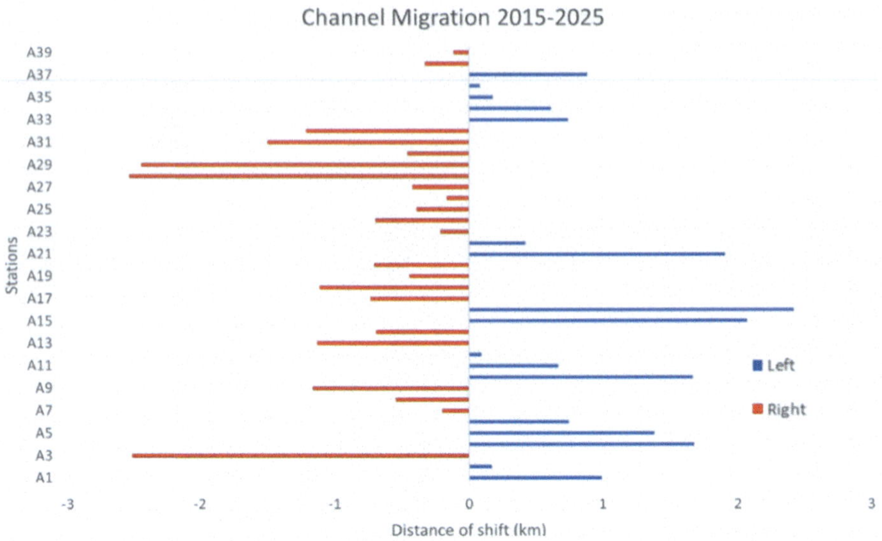

Fig. 11.13 Graph for channel shift of 2015–2025

the middle part of the reach (Chaldhowa Ghat to Na-Ali Koibarta) shows a high sinuosity value with smaller radii of curvature, typical of closely looping meanders developed due to a gentle channel gradient and high lateral expansion. This implies active meander development and development of the floodplain during these years, perhaps under the influences of higher sediment load and flood occurrences in the late 1980s–early 1990s. In later years (2005–2025), sinuosity has a straight trend in some reaches and formed potential cutoffs of older meanders, which are presumably due to avulsions or anthropogenic factors like embankment placement and flood control structures. These flood events also validate the flood modelling.

The braiding index exhibits as a geomorphic indicator in the spatio-temporal trend. The high braiding index value in 1986 was reported in the upstream reach (Lower Dam to Chaldhowa Ghat) due to high sediment deposition after high-discharge episodes and deposition of unconsolidated material from the banks. But in 2025, the most braided reach was the downstream reach (Na-Ali Koibarta to Brahmaputra confluence) with a number of anabranches and mid-channel bars. This change documents the progressive downstream sediment redistribution and channel belt widening close to the junction, where low slopes and extensive floodplains allow for braiding.

Historical lateral migration trends from the channel indicate recurring lateral movements and high bar dynamics over the study duration. Channel migration rates are greater in the mid and lower reaches, where unconsolidated alluvial deposits and broad floodplains accommodate substantial channel mobility. The alternating phases of bank erosion and deposition, with significant channel position changes observed during 1986–1995 and revived instability after 2015, are likely related to catastrophic flood events and enhanced sediment delivery from higher reaches. The

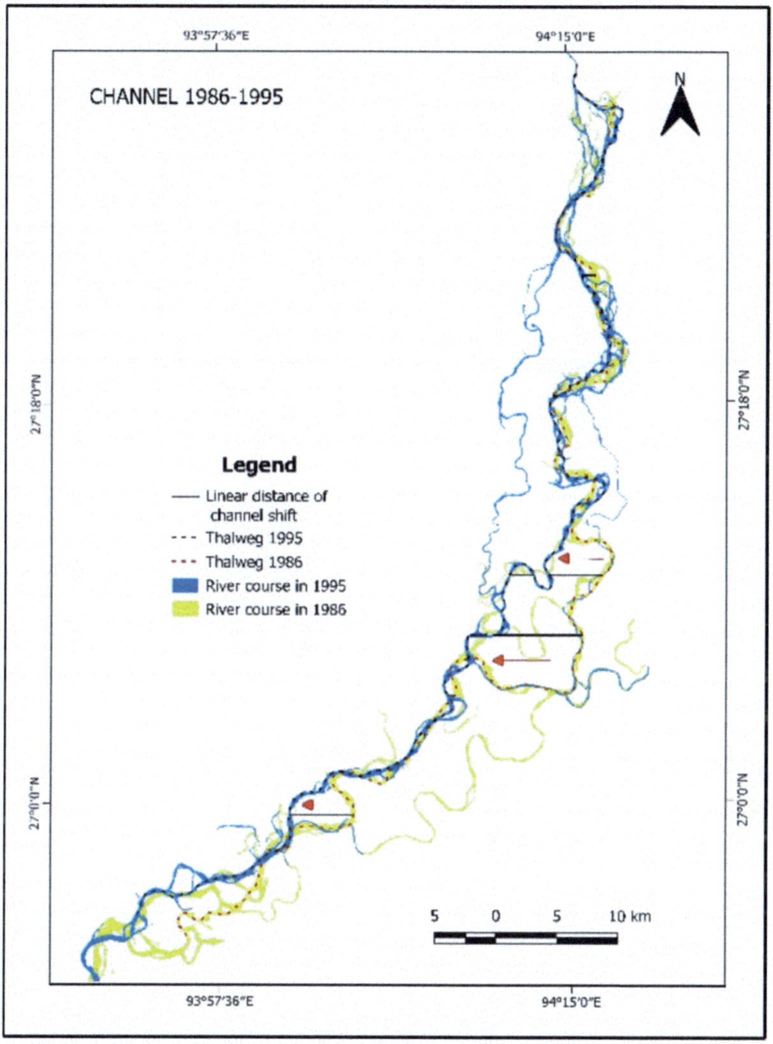

Fig. 11.14 Channel shift of 5 km cross-section section 1986–1995

spatio-temporal variability of river channel migration in the Subansiri River from 1986 to 2025 shows tremendous lateral and sections of Stretch 2 (Chaldhowa Ghat to Na-Ali Koibarta), indicating high-energy lateral shifting pattern variations between different reaches. During the initial phase (1986–1995), the river showed prominent left-bank migration with great shifts of over 6–8 km, especially in the upper reach (A1–A10), and erosion associated with maximum monsoonal discharges during this time. Right-bank migration was relatively localized and less in extent, confined to short reaches (A13–A19). Over the following decade (1995–2005), the direction of migration was reversed, with widespread right-bank erosion controlling the middle

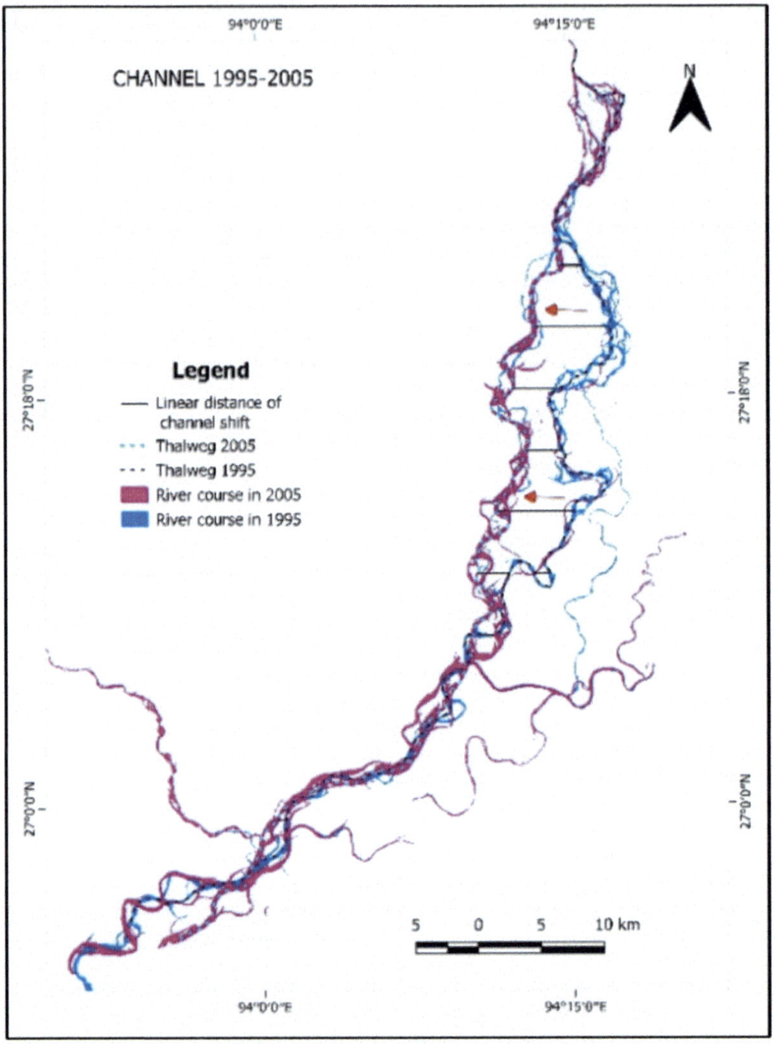

Fig. 11.15 Channel shift of 5 km cross-section section 1995–2005

and downstream parts, where changes of 4–6 km were noted (especially A19–A31). This period is the most extreme channel instability in the record, as extensive avulsion and significant planform realignment are visible in the channel overlays.

The 2005–2015 time interval is characterized by a significant decrease in total migration size, with movements typically limited to 0.5–2 km, reflecting relative stabilization of the channel. In this case, the migration is better left–right bank balanced, with slight right-bank prevalence apparent in downstream sections. The 2015–2025 period, however, exhibits revived instability, with right-bank movements greater than 2–3 km around mid-reaches (A25–A33) corresponding to active bar

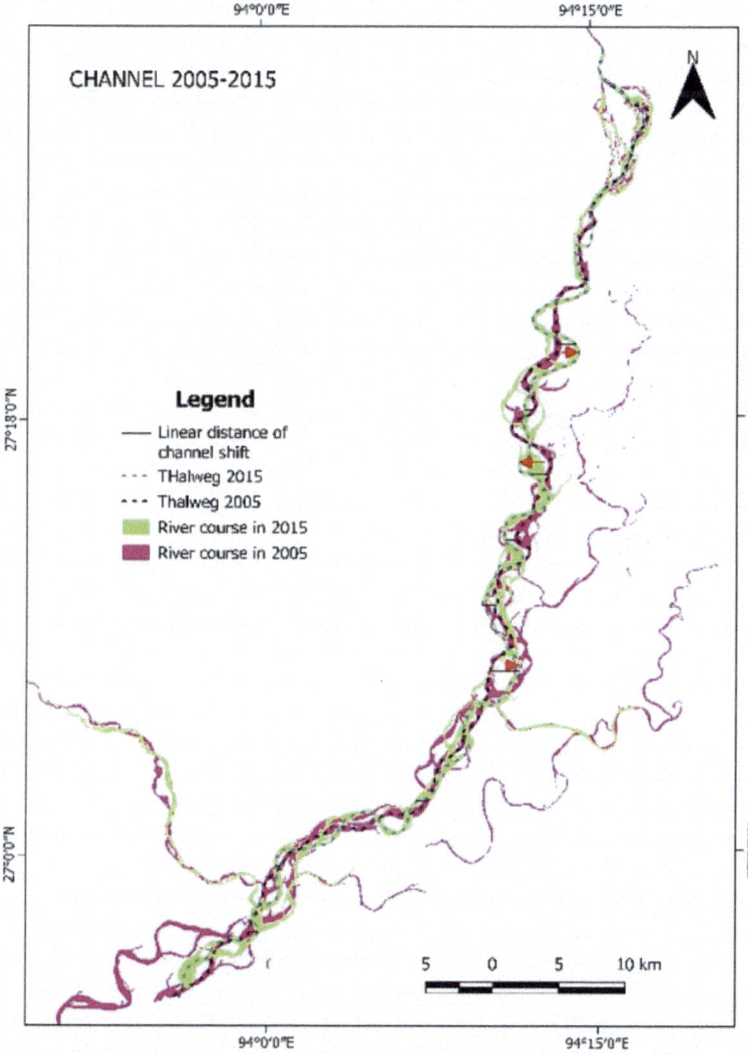

Fig. 11.16 Channel shift of 5 km cross-section section 2005–2015

processes and meander cutoff events. Notably, temporal trends illuminate that Stretch 2 has the highest lateral mobility throughout all time periods at all times, consistent with its broad floodplain, low-gradient, and unconsolidated sediment make-up. In space, migration patterns reflect a westward upgrading of the channel belt, supported by thalweg realignments and confluence position changes with the Brahmaputra.

Overall, the records indicate a shift from extremely dynamic conditions (1986–2005) to partial stabilization (2005–2015), then to resumed but localized changes (2015–2025). Such trends are controlled by hydrological extremes (particularly the

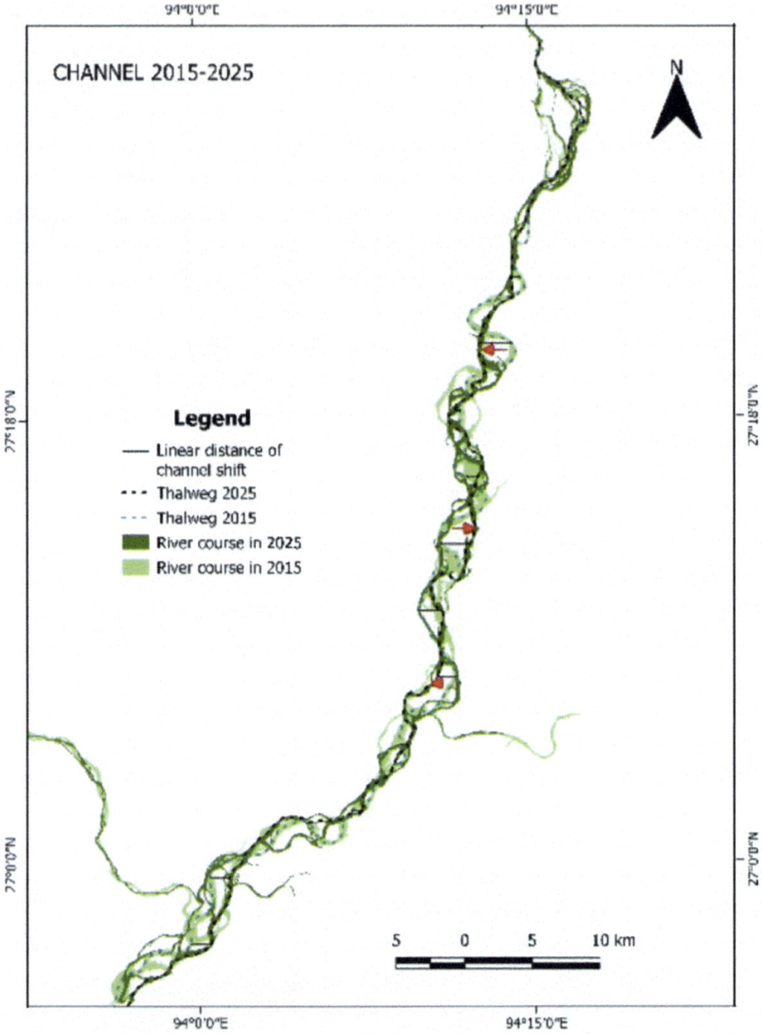

Fig. 11.17 Channel shift of 5 km cross-section section 2015–2025

2000 flood), sediment load fluctuation, and anthropogenic actions like the Dhunaguri Bridge construction, which modified flow allocation and bar construction upstream. The combined temporal (decadal) and spatial (cross-section-wise) analysis underscores the Subansiri's evolving morphodynamics, with implications for flood hazard zonation, bankline management, and LULC planning across the lower basin. LULC change from 1986 to 2025 indicates a large-scale conversion of natural vegetation into agricultural land, indicating increased human use of the floodplain. Urban area increased significantly after 2005, particularly around settlement clusters close to active channel belts, whereas barren land and river deposits varied with channel

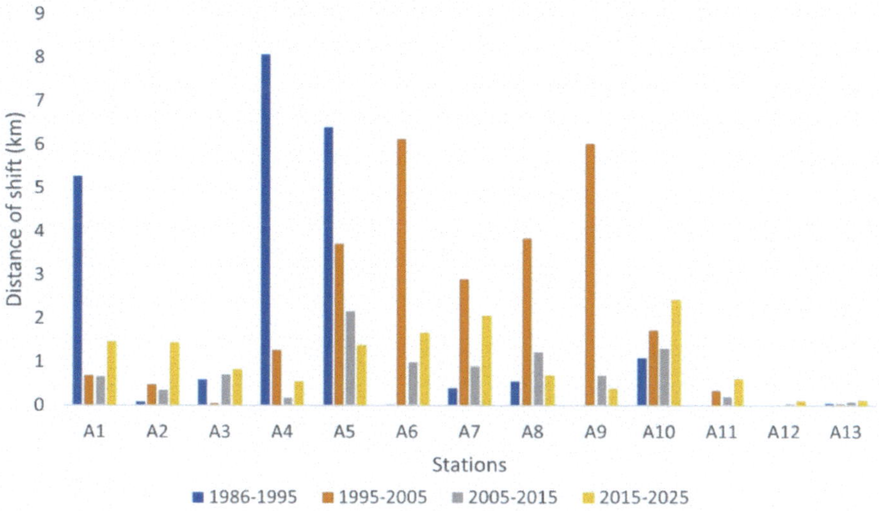

Fig. 11.18 Composite bar graph of the rivers of different timelines

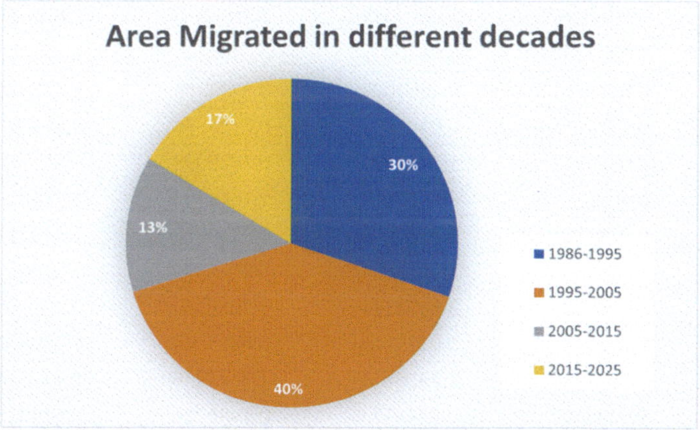

Fig. 11.19 Pie chart for the area migration in different decades

movements and avulsion occurrences. This change has flood vulnerability implications directly, because expansion of agriculture into erosion-prone areas and growth of settlements along migrating channels expose the land to flood risks and bank instability.

Interactions between meander dynamics (sinuosity and radius of curvature) and braiding intensity and channel migration have direct effects on LULC changes in the Subansiri floodplain. Because the river transitions among meandering and braiding forms, periodically forming new floodplain surfaces and abandonment of bars creates cycles of agricultural encroachment, vegetation loss, and settlements. Such dynamic

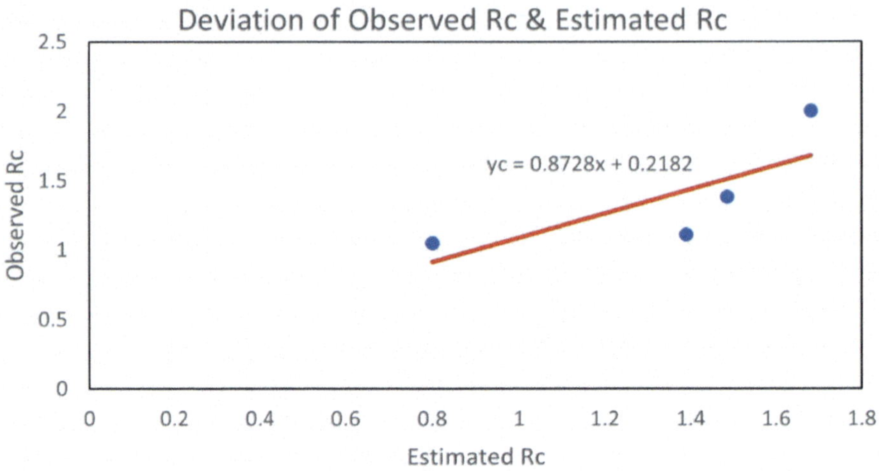

Fig. 11.20 Scatter plot for observed and estimated Rc

Table 11.2 Calculated table for the composite bar graph of rivers

Stations	1986–1995	1995–2005	2005–2015	2015–2025
A1	5.2728232	0.6870637	0.6702499	1.4649915
A2	0.0874761	0.4867078	0.3637755	1.4450224
A3	0.5974835	0.050297	0.7125838	0.838093
A4	8.0723877	1.271659	0.1730093	0.5572789
A5	6.408631	3.7212656	2.1665704	1.3848139
A6	0.0284035	6.1385401	0.9910967	1.67001
A7	0.3989538	2.9063332	0.9026375	2.0706249
A8	0.5530971	3.8438111	1.2286139	0.7029833
A9	0.012718	6.0272166	0.6895762	0.389844
A10	1.0947816	1.7324816	1.3145939	2.4413146
A11	0.0248397	0.3448773	0.1986226	0.6143152
A12	0.015384	0.024231	0.0478917	0.1160627
A13	0.0526445	0.3448773	0.0906121	0.1280399

processes highlight the imperative of ongoing monitoring of morphometry and LULC to guide floodplain management and reduce hazards to livelihoods in this extremely dynamic fluvial system.

The comparative values of June and August unequivocally demonstrate the hydrodynamic variability during the 2000 flood season along the Subansiri River. June's increased depth and span are related to the initial monsoon peak, which possibly results from the combined effect of high-intensity precipitation and upstream catchment snowmelt, producing significant channel floodplain connectivity. The extreme

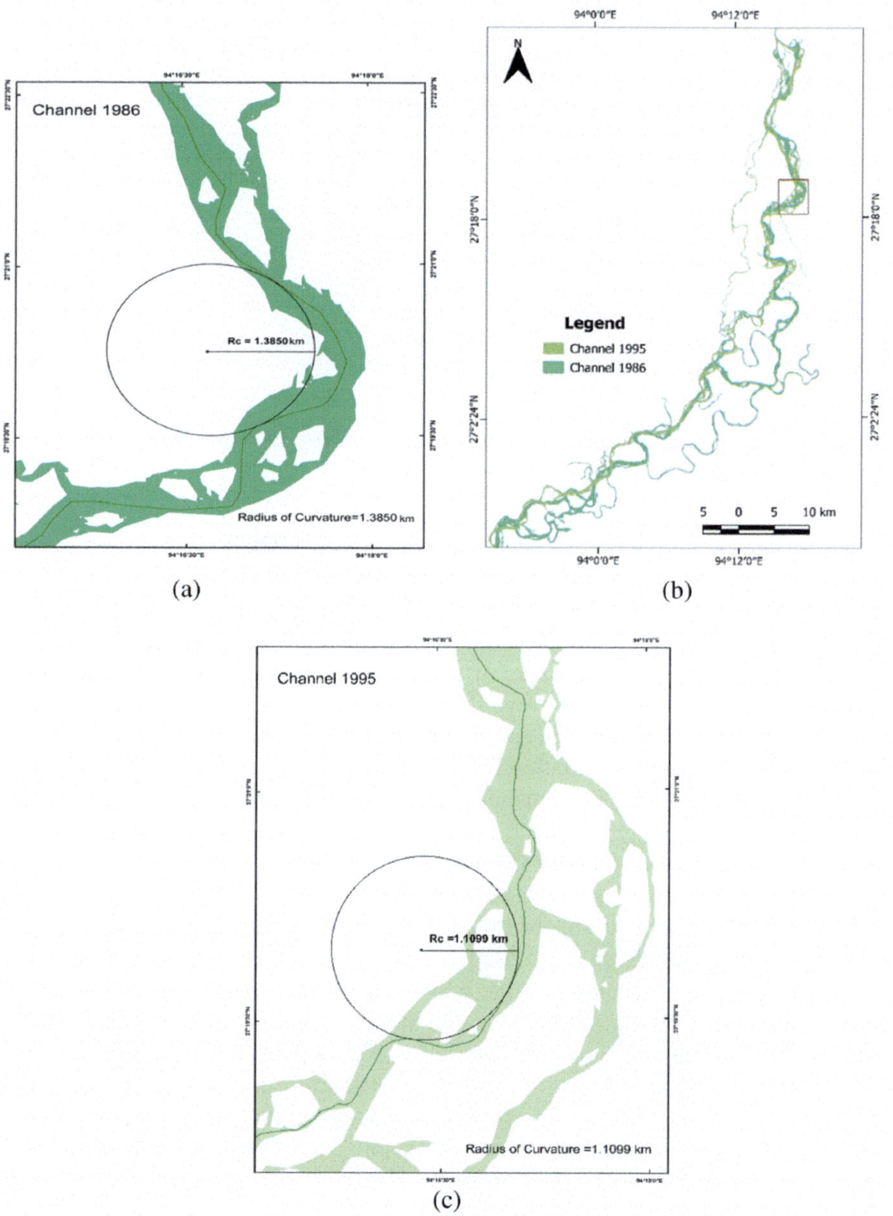

Fig. 11.21 **a** Radius of curvature of 1986. **b** Meandering and channel changes of bend 1 of 1986–1995. **c** Radius of curvature of 1995

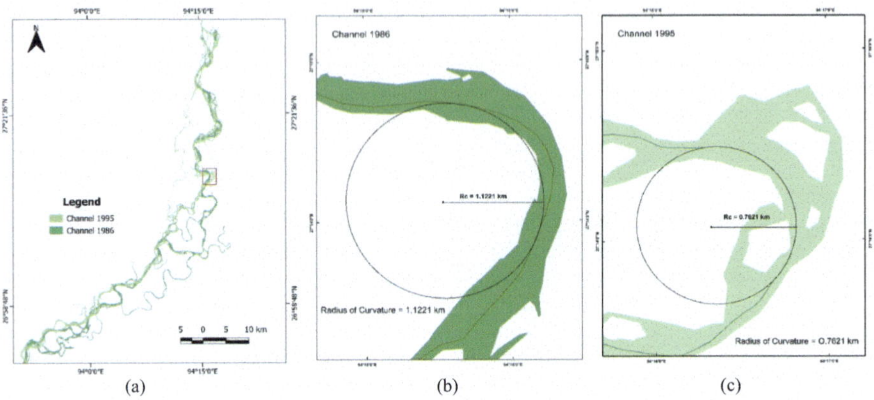

Fig. 11.22 **a** Meandering and channel changes of the 2 of 1986–1995. **b** Radius of curvature of 1986. **c** Radius of curvature of 1995

depths in the main channel and anabranching areas throughout this month indicate the potential for great sediment transport and bank erosion, particularly in the braided reaches where flow energy is localized. By August, while discharge volumes were still high, peak depth reduction and slight floodplain contraction indicate recession development, perhaps because of diminished upstream inflow and sediment deposition changing channel conveyance. For validation of flood water coverage in June and August of 2000, which resulted in channel migration and morphological change, a hydrodynamic simulation by HEC-RAS was conducted for the Subansiri River. The 2000 modelling showed unusually high flood peaks, with discharge values being higher than those experienced during other years, thus making 2000 the most severe flood period over the past decades. This intense hydraulic stress correlates with reported occurrences of floods on sixth and 22nd August 2000. when extensive areas of lower Assam, including the Subansiri basin, were submerged under Brahmaputra flood influences (Arun et al. 2020). The overlap of these waves of floods with maximum discharges accounts for the unprecedented geomorphic energy imposed on the channel. Following surveys revealed that the event initiated enhanced lateral migration, bar formation, and chute cutoffs, speeding up the westward shifting trend of the Subansiri. Abandonment of older channels and formation of new flow paths re-sculpted the floodplain, leading to erosion of agricultural fields and creation of new depositional bars. The 2000 flood is therefore a hydrological extreme and also a turning point in the geomorphic history of the river, as it set channel changes in motion that persisted in subsequent years.

The geomorphic effect of the 2000 flood went beyond direct inundation, as channel planform analysis and historical satellite imagery demonstrate that the record event initiated significant channel migration. Before 2000, the Subansiri in this section had a strongly braided morphology with several active threads. The high-energy and large sediment load of the flood caused rapid bankline erosion, channel avulsion, and filling of secondary channels. Consequently, between 2000 and 2005, the river

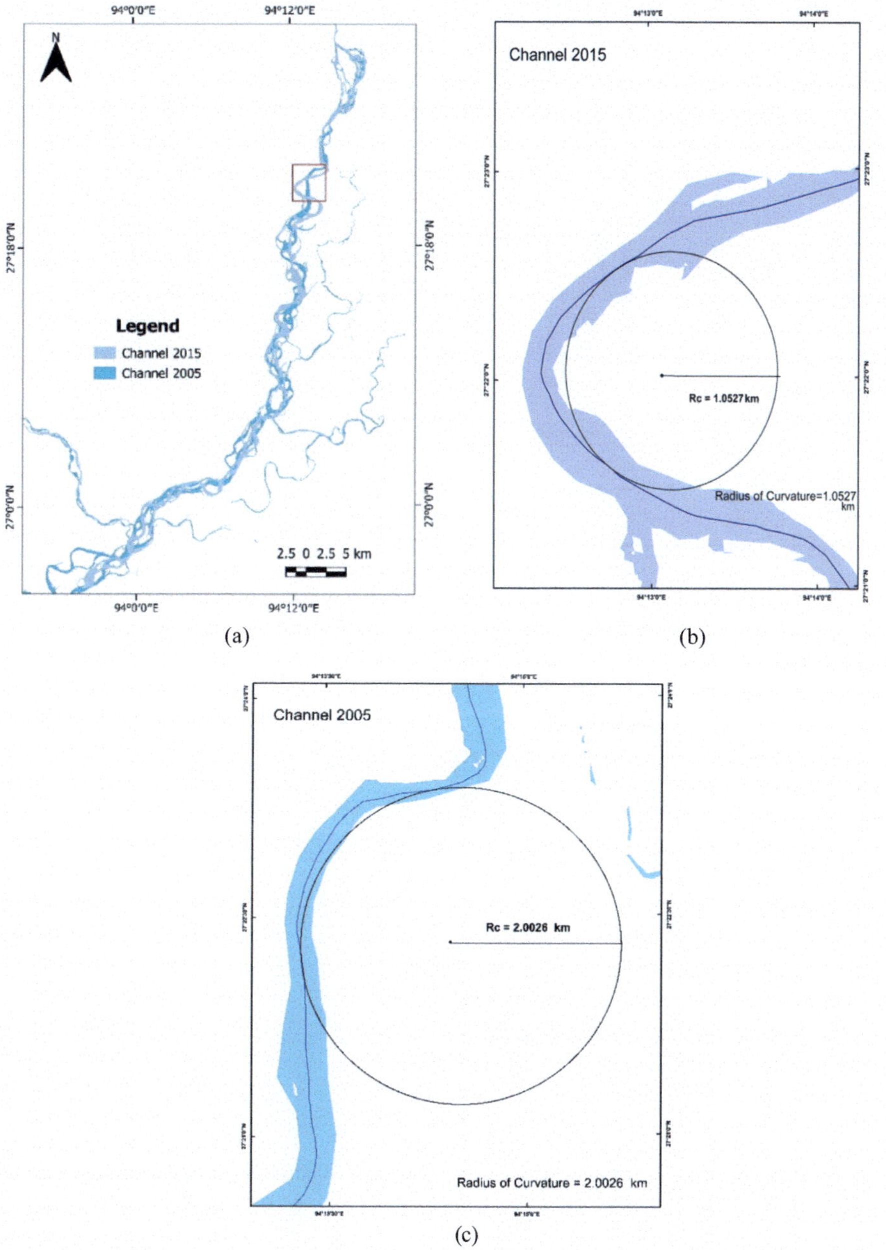

Fig. 11.23 **a** Meandering and channel changes of the 1 of 2005–2025. **b** Radius of curvature of 1986. **c** Radius of curvature of 1995

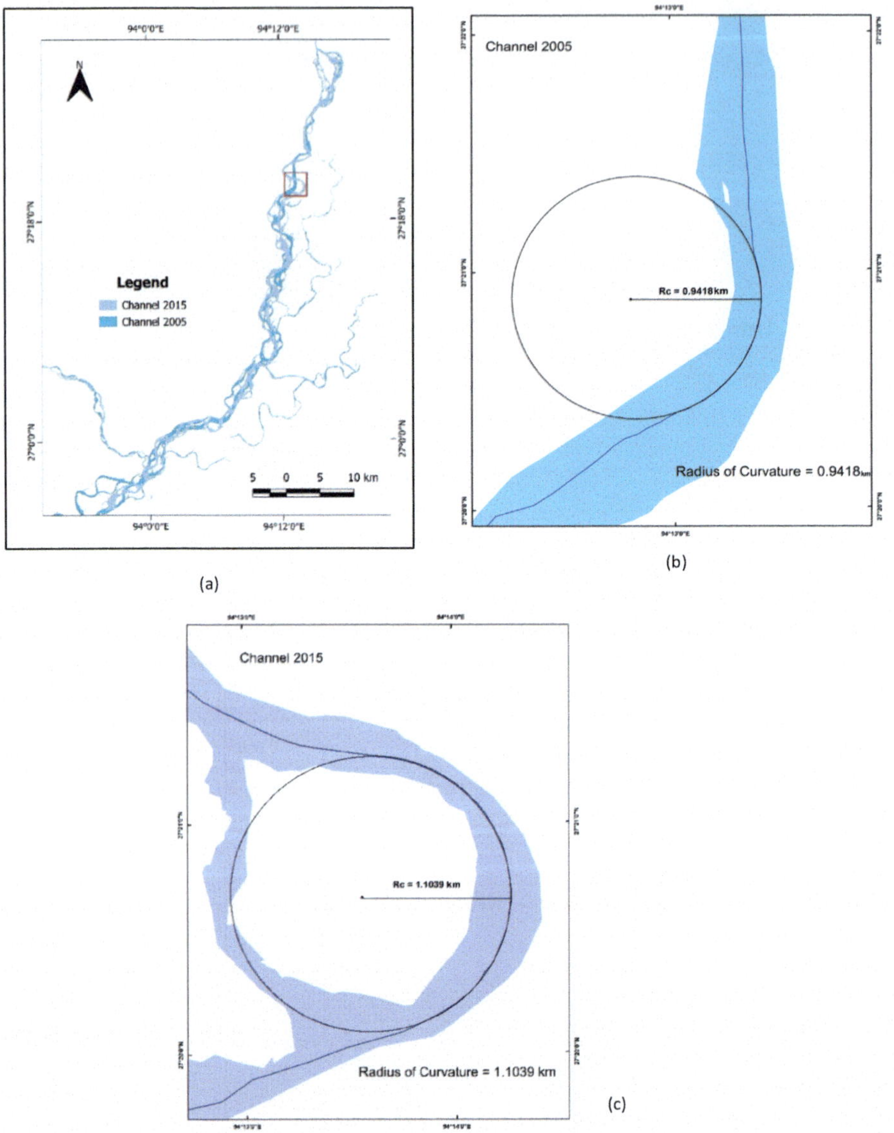

Fig. 11.24 **a** Meandering and channel changes of the 2 of 2005–2025. **b** Radius of curvature of 1986. **c** Radius of curvature of 1995

realigned quite extensively, abandoning several anabranches and taking the main channel to a straightened course. The straightened alignment continued up to 2025, representing a sustained geomorphic change. The flood of 2000 served as a driving force behind this change by overtopping channel stability thresholds, redistributing sediment, and excavating a preferential flow pathway that went on to become the

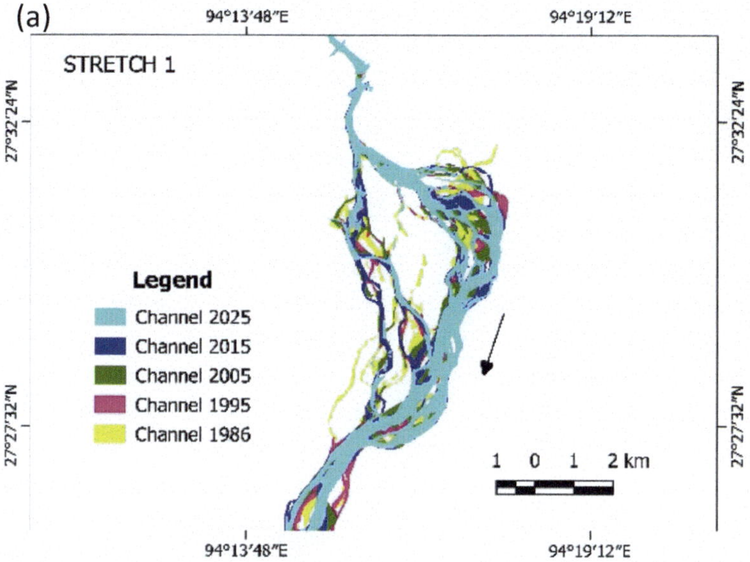

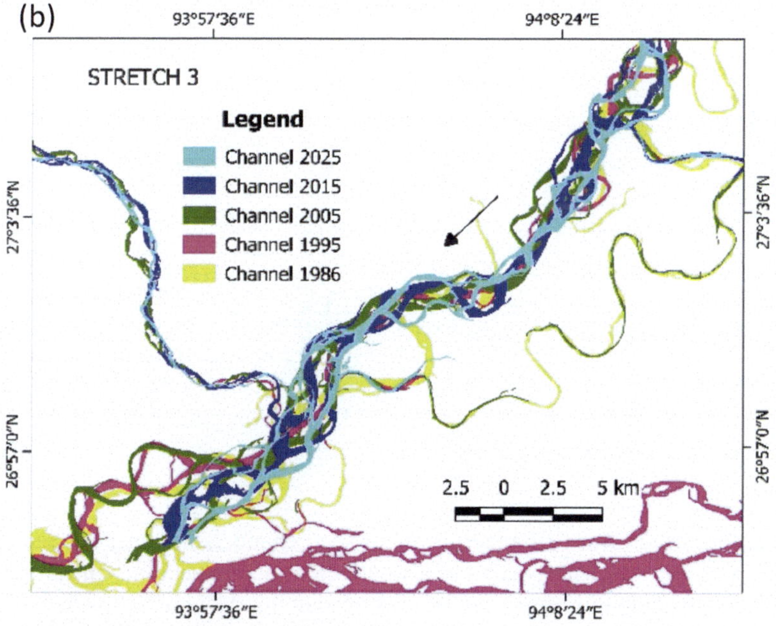

Fig. 11.25 **a**, **b** Braiding pattern temporal channel changes of Stretch 1 and S3

Table 11.3 Calculation table for observed and estimated Rc

Year (Stretch-2)	Area	Observed Rc	Estimated Rc
1986	6.025972	1.385037757	1.4866
1995	3.870263	1.109988034	1.3907
2005	12.598	2.002621433	1.6807
2015	3.481446	1.052756398	0.8015
BEND2			
2005	2.786346	0.941814951	1.1581
2015	2.786346	1.103928782	0.7102
BEND2			
1986	3.955371	1.122126117	0.6869
1995	1.824449	0.76210344	0.4949

Table 11.4 Sinuosity Index for three stretches of the respective rivers

Year	Reach	Thalweg	Straight length	Sinuosity	Total thalweg length in km
1986	S1	15.67	11.291	1.388	127.14
	S2	67.762	39.138	1.731	
	S3	44.402	32.499	1.366	
1995	S1	15.845	11.296	1.403	114.997
	S2	60.067	39.138	1.535	
	S3	41.03	34.592	1.186	
2005	S1	14.91	11.296	1.32	99.825
	S2	47.47	39.204	1.211	
	S3	36.67	32.53	1.127	
2015	S1	14.504	11.296	1.284	111.62
	S2	55.872	39.325	1.428	
	S3	41.337	32.369	1.277	
2025	S1	14.393	11.296	1.274	99.944
	S2	46.9	39.325	1.193	
	S3	37.226	31.663	1.176	

prominent thalweg. This rechannelization of such channels not only changed flood hydraulics in the following years but also affected the connectivity of floodplains, possibly constricting the lateral spread of moderate floods while focusing energy in the main channel. This connection between the 2000 flood and the following channel straightening highlights the role played by extreme events in influencing long-term river morphology change in the Subansiri–Brahmaputra confluence zone.

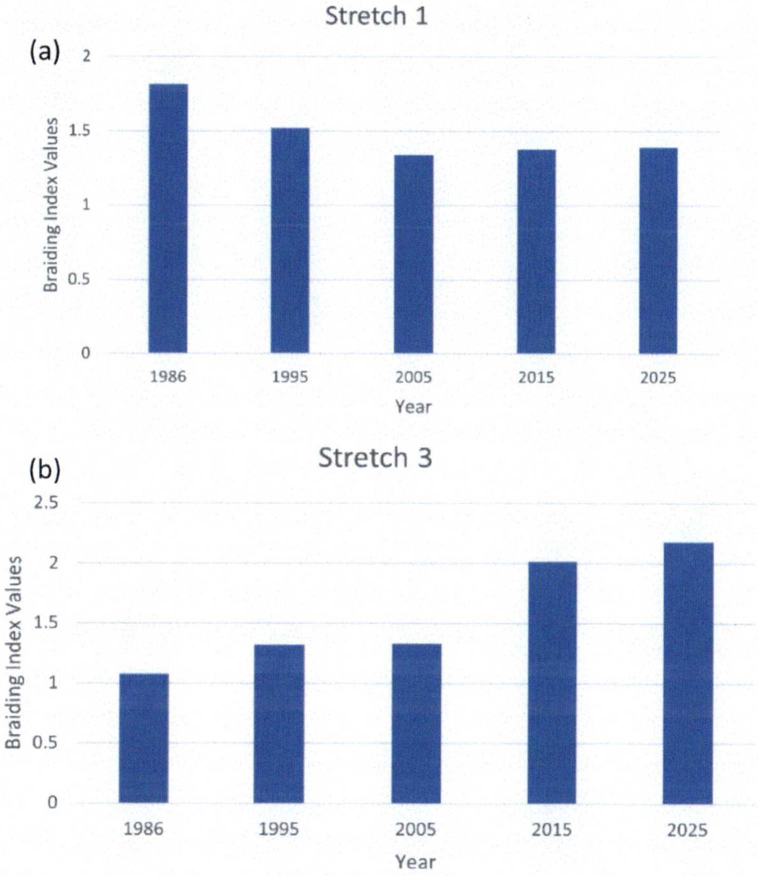

Fig. 11.26 **a**, **b** Bar graph showing the values of Braiding index over time

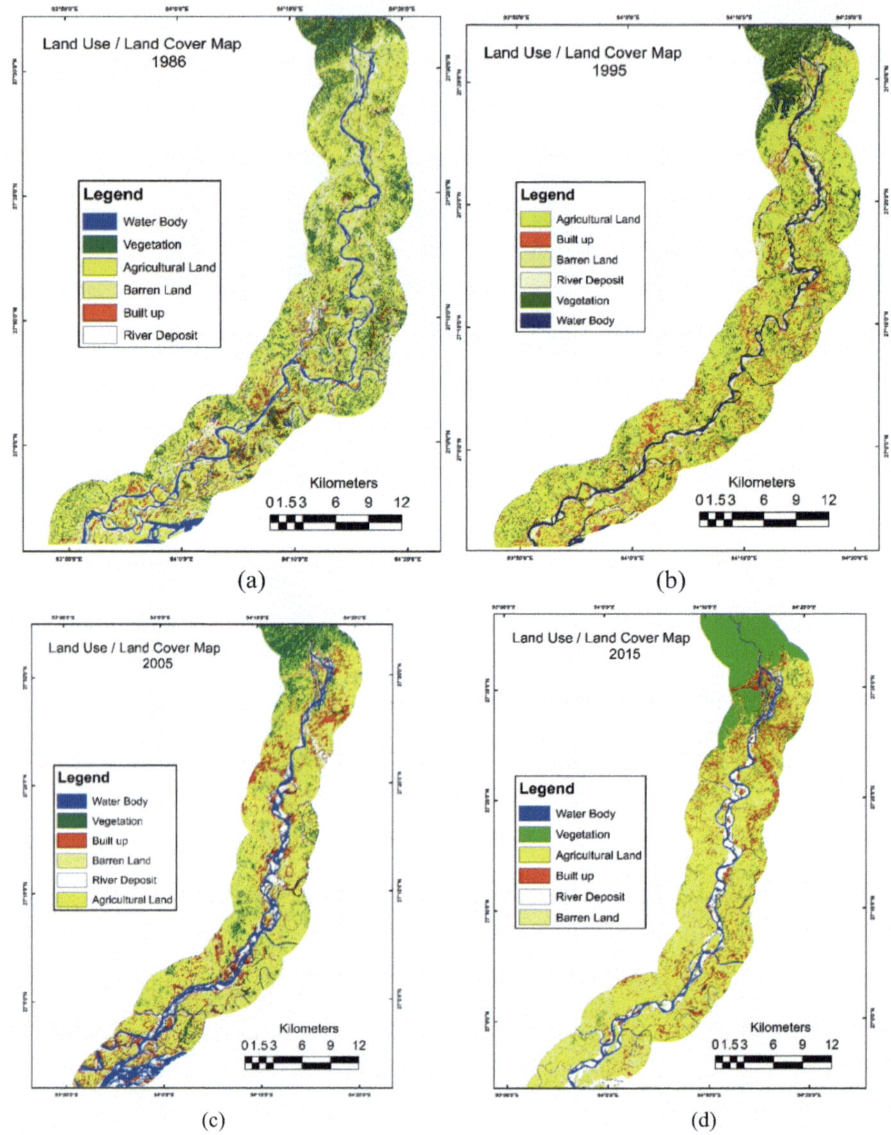

Fig. 11.27 Land use land cover classification **a** 1986, **b** 1995, **c** 2005, **d** 2015, **e** 2025

Fig. 11.27 (continued)

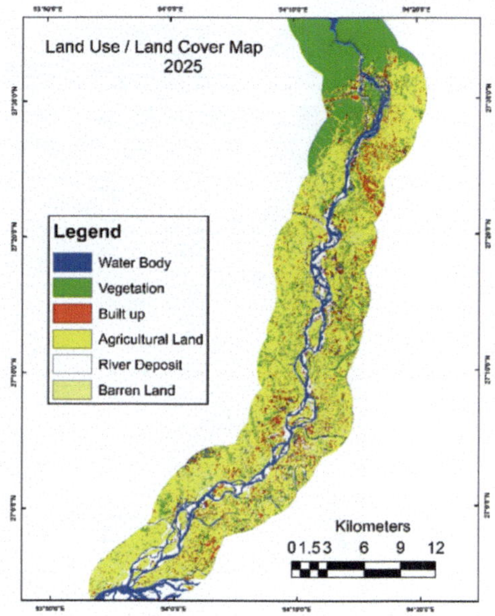

(e)

11.7 Conclusion

The study of river channel dynamics and flood propagation in the lower Subansiri reflects an extreme modification in the morphology and hydrologic regime of the river over the past four decades. Analysis of significant parameters such as radius of curvature, sinuosity, braiding index, channel migration, and LULC alteration confirms how the Subansiri evolved from a broadly meandering channel in the 1980s and 1990s to a comparatively straighter planform by 2025, reflecting significant shifts in fluvial processes and sediment control. A critical turning point in this evolution was the 2000 catastrophic flood, which drastically changed the river's dynamics in all reaches. This flood, fuelled by intense monsoonal precipitation and augmented by heavy sediment input from the upstream Himalayas, caused massive bank erosion, avulsions, and extensive large-scale channel migration. It produced extensive bar reworking and channel widening in Stretch 1 (Subansiri Dam to Chaldhowa Ghat) and initiated meander cutoffs in Stretch 2, laying the foundation for a transition to straighter planforms in the following decades. The flood also left immense sediment loads in the downstream section, altering confluence zone morphology with the Brahmaputra and redefining floodplain hazard zones.

The Subansiri dam to Chaldhowa Ghat section (Stretch 1), which earlier had exhibited a braided morphology, was greatly altered following the construction of the Dhunaguri Bridge, with bar multiplication, channel broadening, and subsequent westward channel migration rewriting its braided morphology. Concurrently, the middle reach (Stretch 2) saw a declining trend in sinuosity and migration

(a)

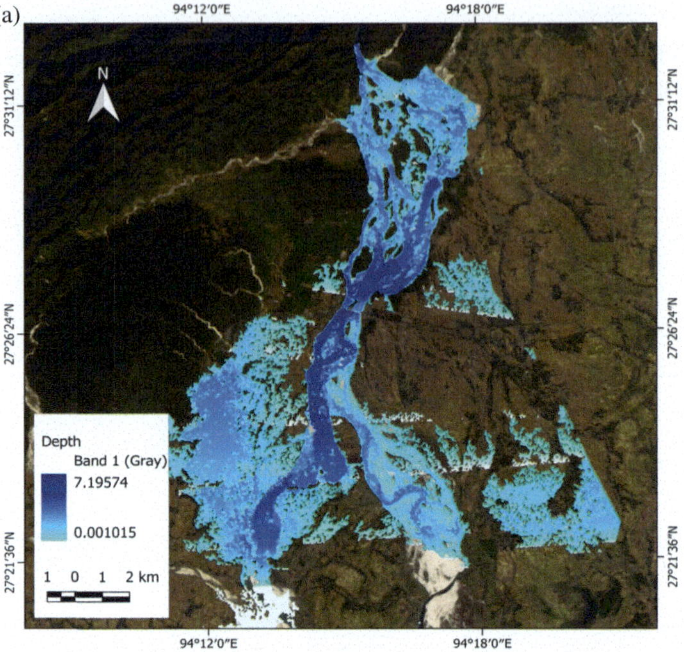

(b)

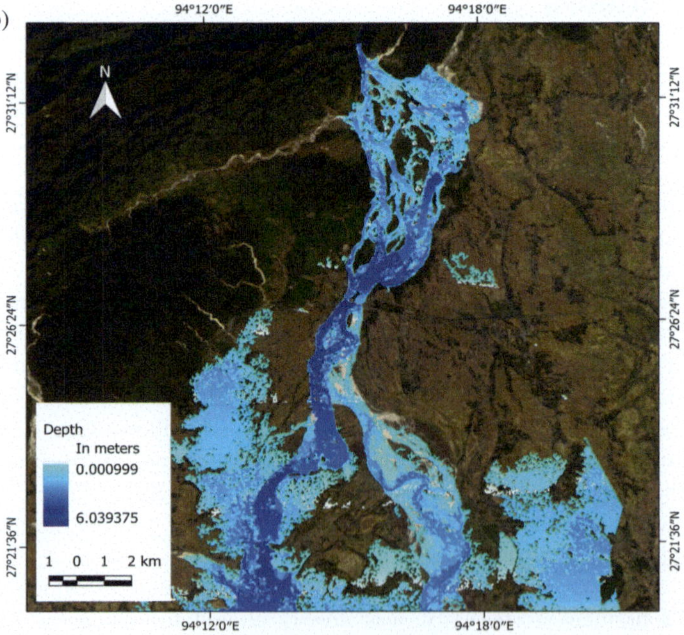

Fig. 11.28 **a** HEC-RAS flood modelling during low flood time, June 2000. **b** HEC-RAS flood modelling during high flood time, June 2000

rates with older meanders truncated and abandoned, while the downstream reach (Na-Ali Koibarta to Brahmaputra confluence) developed towards a more complex anabranching system. By 2025, the Brahmaputra confluence zone had shifted too, resulting from cumulative sedimentation and lateral migration, altering flood routes and areas of hazard. LULC analysis corroborates these changes, indicating a sharp decline in natural vegetation and expansion of agricultural and built-up land, particularly in flood-prone belts, which exacerbates vulnerability to channel shifts and flood inundation. The overall findings underscore the dynamic interplay between hydrological regime, sediment load, tectonic activity, and anthropogenic interventions in shaping the Subansiri's planform evolution. These transformations call for continuous monitoring of flood behaviour and channel morphology, especially in view of infrastructural development and increasing human intrusion onto the active floodplain, in order to formulate effective river management and hazard mitigation measures for the lower Subansiri basin.

References

Acharjee S, Das RT, Shekhar S, Singh N, Goswami U, Dilini MMGS, Negi RS et al (2022) Channel migration and consequential land use land cover changes of Subansiri River, Assam, north-eastern India. IOP Conf Ser: Earth Environ Sci 1032(1):012009. https://doi.org/10.1088/1755-1315/1032/1/012009

Annayat W, et al (2022) Assessment of river channel migration and floodplain dynamics using geospatial techniques. Environ Earth Sci 81(4):1–17

Annayat M, Sil A (2020) Meander parameter estimation and channel shift prediction: a case study of the Barak River, Northeast India. Geocarto Int. (DOI not available)

Arun G, Sharma SK, Tirkey G, Lohani AK (2020) Integrated flood assessment studies using real time satellite precipitation data and flood inundation mapping in Assam—a novel approach. In: Proceedings of the national conference on emerging trends in civil engineering, 26–27 June 2020. National Institute of Hydrology, Guwahati, Assam. https://www.researchgate.net/publication/342850754

Borgohain P, Das D, Sarma JN (2016) Channel dynamics and bank erosion of the Brahmaputra River in Assam, India. Arab J Geosci 9(5):1–14

Brice, J. C. (1964). Channel patterns and terraces of the Loup Rivers in Nebraska. U.S. Geological Survey Professional Paper 422-D. https://doi.org/10.3133/pp422D

Coleman JM (1969) Brahmaputra River: channel processes and sedimentation. Sediment Geol 3:129–239. https://doi.org/10.1016/0037-0738(69)90010-4

Couper PR (2003) Effects of silt–clay content on the susceptibility of river banks to subaerial erosion. Geomorphology 56(1–2):95–108

Das B et al (2014) River bank erosion hazard assessment of the Brahmaputra River using remote sensing and GIS. Nat Hazards 70(3):1829–1841

Das I, Stein A, Kerle N, Dadhwal VK (2012) Landslide susceptibility mapping along road corridors in the Indian Himalayas using Bayesian logistic regression models. Geomorphology 179:116–125.

Debnath K, Das P, Bandyopadhyay S (2016) Morphological changes and channel migration analysis of the Brahmaputra River using geospatial techniques. Spat Inf Res 24(2):145–154

Debnath J, Pan ND, Ahmed I, Bhowmik M (2017) Channel migration and its impact on land use/land cover using RS and GIS: a study on Khowai River, Tripura, Northeast India. Egypt J Remote Sens Space Sci 20:197–210. https://doi.org/10.1016/j.ejrs.2017.07.003

Florsheim JL, Mount JF (2008) Natural bank erosion as ecological disturbance. Bioscience 58(4):305–313
Ghosh S, Mukhopadhyay A (2021) Avulsions and bankline migration dynamics of Dwarkeswar River, Eastern India. Geomorphology
Goswami DC (1985) Brahmaputra River, Assam, India: Physiography, basin denudation, and channel aggradation. Water Resour Res 27(7):959–978
Goswami U, Patgiri AD, Sarma JN (1999) River channel changes of the Subansiri in Assam, India. Geomorphology 30(3–4):227–244. https://doi.org/10.1016/S0169-555X(99)00032-3
Gran KB, Paola C (2001) Riparian vegetation controls braided-to-meandering transitions. Water Resour Res 37(12):3335–3344. https://doi.org/10.1029/2000WR000009
Hazarika N, Das AK, Borah SB (2015) Assessing land-use changes driven by river dynamics in the upper Brahmaputra plains using RS-GIS techniques. Egypt J Remote Sens Space Sci 18:107–118
Howard AD, Hemberger JL (1970) Quantitative characteristics of braided streams. J Geol 78(6):585–594. https://doi.org/10.1086/627557
Imsong W et al (2018) GIS-based analysis of river channel dynamics and flood hazard zonation in northeast India. Appl Geomat 10(4):451–463
Khosronejad A et al (2022) Numerical simulation of braided river morphodynamics and flood propagation. J Hydrol 612:128089
Kotoky P, Bezbaruah D, Sarma JN (2003) Geomorphic adjustments in Majuli Island due to tributary influences. Geomorphology 60:1–19. https://doi.org/10.1016/S0169-555X(03)00266-6
Kotoky P, Bezbaruah D, Baruah J, Sarma JN (2005) Bank erosion and channel migration along the Brahmaputra. J Indian Soc Remote Sens. https://doi.org/10.5958/2277-937X.2017.00024.7
Lahiri SK, Sinha R (2012) Tectonic controls on the morphodynamics of the Brahmaputra River system in the upper Assam valley, India. Geomorphology 169–170:74–85
Lawler DM (1993) A systematic method for measuring lateral migration in meandering rivers. Geomorphology 7(2):115–129
Leopold, L. B., & Langbein, W. B. (1966). River meanders: Theory of minimum variance. U.S. Geological Survey Professional Paper 422-H. https://doi.org/10.3133/pp422H
Leopold LB, Wolman MG (1957) River channel patterns: Braided, meandering, and straight. US Geol Surv Prof Pap 282-B:39–85
Mani P, Bora AK, Kumar R (2003) River channel migration studies using remote sensing and GIS techniques in northeast India. Int J Remote Sens 24(4):833–843
Mukherjee A (2022) Channel planform dynamics of the Ramganga River: bend evolution and bed morphology changes. Geomorphology
Neog P (2018) GIS-based delineation of erosion–deposition zones in northeast Indian rivers. Appl Geomat
Sarma JN, Phukan MK (2004) Origin and geomorphic evolution of Majuli Island, Brahmaputra. Assam Geomorphology 60(1–2):1–19. https://doi.org/10.1016/S0169-555X(03)00266-6
Sarma JN, Phukan MK (2006) Bank erosion and bankline migration of the Brahmaputra River in Assam during the twentieth century. J Geol Soc India 68:1023–1036
Sarthak et al (2015) River channel migration and erosion–deposition analysis using geospatial techniques. Appl Geomat 7(3):145–156
Schumm SA (1963) Sinuosity of alluvial rivers on the Great Plains. Geol Soc Am Bull 74(9):1089–1100
Schumm SA, Lichty RW (1963) Channel widening and flood-plain construction along Cimarron River in southwestern Kansas. US Geol Surv Prof Pap 352-D:71–88
Tal M, Paola C (2007) Dynamic single-thread channels maintained by the interaction of flow and vegetation. Geology 35(4):347–350
Verma S, Maaß F et al (2021) Channel morphology and migration of the Ganga River: predictive modelling using GIS and ARIMA. J Hydrol. (DOI not available)
Wahiduzzaman, M, & Yeasmin, T (2022) Transition of the Ganga–Padma River from meandering to braiding: sinuosity and braiding index analysis. Appl Sci, 12(22):11604. https://doi.org/10.3390/app122211604

Zolezzi G, Tubino M et al (2012) Meandering to braiding threshold: curvature–width interactions and bend stability. Water Resour Res. https://doi.org/10.1029/2012WR012097

The manufacturer's authorised representative in the EU is Springer
Nature Customer Service Centre GmbH, Europaplatz 3, 69115 Heidelberg,
Germany. If you have any concerns regarding our products, please
contact ProductSafety@springernature.com

Printed and bound by CPI Group (UK) Ltd, Croydon, CR0 4YY
07/10/2026
02229350-0004